P9-BID-314

SIGNATURE WOUNDS

Signature Wounds

The Untold Story of the Military's Mental Health Crisis

David Kieran

NEW YORK UNIVERSITY PRESS
New York

NEW YORK UNIVERSITY PRESS
New York
www.nyupress.org

References to Internet websites (URLs) were accurate at the time of writing. Neither the author nor New York University Press is responsible for URLs that may have expired or changed since the manuscript was prepared.

Library of Congress Cataloging-in-Publication Data
Names: Kieran, David, 1978- author.
Title: Signature wounds : the untold story of the military's mental health crisis / David Kieran.
Description: New York : New York University Press, [2019] | Includes bibliographical references and index.
Identifiers: LCCN 2018037659 | ISBN 9781479892365 (cl : alk. paper)
Subjects: LCSH: Soldiers—Mental health—United States. | Soldiers—Mental health services—United States. | Veterans—Mental health—United States. | Veterans—Mental health services—United States. | Combat—Psychological aspects. | Military psychiatry—United States. | Psychology, Military.
Classification: LCC UH629.3 .K54 2019 | DDC 362.2088/35500973—dc23
LC record available at https://urldefense.proofpoint.com/v2/url?u=https-3A__lccn.loc.gov_2018037659&d=DwIFAg&c=slrrB7dE8n7gBJbeOog-IQ&r=gT953V3c8BdcJV4pugGaWueY1IXCnXKbUfEC8Smo3jI&m=MP-GJ562mYu3PDS2bpIlm5Je-F1nOhMD8wY8j_PuQXs&s=JoWU-6XhWTJXgcE4pILvVbUsYQKRY5HRoHJbxKjoo9w&e=

New York University Press books are printed on acid-free paper, and their binding materials are chosen for strength and durability. We strive to use environmentally responsible suppliers and materials to the greatest extent possible in publishing our books.

Manufactured in the United States of America

10 9 8 7 6 5 4 3 2 1

Also available as an ebook

Once Again, for Emma

CONTENTS

Introduction

"These Unseen Wounds Cut Deep"

On November 14, 2004, the *Los Angeles Times* published a troubling headline. "These Unseen Wounds Cut Deep" it read, with the subtitle "A Mental Health Crisis Is Emerging, with One in Six Returning Soldiers Afflicted."[1] A year and a half into the Iraq War, this was perhaps the first article to define the growing number of psychological issues that veterans faced as a "crisis." Six months later, then Sen. Barack Obama (D-IL) took to the Senate floor to tell his colleagues that "many of our injured soldiers are returning from Iraq with a condition known as traumatic brain injury, or TBI," explaining that "some doctors are saying that it could become the 'signature wound' of the Iraq War," the first public use of the term by a U.S. politician.[2] Obama's remarks were inspired by an article that Gregg Zoroya had written a month earlier for *USA Today*, in which he became the first journalist to use the term.[3]

Within two years, the term had become plural. In February 2007, *Washington Post* reporters Dana Priest and Anne Hull published a series of articles exposing the poor conditions under which patients were living at the Walter Reed Army Medical Center, the military's largest and most prominent hospital.[4] The outrage that followed this reporting focused attention on the wider question of whether the George W. Bush administration was doing enough to care for veterans.[5] In an April editorial condemning the administration, the *New York Times* referred to a report that "concluded that the Pentagon is failing to deal adequately with the signature wounds of the Iraq conflict—the thousands of cases of brain damage from roadside bombs and of post-traumatic stress disorder [PTSD]" before concluding that "it is the White House that is in denial. It is past time for it to provide America's soldiers the care they are owed."[6] The term also entered the Department of Defense's vocabulary.[7] Later that year, when Pete Geren—a former Democratic congressman who been Undersecretary of the Army and then assumed the role of Acting Secretary in the scandal's wake—appeared before the Senate Armed Services Committee, which was considering his nomination to permanently assume the position of Secretary, he used the term to mount a qualified defense against these charges: "PTSD as well as TBI, are two of the signature wounds of this conflict," he told Sen. Carl Levin (D-MI), "and we're working them, but we have a lot of work to do."[8]

This media coverage and political rhetoric reveals that, in the course of the U.S. twenty-first-century wars, Americans have become increasingly preoccupied by the mental health and cognitive issues that troops have developed as result of their service, particularly in Iraq and Afghanistan.[9] It demonstrates that in the early twenty-first century, politicians, military leaders, researchers and clinicians, and many in the general public viewed with alarm accounts of service members who returned from war traumatized by what they witnessed, who suffered from brain injuries, and who in some cases went on to take their own lives. These stories raised a host of questions: Why hadn't the military been better prepared to treat post-traumatic stress disorder (PTSD) and mild and moderate traumatic brain injury (mTBI), especially given the purported lessons of the war in Vietnam?[10] Why weren't the Army and the Department of Veterans Affairs (VA) more effectively screening for PTSD and mTBI? Why were troops being denied care, sent back to Iraq despite their mental health needs, or discharged from the military without benefits? Why were so many soldiers and veterans killing themselves? Why wasn't the Army and the VA doing more to address these issues? These examples also illustrate that these conditions took on a range of meanings: They were, in the *Los Angeles Times*'s and Obama's usage, cause for deep concern. For the *New York Times*, they were evidence of the Bush administration's disingenuousness, malfeasance, and indifference. For Geren, they were challenging problems that required research.

Signature Wounds examines how Americans—policy makers, military leaders, caregivers, Army and VA researchers, and the broader public—understood the mental health issues that the United States's twenty-first-century wars generated and how they mobilized those understandings to engage in broader political and cultural debates about the wisdom and management of the Iraq and Afghanistan Wars, the military's place in American culture, and the United States's role in the world. This story embraces anthropologists Sarah J. Hautzinger and Jean Scandlyn's claim that "in the post-9/11 wars psychological injuries have taken center stage in the ways we talk about, digest, and engage with war and its consequences" and Kenneth T. MacLeish's assessment is that "the length, scale, and distinct character of the Iraq War have subjected American soldiers and their families to longer and more frequent deployments, leading to unprecedented rates of postratumatic stress disorder (PTSD) and traumatic brain injury (TBI) and the overburdening of military institutional supports."[11] It unfolds by analyzing the intersecting and often intertwined ways in which these stakeholders understood these issues and debated their significance. In media, popular culture, and political rhetoric, the increasing incidence of post-traumatic stress disorder, traumatic brain injury, and soldier and veteran suicide was evidence that the Iraq War was misbegotten and mismanaged by a callous administration unwilling to acknowledge or address the human consequences of its misguided policies. For researchers and clinicians, these issues were complicated but solvable medical problems, conditions about which there was great uncertainty and that demanded rigorous research around prevention, diagnosis, and treatment. For Army leaders, they were at once tactical dilemmas that required immediate attention amid the demands of two prolonged wars and evidence of a crisis that raised larger, existential questions about military institutions and their place in the nation.

These three discourses were not discrete. In the decade that followed the United States's invasion of Iraq, they often intersected, and the resulting debates were sometimes acrimonious. Members of Congress and senators who seized upon notions of mental health issues to fuel anti-war critiques often looked askance at Army researchers who maintained that combat stress was expected if not inevitable and usually temporary and treatable, raged at Army policies that allowed troops with mental health diagnoses to deploy to Iraq, and portrayed VA officials who maintained that they were making effective inroads in addressing veteran suicide as emblematic of the administration's broader dishonesty. Soldiers, veterans, and their families frequently felt that Army and VA leaders were merely paying lip service when they claimed to be aware of and working to address mental health needs, and they criticized a system that they viewed as bloated with self-interested bureaucrats. Veterans Administration officials grew increasingly frustrated when anecdotal accounts that the failure to treat a particular veteran had resulted in a suicide overwhelmed their claims to be providing the best evidence-based care available. Inside the Army Medical Department, or AMEDD, researchers worked to develop protocols for understanding how deployments affected troops' psychological well-being, to determine the etiology of traumatic brain injury, and to prevent suicide, and they butted heads internally over competing theories and externally with a larger Army that was often resistant to taking mental health seriously. Army leaders worried about their troops' well-being grew frustrated with Army doctors who maintained that applying a treatment that hadn't been fully vetted was unethical. And as the stress of prolonged warfare increasingly took a psychological toll on soldiers and their families, Army researchers and leaders worried that the culture that the Army had built during its post-Vietnam recuperation, if not the All-Volunteer Force itself, was imperiled.

The story of mental health during the Iraq and Afghanistan Wars, then, is multifaceted. Looking back on the Army's efforts to understand and address these issues, Lt. Gen. Eric Schoomaker, who served as Surgeon General from 2008 to 2011, reflected that "I don't think any of us had a complete picture of it at any one time"; instead, he argues, there were "many, many competing viewpoints, agendas, and many multigenerational issues that we were facing."[12] The story of mental health during this war is the story of how different groups—soldiers, veterans, and families; anti-war politicians and the general public; researchers and clinicians; and military leaders—approached these issues from different perspectives, with different agendas, and valuing different kinds of evidence. It is the story of how the advancement of medical knowledge moves at a different pace than the needs of an army at war and of how medical conditions and medical knowledge intersect with larger political questions about militarism and foreign policy.

It is also a story that has thus far been told in fragments. Most Americans who have read about these issues have done so in texts that largely take the perspective of veterans who have been underserved by organizations lacking both resources and empathy and that illustrate the Bush administration's malfeasance or profile a few maverick leaders who have forced hidebound institutions to change. Long-form journalism of the sort written by Ann Jones, Aaron Glantz, Yochi Dreazen, David Finkel, and C. J. Chivers are compelling narratives that offer important insights into the soldiers' and veterans' struggles and military and governmental failures.[13] The popular acclaim that these pieces have earned thus emerges in part from how neatly that narrative comports with widely held, and sometimes well-deserved, skepticism of the military and the federal bureaucracy, opposition to the wars, and disdain for the Bush administration. However, they do not tell a complete story. Neither do medical journal articles that detail particular developments or, in a few cases, offer overviews and lessons learned.[14] In *Signature Wounds*, I tell a broader, more nuanced story. This book joins a body of scholarship that has studied the cultural implications of what war does to the bodies and minds of those exposed to its violence. Scholars including Beth Linker and John M. Kinder, for example, have written deeply researched accounts of how reckoning with veterans' wounded bodies shaped U.S. culture after the First World War.[15] Other historians of the Great War have explored that war's defining injury, shell shock, although this has primarily been done in the context of British experiences.[16] And while some attention has been paid to the psychological struggles of World War II veterans, a considerable body of scholarship has studied veterans of the United States's war in Vietnam.[17]

A growing number of scholars have investigated the medical cultures of the twenty-first-century wars. Jennifer Terry, for example, argues that the medical advancements that war generates "elaborates processes of militarization through which war comes to be tacitly accepted as a necessary condition for human advancement."[18] Allison Howell would agree with this claim, as she views efforts to produce soldier resilience as means of enabling continued warmaking.[19] Similarly, Rebecca A. Adelman argues that military families and veterans diagnosed with PTSD and TBI have become "objects of intense emotional and discursive investment and . . . political subjects that are partially or fully unknowable."[20] The consequence of such constructions, she contends, "promotes a shallow ethics in response to their suffering, the erasure of their political subjectivities, and, ultimately, propogation of the very militarism that begets their victimizations."[21] Other significant research on the intersections of war, culture, and medicine in the twenty-first-century United States has been done by anthropologists including Zoe Wool, who reveals the experiences of soldiers and their families at Walter Reed

Army Medical Center; Erin Finley, who has investigated the culture that surrounds PTSD treatment in both the Army and the VA; Emily Sogn, who has written about the Army's embrace of resiliency; Hautzinger and Scandlyn, who tell the story of a platoon at Fort Carson that became particularly infamous; and Kenneth T. MacLeish, who has deeply studied troops' experiences at the Army's largest post, Fort Hood.[22]

The work of both journalists and academics has largely focused on the experiences of the service members who directly experienced these wars and who live with their physical and psychological consequences. C. J. Chivers, for example, explains that *The Fighters* "rejects many senior officer views" because "the voices of combatants of the lower and middle ranks are more valuable, and more likely to be candid and rooted in battlefield experience."[23] There are valid reasons for taking this approach. As MacLeish explains, "Focusing on the vulnerability of soldiers and their families and communities allows us to ask what is involved in recognizing the harm done to those whose job it is to produce war on the nation's behalf."[24] It is also the case, however, that apprehending the Army as an organization at once embedded in and distinct from the broader culture demands interrogation of how its senior leaders and its researchers thought about and sought to address the psychological challenges that the Afghanistan and Iraq Wars posed for soldiers and their families.

Signature Wounds tells that story. As it does so, it challenges some popular assumptions and scholarly arguments. Contrary to the dominant narrative that surrounds military mental health during the United States's twenty-first-century wars, Army and VA researchers and leaders were deeply concerned about matters of mental health from the wars' first days. This concern, however, manifested as the wars lasted longer and were more brutal than anticipated and amid both profound uncertainty about the nature of the issues service members confronted and a set of cultural dynamics that complicated efforts to identify and treat them. Most psychologists in the post–Cold War Army and VA, for example, had little expertise in treating combat-related post-traumatic stress disorder, most physicians had a limited understanding of brain injury and were reliant on civilian science and protocols for diagnosis and treatment, and there was little certainty about whether an explosion's blast wave alone could damage the brain. Most leaders and medics in the field had little combat experience. By 2001, anyone who had fought in the one hundred hours of ground combat that occurred during the Persian Gulf War of 1990–1991 would have served for more than a decade in an army in which enlistments were usually much shorter and that had seen a considerable reduction in force in the aftermath of the Cold War. Moreover, mental health care was highly stigmatized in the military's hypermasculine culture; many soldiers feared—often with good reason—that they would be disparaged for seeking mental health care or that their careers would be imperiled.[25] Perhaps most important, very few in the United States predicted that the Iraq and Afghanistan Wars would last for a combined quarter century, would degenerate into protracted counterinsurgencies, and would require the equivalent of one and a half million years of service spread over more than one million people.[26] Quite simply, neither the Army nor the nation expected or prepared for conflicts that would place so much strain on so few soldiers and their families.

Army researchers and clinicians thus faced numerous challenges as they struggled to address mental health issues and traumatic brain injuries: identifying exactly how deployment experiences, multiple deployments, and shortened dwell times affected mental health; determining the causes and consequences of traumatic brain injury and distinguishing them from other conditions, like post-traumatic stress disorder, that were often caused by the same event; figuring out under what circumstances a soldier with a mental health diagnosis and perhaps a set of prescriptions could deploy again; identifying which soldiers were prone to suicide and why; cajoling leaders and soldiers into taking mental health seriously; and ensuring that the efforts to maintain a combat-ready force did not mean failing to attend to the mental health needs of families consumed by anxiety over their loved one's well-being and the increased demands that deployments placed on those at home. And all of this was happening as the wars placed increasing stress and strain on the Army, as anti-war sentiment increased, and as senior leaders made more vocal demands for protocols that they could quickly implement on behalf of their troops. Pete Geren described the Army's approach as "changing tires on a bus going sixty miles an hour."[27]

Understanding the centrality of mental health to the culture of the U.S. twenty-first-century wars thus demands understanding the Army as a complicated institution that is at once part of and discrete from

American culture at large. It also requires understanding PTSD, TBI, and suicide—the conflict's signature wounds—as medical conditions and within their historical and cultural context.[28] Recently, scholars have pointed out the culturally constructed nature of the term "signature wound." The sociologist Jerry Lembcke, for example, compellingly explains that "the establishment of TBI as the 'signature wound' of the new wars begs the question of what that means. That a wound 'signs' something implies that it has cultural or social meaning separate from its literal meaning."[29] Similarly, John Kinder points out that "the term 'signature wound' does not describe the most common forms of wartime injury or even the most significant of the time," and it "is applied in such random or haphazard fashion that it betrays its own meaninglessness."[30] Particular conditions, Kinder explains, do not become a conflict's signature wounds because they are new or because they were the "most common forms of wartime injury or even the most significant at the time."[31]

This is not to say, however, that "signature wounds" are not materially real. Of course they are, and this includes psychological conditions and traumatic brain injury.[32] However, they are significant because they are politically and culturally salient.[33] As the disability scholars Nancy J. Hirschmann and Beth Linker explain, defining the significance of a particular health condition occurs both medically and culturally. Although there are "specific instances of physical or intellectual disability, disability itself is not a physical condition pertaining to a 'defective' or 'inferior' or 'abnormal' body but rather a social condition brought about by social norms, practices, and beliefs. It is both socially produced and socially experienced."[34] The same could be true of conditions that become "signature wounds"; they are those that in a particular historical moment take on particular meanings—as Kinder puts it, they are "a form of ideology, a concept"—that in turn shape how individuals and groups think and talk about that condition and its intersections with broader cultural and political problems.[35] When a particular condition dominates Americans' imaginings of a war, it in turn shapes how they evaluate the experiences of those who fought, whether the conflict is worthwhile, whether the government is meeting its obligations to veterans, and what the conflict's legacies will be.[36] This process has occurred throughout the history of American warfare, as Megan Kate Nelson and Kinder have respectively illustrated in their analyses of post–Civil War U.S. culture and the aftermath of the First World War.[37] The wounds of war are thus not only physical conditions and the concerns of medical science; they are also cultural phenomena that shape how Americans think about the nation's wars and foreign policy.

For many Americans, PTSD and TBI became the signature wounds of the twenty-first-century wars because they helped advance anti-war narratives.[38] Popular understandings of these conditions and the notion that many veterans seemed to struggle with them appeared in the media and were discussed on the floors of Congress and in Capitol Hill meeting rooms, fostering and facilitating critiques of the Iraq War by providing tangible evidence of the war's brutality and the George W. Bush administration's callous indifference to the human toll of its misguided, malfeasant foreign policy. Attacking the Department of Veterans Affairs for failing to lower the suicide rate among veterans, for example, became a convenient proxy for attacking the Bush administration as a whole, and expressing outrage over the failure to care for traumatized, concussed, or suicidal soldiers insulated the war's critics from charges that they were insufficiently patriotic. In a climate in which anti-war sentiment was often dismissed with assertions that critics were not supporting the troops, pointing out how the wars were harming those troops facilitated broader policy critiques.

Post-traumatic stress disorder, traumatic brain injury, and suicide are also the signature wounds of these wars because they prompted significant debate and cultural change within the Army itself. *Signature Wounds* tells the story of these debates and that change and thus joins a growing body of scholarship that locates the military as a significant actor in U.S. culture.[39] The dominant popular image of the Army emphasizes the combat arms and its warrior culture, but it is, of course, more complicated than that. One researcher whom I interviewed might have overstated the point when he guessed that "only about twenty percent of Americans realize that there even is a military health system," but it is nonetheless true that the Medical Department does not dominate popular imaginings of the Army.[40] And yet it is central to the cultural debates that played out both within the Army and between it and the larger culture in this moment. If AMEDD scientists and doctors struggled to understand PTSD and TBI and often disagreed about etiology and treatments, they also contended with other aspects of the Army that resisted both their commitment to the scientific method and their efforts to improve care and make it more accessible. Some senior leaders, for example, were frequently frustrated that the AMEDD's insistence on a rigorous research process meant that applicable knowledge could be slow to emerge when troops in the field were in danger or suffering. For action-oriented leaders who were worried about protecting forward-deployed troops, the notion of a three-year trial for a concussion treatment was maddening; for researchers, using an unvalidated treatment was unethical. Army Medical Department staff also faced resistance from soldiers and leaders who disparaged mental health issues as signs of weakness. For those leaders and soldiers, to admit to a mental health issue was to risk embarrassment and censure; for AMEDD clinicians, breaking down these barriers was a key means of ensuring that soldiers got the care they needed.

These internal debates and efforts at change intersected with broader concerns about what kind of institution the Army was and what impact the wars had had on it. When the Army contemplated the psychological impact of multiple tours, the increasing strain on families, and the rising suicide rate, it also debated whether the culture and reputation that it had carefully restored after Vietnam was being imperiled by the wars. At times, these anxieties produced subtle critiques of the wars themselves, the Army's political leadership, and a culture that had sent them to fight but seemed detached from their experiences. In telling these stories, then, *Signature Wounds* is a story of culture and cultural change.[41] It is a story about how the Army responded to the unanticipated realities of the Iraq and Afghanistan Wars, how those conditions prompted debate about what kind of culture the Army has and its place in American culture, and what the United States's role in the world should be.

To tell this story, I rely on a wide variety of sources, including publicly available documents produced by the Army, the VA, and the U.S. Congress, as well as media coverage in U.S. newspapers and broadcast journalism and popular texts like memoirs and novels. To describe the Army's and VA's internal efforts and debates, I turn to published medical literature, thousands of pages of documents received through the Freedom of Information Act, and more than three dozen interviews with Army leaders, clinicians, and researchers, as well as several sets of documents that some of those individuals provided.[42]

Together, these documents provide a comprehensive, but not a complete, account. Some important initiatives—most notably the Army Study to Assess Risk and Resilience in Servicemembers (Army STARRS) program, which sought to capture long-term longitudinal data on mental health—are only briefly mentioned, as are the development and use of particular treatments for treating PTSD.[43] The important issue of military sexual trauma also receives regrettably little attention in what follows. This complicated topic emerged as an issue of significant concern relatively late in the wars through increased media attention, the embrace of the issue by political leaders like Sen. Kirsten Gillibrand (D-NY), and the appearance of popular texts like Kirby Dick's 2012 documentary *The Invisible War*.[44] An issue of this complexity warrants sustained, book-length inquiry, particularly as sexual assault has become a more prominent issue in American life in the era of the #MeToo movement. As well, although I do mention some memoirs, novels, and films in the chapters that follow, *Signature Wounds* does not interrogate the range of cultural products about these wars that touch on mental health issues. Given the centrality of fiction and film to Americans' understandings of war, a comprehensive, book-length study of those cultural products is likewise necessary.

What follows here, though, is a comprehensive account of the political, public, and institutional debates over PTSD, TBI, and suicide during the Iraq and Afghanistan Wars, one that reveals how mental health became a significant terrain for debating the impact of these conflicts on U.S. culture. In chapter 1, I begin with the generation before the September 11 attacks, when the behavioral health doctrines that the Army would carry into Iraq and Afghanistan were first developed. Imagining that the next war would be a conventional conflict fought in Central Europe against the Soviet Union and would last only a few days, the Army imagined that most psychological casualties would result from battle fatigue, a condition that they understood as temporary and treatable. This vision informed the Army's development of the Combat Stress Control doctrine that shaped the Army Medical Department's approach to the 1990–1991 Gulf War and the twenty-first-century conflicts. Although this approach was well suited to treating acute combat-related mental health issues, the Army was unprepared for the psychological consequences of prolonged, asymmetrical warfare. The challenge that the Afghanistan and Iraq conflicts posed were compounded by a relative paucity of expertise in combat-related post-traumatic stress disorder inside the Army Medical Department.

The Army's and the nation's growing awareness that the twenty-first-century conflicts were creating significantly greater challenges than anticipated is the subject of chapter 2. I describe how Army researchers drew upon research capacities developed during peacekeeping deployments in the 1990s to organize and deploy Mental Health Advisory Teams (MHATs) to Iraq. The reports that these teams generated alerted both Army leaders and the public to the magnitude of soldiers' struggles and increasingly raised questions about the nature, pace, and frequency of deployments. By 2006, the MHAT reports had become documents that implicitly critiqued aspects of American military strategy and public attitudes toward the war. The increasing awareness of mental health issues in the force, however, also pitted Army researchers, clinicians, and leaders against politicians and a broader public increasingly restive about the Iraq war. While Medical Department researchers insisted that mental health issues were normal, temporary, and treatable, cultural understandings of post-traumatic stress disorder as permanently disabling and growing concern over the deployment of troops with behavioral health diagnoses fueled critiques that the wars were mismanaged and descending into chaos.

The intersection of debates over mental health and growing anti-war sentiment is likewise the subject of chapter 3. As Democrats took control on Congress in 2007, the Iraq War's critics increasingly seized on veteran suicide to support their opposition. Against VA officials' assertions that veteran suicide was not an epidemic and that the organization had the capacity to effectively treat veterans, accounts from anti-war legislators, grieving parents, and journalists presented the Iraq War as a disaster created by a disingenuous Bush administration that had forced idealistic young men to witness and commit atrocities and then failed to care for them. For the war's critics, the rising suicide rate and the perceived disregard for those who had taken their own lives was evidence of the Bush administration's coldhearted disregard for those it had sent to fight and evidence as to why the war must end.

Chapters 4 and 5 return attention to the Army. Beginning in 2007, a new generation of officers who had served in Iraq became the Army's senior leaders. Combined with the fallout from the scandal at the Walter Reed Army Medical Center, this created an increased focus on mental health and an awareness that the Army's culture needed to change. Chapter 4 describes how the Army worked to understand why soldiers were reluctant to seek care, to reduce the stigma that surrounded doing so, and to make care more available. In focusing on efforts to revise the process for discharging soldiers with mental health diagnoses, studying the links between mental health diagnoses and criminal behavior, and creating programs to embed mental health care in individual units and primary care settings, I illustrate that, while the Army's top-down efforts to prompt immediate change had inconsistent results, the grassroots efforts that changed individual leaders' and soldiers' attitudes produced slower-paced but more enduring cultural change.

The Army also struggled to meet the psychological needs of soldier families. The All-Volunteer Force has a significant number of dependents for whom the Army is responsible for providing care. The Iraq War strained those resources, and military families began reporting increased mental health issues including depression, substance abuse, and suicide attempts. After examining how these issues were presented in the media and by military partners on social media, in chapter 5 I contrast the efforts under Chief of Staff George Casey and Secretary Pete Geren to better serve families through the Army Family Covenant with the increasing agitation by activists and families, who claimed that the increased resources and the turn to resilience was too little, too late, given the strain that the wars had produced. Ultimately, this chapter illustrates how the army's efforts fell short of ensuring better mental health care for all families.

In chapter 6, I discuss accelerated efforts to understand traumatic brain injury. Although the Army had long studied concussion, before 2006 clinicians had generally followed civilian protocols, and very little research had explored whether the pressure waves that an explosion produced had the capacity to damage the brain. The uncertainty surrounding the causes and consequences of mild traumatic brain injury (mTBI) created significant controversy both among Army researchers and between the Medical Department and Army leadership. Some psychologists and psychiatrists pointed to research that concluded that post-traumatic stress symptoms were the most effective indicator of whether a soldier would have long-term issues. Army neurologists, in contrast, pointed to a growing body of research that suggested that blast waves did in fact damage the brain at a cellular level. Untangling when symptoms were attributable to PTSD and when they were the result of mTBI, particularly when an event like hitting an improvised explosive device might simultaneously precipitate both conditions, thus posed a major challenge.

Army researchers also struggled to understand how severely a blast would damage the brain, what protocols would protect soldiers, and which treatments were effective. These efforts sometimes generated acrimonious debates. So, too, did the Medical Department's insistence that any protocol or treatment be validated through rigorous testing before implementation increasingly frustrate some Army leaders, who criticized the Medical Department's academic approach as they sought ways to protect troops in the field. In this debate, the limited understanding of the brain pitted not only different academic approaches but also the pace of academic research and the demands of the wars against one another.

Chapter 7 focuses on the Army's efforts to address the rise in active-duty suicides. Although the Army had long prided itself on having a lower suicide rate than the country as a whole, in 2007 the number of soldiers taking their lives tipped above the national average.[45] The Army's efforts to understand why and lower the rate became, in effect, a forum on how the Iraq and Afghanistan Wars were challenging the Army's identity as an institution. In reports and briefings, career Army officers who had come of age in the post-Vietnam All-Volunteer Force expressed increasing anxiety that the nature of the twenty-first-century wars had changed the Army's culture, and not for the better. Lamenting that the new generation of leaders was so focused on combat deployments that they no longer fulfilled the previous generation's role of mentoring young soldiers, Army officials who established Army suicide prevention efforts implicitly critiqued the wars, and the country that demanded they fight them, as they worried that the Army that leaders had worked so hard to create after Vietnam had been irrevocably damaged by the wars. Suicide prevention discourse was thus a venue in which some Army officials came as close as they could to publicly critiquing the wars and the demands that a disengaged nation had put on the All-Volunteer Force.

Chapter 8 maintains a focus on suicide, asking how the Department of Veterans Affairs worked to address the rising number of veterans taking their own lives. In challenging conventional notions that the VA was a disengaged bureaucracy, I detail the extraordinary efforts of VA officials to adopt a public health approach to understanding and preventing veteran suicide. Following the passage of the 2007 Joshua Omvig Veterans Suicide Prevention Act—the result of the activism described in chapter 3—the VA embarked on an ambitious epidemiological research project, through which it determined that twenty-two veterans took their lives each day, and then the VA focused on addressing suicide risk factors by creating the Veterans' Suicide Prevention Hotline and on instigating measures to prevent suicidal veterans from accessing firearms. These efforts proved controversial. Although they succeeded in saving the lives of many veterans, the VA continued to receive criticism. Individual cases of failures to successfully save a veteran's life contributed to an enduring public narrative that the VA was not doing enough to stem what many Americans increasingly perceived to be an epidemic. More important, though, the VA's public health efforts, particularly its efforts to help veterans secure their guns, were criticized by conservative politicians and pundits who accused the organization of abridging veterans' constitutional rights.

Signature Wounds concludes with an analysis of the Army's turn to resilience and the Comprehensive Soldier Fitness Program. This program, like every other, proved controversial. If some Army leaders and researchers were convinced that mental health issues could be reduced if the force became more resilient, others worried that the emphasis on resilience took attention away from the continuing need to treat soldiers' and families' mental health issues and that it unnecessarily sacrificed the successes of programs that had already proven their effectiveness.

* * *

The story of how researchers, soldiers, military leaders, legislators, and ordinary Americans thought about mental health amid the United States' twenty-first-century wars is a story of institutional and cultural change. Throughout this book, two overarching themes emerge. The first is that, when faced with wars that were longer and more violent than the they had anticipated, that put unprecedented demands on the

All-Volunteer Force, and that introduced new medical challenges about which there was profound uncertainty, both the Army and the VA were more aggressive and more progressive in attempting to address mental health issues than popular discourse has generally recognized. Although there were vigorous, and at times heated, debates over how best to care for soldiers and veterans, and while there were some profound failures, both institutions adamantly sought to better prevent, diagnose, and treat mental health issues and to change the culture surrounding mental health in the Army and the nation. Second, the debates over mental health during these wars were as much cultural as they were medical. Americans both within and outside of the military used their understandings of mental health issues like PTSD, TBI, and suicide to ask questions about the conflicts themselves, the experiences of the men and women fighting them and their families, the relationship between the military and the nation, and the United States' place in the world.

The story of those efforts and debates matters as we look back at more than fifteen years of conflict; as U.S. troops remain deployed in combat roles in the Middle East, in Afghanistan, in Africa, and elsewhere; as their families struggle with the stress and anxiety of their loved ones' absences; as veterans continue to take their lives at an alarming rate; and as new military threats loom on the horizon. Understanding the psychological impact of the nation's longest wars prompts us to ask not only what resources we must commit and what capacities we must build to ensure the well-being of those who serve but also under what circumstances we should expose human beings to circumstances that place their bodies and minds at such great risk.

1

"At the Time People Hadn't Been Asking Those Sorts of Questions"

Army Mental Health Research between Vietnam and Iraq

On September 3, 1981, twenty-five Army psychiatrists and psychologists settled into a classroom at Fort Sam Houston's Academy of Health Sciences in San Antonio, Texas, to chart the future of the Army mental health. Most of the seven majors and eight captains there had likely not been in the Army long enough to have served in the war in Vietnam, from which the United States had withdrawn combat forces in 1973.[1] The ten colonels and lieutenant colonels there might have been among the relatively few psychiatrists who deployed to Southeast Asia, but even if they had, they likely spent the bulk of their time treating the rampant substance abuse issues that troops faced.[2] Very few had significant experience with combat stress, much less post-traumatic stress disorder, a condition that had been included for the first time in the previous year's revision of the American Psychiatric Association's *Diagnostic and Statistical Manual of Mental Disorders* (*DSM-III*).[3]

They were all, however, part of an institution still recuperating from the war. In the late 1970s, Americans worried about a "hollow Army" populated by soldiers who were unprepared for modern warfare.[4] That summer, the *New York Times* had reported that the Army was struggling to recruit qualified soldiers and that officers in the U.S. Seventh Army in West Germany were finding that "many of the soldiers coming . . . from training in the United States cannot read or write or do simple arithmetic."[5] For those ill-prepared soldiers, a tour of duty in Germany was hardly an adventure.[6] During the day, they trained on outdated equipment; at night, they slept in dilapidated barracks. "In most American prisons," the *Times* wrote in May, "such working and living conditions would probably lead to riots."[7] Morale, unsurprisingly, was low.[8]

These issues mattered greatly. After Vietnam, the U.S. military turned away from fighting insurgencies in the developing world and focused on building capacity to fight a conventional war to counter potential Soviet aggression in Europe, and the Seventh Army was the bellwether of the United States's ability to meet that challenge.[9] For Ronald Reagan, who would oversee a massive increase in defense spending and the recuperation of the U.S. armed forces, that threat was not notional. Telling the West German Bundestag in 1982 that "we're menaced by a power that openly condemns our values and answers our restraint with a relentless military buildup," Reagan called for "the presence of well-equipped and trained forces on Europe" and promised "a national effort . . . to make long-overdue improvements in our military posture."[10]

This anxiety and these promises benefited the U.S. Army. As Andrew Bacevich explains, "Preparing to fight Russians was . . . the ready-made answer to every question essential to institutional recovery and continual health."[11] Reagan's increased defense spending led to massive modernization of equipment, and the Army finally figured out how to recruit quality soldiers.[12] The Seventh Army alone saw an increase of $239 million in its budget in 1983, and new weapons like the M-1 Abrams tank and the Bradley fighting vehicle poured into its arsenal.[13] By the fall of 1984, soldiers in the Seventh Army were boasting of the unit's morale and capabilities, and simulations showed that the Army was capable of defeating a Soviet advance.[14] An entire generation of Army armor and infantry officers—those soldiers who became senior leaders in the twenty-first century—made their careers by demonstrating their ability to lead troops in an imagined battle on the plains of Central Europe.[15]

So it was in the Army Medical Department, or AMEDD, as well. Medical Department researchers contemplated the likely psychological consequences of the brutal, brief war with the Soviet Union that the larger Army imagined and studied how they could be prevented, mitigated, and treated. Two major schools of thought approached this question. At the Walter Reed Army Institute of Research (WRAIR) outside of Washington, DC, a group of researchers led by Dave Marlowe investigated how soldiers could be made more resilient, focusing on issues including the creation of cohesive units, the impact of strong leadership, and the effects of sleep deprivation. Marlowe was a World War II infantry veteran and Harvard-trained anthropologist who was, one colleague explained, the kind of man who "could spend hours watching you make soup, just to understand something about you."[16] The WRAIR faction thus embraced his anthropological approach and viewed combat stress somewhat differently than a group at the AMEDD Center and School in San Antonio. This was led by A. David Mangelsdorff, a reservist and civilian psychologist for the Army's Medical School who had taken an interest in how soldiers could become better able to cope with adversity, and Col. James Stokes, an Army psychiatrist. This group focused primarily on combat stress and battle fatigue—a temporary but treatable ailment brought about by the overwhelming violence and chaos of combat—and took a traditional psychomedical approach.[17] According to Col. James Martin, who served under Marlowe and led WRAIR's research unit in Heidelberg, Germany, the two groups were akin to "people with different religious beliefs," which meant that "there were profound differences in their approach" because "we each had a different view of the elephant."[18]

The research conducted by each group and its implementation in the 1980s and 1990s offers an important window into the assumptions and capabilities regarding mental health that the Army brought into the twenty-first century. In particular, the belief that the Army would henceforth fight short, intense conflicts led to a focus on combat stress and battle fatigue that, while clinically valid, did not account for the possibility of protracted irregular warfare that would require individual soldiers to deploy multiple times. Perhaps more important, when the Army did fight in the 1980s and 1990s, the Army persistently struggled to effectively implement its Combat Stress Control program. The combined failure to prepare for lengthy conflicts and to effectively apply the doctrine that it was developing was joined by both a shrinking of the AMEDD's mental health capabilities after the Cold War and a wider Army culture that disdained mental health care. Despite important successes, like the creation of a combat stress control doctrine and the development of a capacity to survey the well-being of deployed troops, these factors left the AMEDD largely unprepared for, and having to learn in real time how to address, many of the challenges that the twenty-first-century wars would present.

* * *

Four decades later, it is perhaps difficult to recall the seriousness with which U.S. and NATO military planners accorded a potential Warsaw

Pact invasion of Western Europe. In 1981, the RAND Corporation, the most prominent defense think tank, issued a white paper that tallied the number of Soviet divisions in East Germany, Poland, and Czechoslovakia at thirty-one, with another three dozen in the western Soviet Union.[19] Although not all of these were combat ready, the balance of forces between the two sides was grim; the Warsaw Pact had more troops, more tanks, more artillery, and more aircraft.[20] The report argued that, "in an attack against NATO, 36 divisions could be used in the first wave with the remaining 18 in a second echelon about 72 hours later."[21] That meant that about half a million Soviet troops would pour in Western Europe in an invasion's first days.[22]

Given these figures, U.S. Army leaders envisioned a nightmare scenario if war came. At the first Users' Workshop on Combat Stress, Maj. Raymond Keller told the assembled psychiatrists and psychologists that such a war would be "continuous battle against a seemingly unending stream of fresh enemy forces . . . [that will] provide for no respite from combat" and that "tactical units will be walled off from their support base . . . by conventional and nuclear munitions as well as chemical agents, surrounded and then destroyed in detail."[23] A year later, Maj. William H. Thornton offered a similarly bleak assessment: a Soviet invasion of Central Europe promised "perhaps the most difficult conditions ever faced by American soldiers" and that "there will be neither rest nor respite from the terrors of battle."[24]

However terrible they assumed the war would be, though, these presenters also assumed that it would be brief. Assessments of the Soviet doctrine assumed that "the Soviets hope to produce a swift and sudden collapse of NATO, conquer Europe, and sue for peace" and that they would use nuclear weapons at most "after 1–2 days."[25] Another prediction, based on the assumed similarities between the 1973 Arab-Israeli War and a potential war against the Soviet Union, estimated that the conflict would last three or four days.[26]

These assumptions shaped the Army's approach to mental health. Army psychologists and psychiatrists maintained that combat inevitably produces psychological changes and that "combat stress reactions" could be either beneficial or debilitating. A solider, for example, might do something uncharacteristically heroic; by the same token, he might suffer a psychotic break.[27] Of course, it was combat's "uncomfortable or performance degrading" effects, which they grouped under the term "battle fatigue," that worried them.[28] Combating and treating battle fatigue thus became Army mental health's primary preoccupation in the Cold War's final years. Foremost, the Army sought to minimize battle fatigue casualties. Under the leadership of Gen. Maxwell Thurman, the demanding workaholic leader who as Army Vice Chief of Staff played a defining role in the organization's post-Vietnam reccuperation, the Army focused on building healthy units—ones in which quality, well-trained soldiers trusted their leaders and colleagues, had stable relationships outside of their units, and felt prepared for their mission.[29] Throughout the 1980s, *Military Review: The Professional Journal of the US Army*, which the Command and General Staff College at Fort Leavenworth published, featured articles with titles like "Soldiers: They Deserve Good Leadership" and "Leadership Challenges on the Nuclear Battlefield."[30] Imagining a battlefield in which the Soviets would deploy tactical nuclear weapons, Lt. Col. Jeffrey L. House, the author of the latter article, argued that "an additional decrease in combat effectiveness can be anticipated from the psychological changes in soldiers" and thus that "we will need leaders who have the thinking skills, moral courage and initiative to form disciplined, cohesive units that are capable of fighting and winning in such an environment."[31] Another piece argued for training under realistically stressful conditions.[32] In the field, ensuring that soldiers were well rested was critical because studies showed that soldiers quickly became unable to perform their missions if they got less than four hours of sleep.[33] A 1986 "Leader Actions for Battle Fatigue" graphical training aid counseled, "Never waste a chance for sleep."[34]

Army researchers knew, however, that battle fatigue was not entirely preventable. Rather, given the war that they expected to fight, one presenter anticipated that over half of the casualties "will be psychiatric casualties within the FIRST 48 hours," while another predicted that "high stress conditions will prevail and make <u>every</u> soldier susceptible of becoming a psychological casualty," to the point that "medical facilities will be overwhelmed and engaged forces depleted."[35]

These predictions were not, however, as dire as they might have seemed, for the prevailing medical opinion was that the condition was both short-lived and easily treatable. Central to the Army's treatment protocol were the principles of proximity, immediacy, and expectation, or PIE, which held that battle fatigue casualties stood the best chance of recovery if they were treated promptly near the front lines, rather than being evacuated to the rear, and were told that they would return to their units shortly.[36] This concept, of course, was hardly original to the 1980s; it was, as Col. James Stokes, an AMEDD psychiatrist deeply involved in the creation of the Combat Stress Control doctrine, stated, "an understood and well-proved treatment" that had been combat psychiatry's driving assumption throughout the twentieth century.[37] The Army's protocols in the 1980s thus followed six decades of thinking on matters of military psychiatry. The correct prescription for a soldier "staring into space . . . and unable to carry out his duties or sleep [because] one of his friends had been killed" was "sleep (sedate if necessary). Food and liquid, shower and shave, stress normalcy not illness, and do not evacuate to rear, and return to full duty."[38]

The Army sought to communicate this view beyond mental health providers and to train soldiers that battle fatigue was inevitable but not permanently debilitating. In 1985, for example, the First Cavalry Division's Combat Stress Course was promulgated on the idea that "any 'normal' soldier could be expected to become a battle fatigue casualty . . . if exposed to enough stresses in combat," but the course explained that it was "a temporary break down . . . that improved in one to three days with adequate rest."[39] A year later, Thurman issued a memo standardizing combat stress reaction training and tasked Stokes with creating a set of pocket-sized cards detailing facts about battle fatigue, with the intention that unit leaders would review them with their men during breaks in field training exercises.[40] The cards featured Stokes's pen-and-ink drawing of a clearly fatigued soldier and taught leaders, for example, that battle fatigue was to be expected but that "BFCs [Battle Fatigue Casualties] can be restored to duty quickly if rested close to their units and treated positively" and that "a good soldier will be a good soldier again."[41] Enlisted men were similarly told that, "although battle fatigue can take different forms, it is still only battle fatigue"; that "it gets better quickly with rest"; and that, if a colleague were evacuated, he or she should be reminded "that the team counts on them to come back quickly."[42] Another set of cards offered scenarios that soldiers could use to role-play combat stress reactions.[43] The cards offered a range of scenarios, including "Anxiety, Used Up Luck," in which a soldier was instructed to "act Tense and Jittery, but Tired" and "Talk endlessly about the bad events in the exercise's last minutes/hours/days, like FRIENDS or LEADERS KIA or WIA."[44] Another card instructed a soldier to "tell the medic or sergeant your stomach hurts," to similarly describe horrors that the soldier had seen, and to "worry about someone back home (pick one: sick parent or child, pregnant wife, girlfriend, newborn baby)."[45] Soldiers were told to "Improve only if buddy, medic, leader say you have battle fatigue" and follow the proper protocols.[46]

What is striking, given the form that the twenty-first-century wars would take, is that the Army did not differentiate between the impact that short deployments and lengthier ones were likely to have, nor did it differentiate in the treatment that they required. Rather, in a paper distributed at the 1983 User's Workshop, Stokes compared battle fatigue to physical exhaustion, noting that, just as "runners can be temporarily 'exhausted' pushing too hard in a 100-yard dash or in a 27-mile marathon" and that "both cases look alike, lying on the ground gasping for breath" and that "the treatment is the same: get them up, walking around; cool off; and replenish fluids," soldiers suffering from battle fatigue had the same symptoms regardless of how long they had been in combat and required the same treatment. "Soldiers can be temporarily emotionally overloaded in a few seconds of horrifying combat or in days or weeks of less intensive experience," he wrote, adding that, "in most cases, the treatment is the same: sleep, replenishment, hygiene, structured military activities, supportive psychotherapy or counseling and positive suggestion."[47] Army psychiatrists were confident that this treatment protocol would be effective and that about three-quarters of soldiers who received it would return to effectiveness promptly.[48]

There was little consideration that repeated deployments to combat were likely to result in significant negative outcomes. Nor was there much attention to post-traumatic stress disorder at these workshops. In 1981—a year after the American Psychiatric Association had recognized PTSD—one colonel offered a "historical review" of combat stress and asked the audience, "What is 'Viet Nam Veterans Syndrome'?"[49] It was not until 1985 that the workshop included a paper on the topic.[50]

This omission likely reflected Army psychologists' belief that that effectively treating combat stress would prevent long-term psychological problems. A consistent idea across the workshops was that making too much of battle fatigue might precipitate permanent disability. In the first workshop, for example, one paper specified the familiar prescription for adverse combat but warned that, "if effective treatment is not applied, then the anxiety will crystalize along the lines of a premorbid personality and there is a high likelihood that the soldier will become permanently disabled."[51] A year later, a training protocol presented the example of a catatonic nineteen-year-old infantryman who "should be returned to duty at his unit immediately" because "if the seed of suspicion of mental illness is allowed to germinate . . . [he] is quite likely to become permanently impaired by anxiety and self-doubt."[52]

* * *

From the distance of thirty years, half of which have seen the U.S. military immersed in prolonged, asymmetrical conflicts, these assumptions might seem overly simplistic. However, it is important to note that they were in keeping with the best research of the moment as well as with a long history of wartime psychiatry. The Army, like the rest of U.S. culture, had an incomplete understanding of conditions like the newly recognized post-traumatic stress disorder.[53] Similarly, it was not until the 1990s that the National Vietnam Veterans' Readjustment Study provided significant insight into the long-term psychological impact of combat exposure.[54] It thus remained unsettled exactly how combat exposure and PTSD were related, and in the absence of a consensus, the Army chose a protocol that they believed would eliminate many of combat's long-term effects.

Embracing combat stress control was one thing. Implementing it was another. To Stokes's frustration, the AMEDD was not quick to elevate "Combat Stress Control"—a name that he and his colleagues chose to emphasize that preventing battle fatigue was "a leadership function"—to the level of Army doctrine.[55] Although he and others had been talking about combat stress control throughout the 1980s, it wasn't until 1994 that the Army issued Field Manual 22-51, *Leaders' Manual for Combat Stress*. Throughout the 1980s, Stokes explains, efforts to organize behavioral health units were largely stymied. In his view, this was because the AMEDD didn't value mental health or want to commit resources to it. "We had the lowest priority," he recalls, referring to the competition over resources inside the AMEDD.[56] "Everything that went to combat stress," he explains, "was not going to go to . . . whatever they regarded as more important than this psychobabble."[57] This was perhaps because, as Martin puts it, the WRAIR researchers and those at San Antonio were "loyal to different patrons": Stokes's group was solely dependent on getting money from the AMEDD budget while the WRAIR scientists, Martin explains, received considerable additional money by contracting with Army commands interested in studying particular issues.[58] This meant that, while Stokes was "in a staff office in San Antonio, siloed in the Surgeon General's office," Dave Marlowe had "unparalleled access to the operational Army as well as important [relationships] with multiple generations of senior Army leaders."[59] As Martin recalls, "The part that would drive the folks in San Antonio nuts was that they were dealing with the Surgeon General, and Dave Marlowe would be in with the Army's Chief of Staff, people with real political power."[60]

As a result, combat stress control was never effectively implemented in Grenada, Panama, or Operation Desert Shield and Operation Desert Storm. In the 1983 invasion of Grenada, the Army did not deploy mental health providers—according to Stokes they were on the tarmac ready to board a plane, but they were not on the manifest and were left behind.[61] Moreover, the training that medics had received seemed not to have had much impact. Army medics, one study found, "were familiar with the term ["combat stress reaction"], but they didn't really know the symptoms or the treatment" and hadn't received much training on it; in fact, "only one medic had an accurate idea of how to diagnose and treat psychiatric casualties."[62] These after-action assessments pointed to an issue that the Army would confront throughout the rest of the twentieth century and into the twenty-first. They may have developed a doctrine that research suggested would be effective, but implementing it remained a challenge.

Even more discouraging was the 1991 Gulf War, which was about as close to the imagined war with the Soviets as the Army could have hoped.[63] Before the war began, WRAIR social scientists led by Marlowe had deployed to the Persian Gulf to assess the well-being of troops deployed for Operation Desert Shield. During the last week of September and the first week of October 1990, Marlowe, Martin, and Robert K. Gifford, another WRAIR psychiatrist, surveyed hundreds of soldiers and found that, for all of the rhetoric about combat or Saddam Hussein's chemical and biological weapons, "the greatest stressor cited by soldiers is the lack of a definite tour length."[64] They also found that soldiers were frustrated by the close quarters in which they lived, by limited access to the outside world in the form of both phone calls home and news, and by what Marlowe's report pointed out was a "cultural and social isolation of soldiers in Operation Desert Shield [that] is as extreme as for any deployment in American military history" and where "there is no place where they can dress and act like Americans."[65] These issues became more significant "the further forward one went," where soldiers consistently complained about the comforts that they knew or imagined the "REMFs [Rear-Echelon Mother Fuckers]" were getting.[66]

Soldiers were also frustrated by some "overzealous interpretations of guidelines" intended to keep troops from breaking Saudi cultural norms, including, in one case, a battalion in which "soldiers were required to turn in or destroy any pictures of their girlfriends that showed arms or legs," an act that Marlowe pointed out "is a clear overreaction to Saudi standards" because "the Saudis do not much care what pictures soldiers have in their wallets or by their cots."[67] In their view, some leaders were too worried that soldier behavior would devolve as it had in Vietnam, and they were at pains to point out that "the challenge for leaders in the Desert Shield theater is to find ways to take care of soldiers that take advantage of the improved quality and motivation of the Army."[68] Indeed, the report's primary conclusion was that effective leaders who sought to improve soldiers' living conditions was a key to soldier well-being.[69]

Marlowe presented the report to Chief of Staff Gen. Carl Vuono and Vice Chief Gen. Gordon Sullivan on December 17, about a month before Operation Desert Storm began. At the conclusion of the briefing, Vuono asked Marlowe to produce another report, this one on post-deployment issues in the event of combat.[70] Marlowe and his staff largely echoed the conventional thinking on combat stress, pointing out the need to help "soldiers who have participated in combat, and who have been exposed to the trauma of war . . . be aware of the possible normal responses to such exposure," adding that after Operation Just Cause in Panama many soldiers lacked this knowledge and thus "were unnecessarily in pain because they did not know that recurrent nightmares, occasional awakening in cold sweats, and recurrent images of horror and death were normal consequences of combat participation, not signs of incipient breakdown in otherwise effectively functioning human beings."[71]

In keeping with Marlowe's anthropological perspective, however, the report went beyond the principles of combat stress that were rooted in psychiatry to suggest some ways in which post-war symptoms could be ameliorated by creating appropriate public opportunities for soldiers to process what they had experienced. They explained that units that had been in combat or in which soldiers had died needed debriefings, soldiers needed information about friends who had been hurt, and that injured soldiers needed to return to their units "to complete the appropriate psychological transition" even if they were going to be discharged.[72] All of these efforts aimed to reduce the likelihood that soldiers would suffer enduring trauma, but Marlowe and his team nonetheless argued that leaders must become more open to the reality that some troops would need post-deployment counseling. In the aftermath of the Panama invasion, they wrote, "A few commanders tried to prevent their soldiers from exposure to mental health and helping personnel," which "led to prolonging normal post combat symptoms for some soldiers who feared that they were going crazy when they had nightmares."[73]

An implicit effort to prevent a repetition of the perceived cultural failure to recognize the service and sacrifice of Vietnam veterans—and thus, perhaps, to prevent the enduring trauma understood to have come from that lack of recognition—also runs through the report.[74] The debriefings, it states, should emulate those conducted by S. L. A. Marshall after the Second World War, and soldiers should be given "valid and truthful information that will help them cope with" comments from "far fringe 'anti-war' groups" that might accuse them of "'baby killing.'"[75] They counseled "keep[ing] the soldiers together for as long as possible" in order to "accomplish the (psychosocial) work required to achieve a recovery context for the restoration of soldiers," an idea that sought to replicate soldiers' return from World War II and reject the way soldiers came home from Vietnam.[76]

Marlowe and his staff also recommended that all soldiers, not only those who had been in combat, needed to have their contributions recognized, and they proposed an award akin to the Combat Action Badge that the Army would create in 2005 that would "signify the soldier's special value in his or her performance in the combat environment."[77] They also counseled that leaders ensure that soldiers felt that their efforts and sacrifices were worthwhile and appreciated, and they suggested that "some effort should be made, at levels above the Army, to get the Saudis and Kuwaitis to publicly show their appreciation to our soldiers for what they have done."[78] They also suggested that "each welcome home parade should involve the entire community and, wherever possible, end with a community-wide picnic and celebration . . . in the manner of the massive community block parties that took place after WWII."[79] When it came to ensuring soldiers' healthy readjustment, Marlowe and his colleagues asserted, anything that replicated World War II was good; anything that echoed Vietnam must not be repeated.

Many of these points were reiterated in other memoranda that Marlowe prepared for commanders in the Middle East during the war, which indicates the degree to which the Army entered Desert Storm thinking about the potential mental health outcomes.[80] Mobilizing personnel to address them, however, turned out to be a challenge. The 528th CSC detachment at Fort Bragg, for example, was tasked to deploy with the 82nd Airborne, but Stokes found them in abysmal shape. Their vehicles hadn't been maintained and all had flat tires, and they lacked a commander.[81]

Stokes maintains that these units eventually reached the point that they could deploy successfully and that "Combat Stress Control was one of the great success stories" of the war, an assessment with which Martin—who notes, "I was there, and he wasn't"—takes issue.[82] When Martin arrived in Saudi Arabia, he discovered that "most of these folks had limited military training" and that they had neither "credibility in the units they were serving" nor "a lot of self-confidence."[83] Planning was poor, mental health providers were stretched thin and often undertrained and often did not or could not make an effort to reach out to units to provide care.[84] In one case, Martin arrived at a unit that had suffered friendly fire casualties two days earlier and found that "no one with a mental health background had been there to talk to them."[85] Just before the ground war began, he and WRAIR psychiatrist Greg Belenky discovered that a 13,000-person Cavalry unit "preparing to spearhead the advance into Iraq" had "no organic mental health resources."[86] He and Belenky cobbled together a unit of their own and accompanied them during the ground war to provide mental health support.[87]

Documents from WRAIR echo Martin's criticisms. One report begins by promising "lessons learned" that would "focus on structural and operational issues that need correction" before archly adding, "The list is long."[88] The report concluded that "we obviously have a psychologically healthy Army" but also offered a blistering assessment of mental health services during the war.[89] "The organization, staffing, equipment, and training of mental health units deployed to Southwest Asia were severely inadequate," it explained, and "many division mental health personnel who deployed to ODS [Operation Desert Storm] lacked 'credibility' with battalion and brigade commanders."[90] The report noted that mental health providers "often lack basic soldier skills" and are thus not taken seriously by commanders, and it specifically pointed to "two reserve psychiatrists whose poor physical conditions and ages (58 and 62), and lack of military experience and training made them unsuitable."[91] The report also noted that some positions were filled by reservists "who lacked clinical credentials and had not practiced in their respective fields for many years."[92] In WRAIR's assessment, mental health care in the Gulf War was anything but a success story.

At the final Users Workshop on Combat Stress, Martin offered a withering critique of the Army's mental health treatment, noting the good fortune that the Iraqi army had not fought the promised Mother of All Battles, for "if Saddam Hussein had been able to deliver on his chemical and biological warfare threats . . . the total number of combat stress casualties could have easily overwhelmed the division and corps-level mental health resources."[93] Martin's conclusion was no less grim and equally damning. "If significant casualties had occurred," he warned, "these teams would have found it very difficult to carry out their mission. They were not adequately staffed, equipped, or trained in peacetime to perform their wartime role."[94]

This troubling assessment of behavioral health care on the ground was joined in the next few years by reports that the Persian Gulf War had left many veterans struggling with mental health issues. Martin's initial research showed that "about 10%" of the soldiers that he had surveyed "'often' experienced one or more signs of psychological stress related to their combat experiences."[95] In 1992, the VA found that 9 percent of veterans qualified for a PTSD diagnosis and that a third "appear to have experienced other forms of psychological distress during the months after their return from the Middle East."[96] If 9 percent was much lower than the numbers associated with Vietnam veterans, it still alarmed the study's authors, who described "a specific relationship between service in the Persian Gulf war and psychological distress."[97] Five years later, another study reported that an increasing number of veterans showed signs of PTSD.[98] These statistics suggested that, whether the Army's doctrine was mistaken or whether it hadn't committed enough resources to the problem, it had failed to protect a significant number of soldiers from lingering mental health issues.

As the WRAIR reports show, the failure to successfully operationalize a combat stress control doctrine was partly the result of poor implementation on the AMEDD's part and partly the result of leaders' failures to take mental health—and mental health providers—seriously. Indeed, for many Army officers coming of age in the post-Vietnam era, mental health was hardly a thought and rarely a priority. In this sense, the Army is not different from the rest of U.S. culture, where the stigmatization of mental health issues has always been significant. Researchers in 1997 determined that 70 percent of people diagnosed with a mental illness found that "they had at least sometimes been treated as less competent by others once their illness was known," while a third found that "co-workers and supervisors were seldom or never supportive or accommodating when they learned about the respondent's mental illness."[99] In the Army's competitive, Type A, and hypermasculine culture, these issues were likely more prevalent.[100]

Indeed, several soldiers who became prominent advocates of improved mental health care recall how little they knew about it or how little regard it enjoyed early in their careers. When George Casey, who in 2007 would become Army Chief of Staff, arrived at his first post in West Germany in 1972 as a freshly commissioned second lieutenant, his primary problem was that one of his men consistently showed up to formation drunk, but he never considered that the sergeant's combat experiences might have been caused by post-traumatic stress.[101] Michael S. Tucker, who as a brigadier general would organize the Army's Warrior Transition Units in 2007 and who joined the Army as an enlisted soldier in 1972, recalls that, in the quarter century after Vietnam, "I don't know if we were even conscientious about mental health regarding battlefield injury."[102] While troubled soldiers were referred to mental health care, "We never considered that mental health was an illness that was the result of exposure to combat situations. And to be honest with you, it was seen as a weakness, not an illness. It wasn't macho. It wasn't 'Hooah.' Nobody ever wanted to admit it for those types of reasons."[103]

Rebecca Porter had more direct experiences with commanders' disdain for mental health. A former military police officer who had returned to the reserves in order to earn a doctorate in clinical psychology, Porter would eventually become the Army's Director of Psychological Health. But in the late 1990s, she was Chief of Community Mental Health at Fort Bliss, Texas, where soldiers were rotating in and out of Saudi Arabia to enforce the No-Fly Zones in Iraq. Commanders there, she recalls, had little use for her:

> When I first started as a psychologist, in '95, we could go to a war game and there would be no mental health play in the whole thing. You could sit in your tent for a couple of days waiting for somebody to show up with shell shock or something. And when I would try and talk to commanders it was kind of like, "Hey, you know, unless he's going to kill himself or somebody else, your services are not required, nor is your opinion or recommendation."[104]

In one instance, Porter called a commander to tell him that she had determined that a woman suffering from PTSD after a car accident was not deployable to Saudi Arabia, and the commander hung up on her.[105]

To make matters worse, the AMEDD as a whole, and its mental health component in particular, had gotten smaller, and less attention was being paid to preparing for deployment-related health issues that soldiers were likely to face because of a deployment. As the historian John Sloan Brown explains, in the immediate aftermath of Operation Desert Storm, "Budget cuts, manpower drawdowns, and congressional pursuit of a 'peace dividend' were not forecasts; they were crippling contemporary realities."[106] Those cuts disproportionately hit the Medical Department. "The combat arms draw down after the Cold War was not nearly as severe as all of the service support units," Army psychiatrist Carl Castro recalls, pointing out that medical care was often pushed into reserve units.[107] Mental health took particularly severe cuts, to the point, Castro recalls, that "we didn't even have enough psychologists and social workers to meet the needs of our service members."[108]

Moreover, the psychological care that the AMEDD was providing had little to with deployment- or combat-related stress. Col. Elspeth Cameron "Cam" Ritchie describes the work she did in the 1990s as "a lot of bread-and-butter psychiatry. . . . Most of the time you had adjustment disorders, marital issues, and occasional bipolar or psychotic person . . . and so you had some PTSD, but relatively little."[109] By the time the Iraq War started in 2003, most veterans of earlier wars who had a psychological issue had left the Army and entered the VA medical system, and the Army was neither prepared to treat nor interested in treating PTSD.[110] As Carl Castro asserts, the attitude inside the AMEDD was "if someone has PTSD, we will just kick them out of the military. They can go to the VA. . . . [The plan] wasn't for you to stay, to treat you and return you to duty. It was to separate you out and get somebody else."[111] As a result, treating PTSD was not an AMEDD priority. Rebecca Porter recalls that the attitude was that "PTSD was something Vietnam vets had, and the VA dealt with that" and that, as a result, "it wasn't something that we felt had to be at the center of our tool box."[112] Charles Engel, a WRAIR researcher who would later develop a method for screening soldiers for mental health issues during primary care visits, makes a similar point, recalling that, in the first few years of the Afghanistan War, "something like only 2–3 percent of primary care clinics in the military were screening for depression. And of course none of them really were screening for PTSD."[113] Castro concurs, arguing that, because there were essentially "no PTSD [cases], clinicians in the military didn't know how to treat people with PTSD. . . . [On] 9/11, maybe 10, 15 percent [of clinicians] knew how to do it."[114]

* * *

Army researchers did take combat stress seriously in the 1980s and early 1990s, but their efforts were neither entirely effective nor sufficient preparation for the twenty-first-century conflicts. Army researchers developed combat stress protocols modeled on notions of battle fatigue that originated in the early twentieth century, largely forgot any lessons that Vietnam might have offered, and focused instead on the war that the United States thought—or perhaps hoped—that it would fight. Nowhere in the literature is there any meaningful consideration of the trauma that a lengthy war, a counterinsurgency, or a non-linear conflict might produce. The Army also did not effectively operationalize combat stress treatment, leaving units that were equipped to treat soldiers sidelined and deploying poorly prepared clinicians. Finally, the culture of the Army neither appreciated mental health care nor committed resources to mental health issues in the force. Together, these factors left the AMEDD unprepared for the kinds of wars the United States would fight in Iraq and Afghanistan.

* * *

When Amy Adler was a doctoral student in clinical psychology at the University of Kansas, the last place she probably imagined sleeping was on the runway of Croatia's Zagreb airport. Nonetheless, there she was, in the middle of a United Nations tent city, in mid-December 1992. A month earlier, the Army's 212th Mobile Army Surgical Hospital had set up a clinic to treat UN peacekeepers from five other countries who were tasked with alleviating the suffering of Bosnians trapped in the ethnic conflict that had followed the collapse of the former Yugoslavia.[115] This deployment would be the first of many in the 1990s in which the Army, transitioning from Cold War missions that anticipated a war with the Soviet Union, increasingly committed troops to humanitarian missions aimed at peacekeeping, democracy promotion, and famine and disaster relief.

Adler was not in Croatia to care for UN peacekeepers. She had moved to Germany in the early 1990s and worked as a clinical psychologist at the Army's 97th General Hospital in Frankfurt, and when that hospital closed James Martin hired her as a civilian researcher at the Army's Medical Research Unit–Europe, a division of the Walter Reed Army Institute of Research.[116] Because units from Germany were the primary ones deploying to the Balkans, conducting research that would help commanders in the field and at the Army's senior levels better understand how deployments affected troops became one of her unit's primary missions.

Throughout the 1990s, WRAIR researchers deployed Human Dimensions Research Teams of psychologists, social workers, and other researchers like the one that Adler joined in Croatia to the Balkans, Somalia, Haiti, and the Middle East to survey soldier well-being. These missions were important to behavioral health research in the twenty-first century for two reasons. First, WRAIR researchers developed a methodology for real-time research that would shape their approach to the Iraq and Afghanistan Wars. Second, they gained a reputation and built relationships with particular commanders in the 1990s that led to their recommendations being taken seriously when they briefed leaders in the 2000s.

Sending researchers into the field was Dave Marlowe's idea—in fact, he, James Martin, and Robert Gifford had deployed on the first one, to the Middle East, in the fall of 1990—and his training and belief in observational fieldwork gave the deployments the flavor of anthropological expeditions. The model, Adler explains, was rooted in a simple assumption that, "if you want to find out what is happening with soldiers, you actually have to go and ask them."[117] Small teams of WRAIR researchers would deploy for short periods—in Croatia, two people made four visits of about a week, and the three-person team that deployed to Operation Restore Hope in Somalia in 1993 was in country for just over a month—and they would sleep in the same tents, eat the same food, and brave the same elements as the soldiers they studied.[118] Often the conditions were less than ideal, and certainly a departure from what the research psychologists were used to. Carl Castro, who commanded WRAIR's medical research unit in Germany in the late 1990s and deployed on four such research teams, recalls his Kosovo deployments as "so goddamned cold it was unbelievable," to the point that he was thankful that the Army required that even its psychologists stay in good enough shape to withstand an austere environment.[119]

The conditions were not the only challenge; the research that the psychologists were conducting was entirely new. Although the Army had conducted some surveys of deployed troops during the Second World War, no research on the psychological well-being of deployed servicepeople had occurred during Vietnam or afterward.[120] In the 1990s, however, occupational psychologists, employers, and the general public became increasingly concerned about work-related stress and were asking what factors aggravated workers and how employers could produce more content—and more productive—employees.[121] "Symptoms of stress in the workplace have increased in recent years," one 1989 *New York Times* article began, pointing out that with longer hours and more travel, many workers were not only on edge by also more likely to become ill.[122] Two years later, the *Times* published a scorecard so workers could determine how stressful their workplace was.[123]

Serving on a peacekeeping mission in Croatia is different from, say, working in a cubicle at AT&T, and the Army's workplaces and demands were quite different from those of the civilian world. Nonetheless, civilian psychology, as it always had and would continue to do, shaped the Army's approach to mental health, and WRAIR psychologists also became interested in what factors created stress for soldiers. For the Army, this was new research. "None of us really had a system for doing this kind of work," Adler remembers "None of us had a way of thinking about it."[124] There were no templates for surveys, no standard questions, not even a sense of what needed to be measured.[125] As a result, their questions seem basic a quarter century later: What was the soldiers' mental health status, and what were the factors that were detrimental to it? "It seems so obvious now," Adler explains, "but at the time people hadn't been asking those sorts of questions."[126]

In general, the Human Dimensions Research Teams found that soldiers were doing well and that they had adequate mental health resources. In Somalia, they found that "soldiers did not suffer major negative effects from the stressors that had, before the operation began, been of most concern as potential mental health threats," and they "detected no problems that were severe enough to threaten to compromise the Army's ability to perform its [Operation Restore Hope] mission, or that would portend major mental health problems in the future."[127] In Bosnia five years later, they concluded that "only 10% of soldiers reported that they were coping poorly with the stress they experienced" and, in fact, that "Operation Joint Guard was more similar to garrison levels of overall distress than the levels reported during other deployments."[128]

The researchers also learned that the Army's focus on combat-related trauma was important but hardly captured all of the potential sources of soldier stress. As it turned out, soldiers often worried less about the potential dangers of a deployment than they did about their daily lives and their families at home.[129] In Croatia in 1992, boredom was the primary issue, to the point that "the surgeons . . . performed elective surgeries such as vasectomies and tattoo removal to find something to do," while one anesthetist had so little work that when he walked into an operating room, a member of his staff didn't recognize him.[130] They also found that issues like a confusing chain of command and inconsistent communications with their families were significant sources of stress, as were overcrowded tents and a lack of privacy—which led, in one instance, to soldiers appropriating hospital bed curtains to hang around their own bunks.[131]

Moreover, the researchers found that the potential for injury hardly affected soldiers' stress levels. In Somalia in 1992, "Small arms fire tended to come out in interviews almost as an afterthought, and was discussed less emotionally than were such issues as the uncertainty of their return to CONUS [the continental United States] or the problems with mail or telephone."[132] In Bosnia, "72% of soldiers reported moderate or extreme stress from the lack of personal privacy" but "reported little worry about their physical safety."[133] Although the Serbians could have easily shelled the runway on which they were encamped, "no one seemed concerned about it," Staff Sgt. Bradley F. Powers told an audience at a 1993 conference, and soldiers instead "joked about going to the Norwegian bar which was in an old bunker" if they were attacked.[134]

This research revealed that the Army's focus on combat stress hardly addressed all of the issues that soldiers were likely to face in post–Cold War deployments. "Talking to soldiers," Adler explains, "really begin[s] to inform [us that] we need to talk not only about traumatic experiences, but also . . . about their work experiences, about their unit climate, about how their leaders are treating them."[135] As in Marlowe's research during the Gulf War, researchers found that soldiers were happier and performed better when they understood their mission and how long they would be deployed, had commanders whom they trusted and thought cared about them, and when they had some modicum of comfort.[136] These were issues that, though not entirely absent, had not always been central to the development of a combat stress control doctrine for an imagined war with the Soviets.

Under Castro's direction, the behavioral health teams instituted a practice in which they would present their results and recommendations to the commanders on the ground before they left the theater. Usually, this information was well received, Adler explains, largely because it cohered with an Army culture that held that "you are supposed to develop lessons learned and you are supposed to be improving all of the time."[137] These briefings were important for three reasons. First, they made the research matter to those commanders who could take immediate action to address issues that left their soldiers unhappy or struggling.[138] Second, the Human Dimensions Research Teams built goodwill among commanders who understood not only what the teams had found but what they were likely to hear from their superiors when they were briefed on the research.[139] And third, the officers who had led troops in the Balkans and elsewhere in the 1990s and who accepted the notion that surveying their troops' well-being yielded useful insights had by the time the twenty-first-century wars began been promoted to more senior positions, with oversight of a greater number of soldiers and junior officers.[140] This meant that when the Army's Mental Health Advisory Teams (MHAT) arrived in Iraq in 2003, officers were likely to understand their importance and facilitate their research.[141]

That commanders reacted so positively to the recommendations of Adler, Castro, and Bliese while Rebecca Porter, who was serving at Fort Bliss around the same time, felt that her expertise was often ignored reveals something significant about how the broader Army approached behavioral health in the 1990s. Army commanders were neither passive nor uncritical consumers of psychology. Rather, they tended to be willing to listen to information that helped them—that made their men happier, made it easier to complete their mission, or made them look good to their superiors. In contrast, they tended to dismiss or ignore any information that might have the opposite effect. As Porter puts it, "At one point they didn't even want to know we existed because we could make a recommendation that would decrease their perceived readiness."[142] This selective embrace of psychology would bedevil the Army's attempts to address mental health issues throughout the twenty-first century.

Walter Reed researchers themselves also took important lessons from the Human Dimensions Research Teams, realizing that they had developed a capability that could easily be adapted to address the more pressing mental health questions during the Iraq and Afghanistan Wars. "I think the Army did a very smart thing, whether they did it intentionally or not, by keeping and maintaining these teams in peacetime scenarios,"

Paul Bliese reflected two decades after he had deployed on the research team that went to Somalia, "because when the wars rolled around we really didn't have to do a whole lot to have the core capability to go back into Iraq and Afghanistan to use surveys and to collect data and to feed it back."[143] Carl Castro agrees. When he went to Iraq in 2004 as the leader of the first MHAT team, he knew that "we were going to look at the mental health measures by PTSD, measure depression and anxiety, we were going to look at leadership issues, we were going to look at deployment-related stressors. These are things that we knew existed from what we had done in Kosovo and Bosnia."[144]

The Human Dimension Research Teams were not the only initiatives that would inform the Army's approach to the post-2001 wars. In 1994, James Stokes's long-held goal of a revised field manual for combat stress control finally came to fruition with the publication of FM 22–51, *Leaders Manual for Combat Stress Control.* Its contents were repackaged in short paperback handbooks with titles like "Combat Stress Behaviors," "Leader Actions to Offset Battle Fatigue Risk Factors," and "Stress Dimensions in Operations Other than War," each of which featured Stokes' pen-and-ink drawings of troubled soldiers. If these handbooks acknowledged a more diverse range of potential battlefields, drew upon the work of the Human Dimension Research Teams, and included more information about PTSD, they largely reiterated what Stokes, Martin, and others had been preaching since the 1980s: that combat stress reactions were normal and usually treatable, and that good leadership that attended to soldiers needs could often ameliorate soldier stress.[145] Leaders were taught, for example, that "it is normal for the survivor of one or more horrible events to have painful memories, to have anxiety (perhaps with jumpiness or being on guard); to feel guilt (over surviving or for real acts of omission or commission); and to dream unpleasant dreams about it" and that "fifty to eighty-five percent of battle fatigue cases . . . returned to duty following 1 to 3 days of restoration treatment, provided they are kept in the vicinity of their units."[146] The Army would take this doctrine into combat in 2001, where in many cases it proved successful in allowing soldiers to move past difficult experiences.

Back in Germany, Adler and Castro were instrumental in research that linked deployment length and mental health issues. This occurred because of the uncertainty that surrounded Gulf War Syndrome, the mysterious constellation of symptoms—"fatigue, muscle or joint pains, gastrointestinal complaints, skin rashes, poor sleep, memory problems and difficulties concentrating"—that veterans of the 1991 war began complaining of in the 1990s but whose origins remained a mystery.[147] There was robust debate over the syndrome—or, more properly, the symptoms attributed to it—with researchers advancing explanations that included the inhalation of toxic fumes from oil well fires, the side effects of pre-deployment vaccines, and post-traumatic stress disorder.[148] However, as a 1998 *Washington Post* feature story on the efforts of clinicians at Walter Reed Army Medical Center to treat it pointed out, "People with it share little more than participation in the Gulf War. . . . They suffered no unique exposures, their illness has no cardinal feature, and their symptoms follow no predictable course."[149] Some researchers thought the conditions might be psychosomatic.[150]

Whatever the etiology, the condition so concerned the Department of Defense that, when U.S. troops deployed to the Balkans in 1996, it worried that service members might develop similar symptoms and stipulated that every soldier who deployed for more than a month undergo a mental health screening upon redeployment.[151] Because the deployments were happening in Europe and were mostly affecting soldiers already based there, the task of implementing that directive for the Army fell to Castro and Adler.[152] The Post-Deployment Health Assessment (PDHA) that they developed was a simple document adapted from a self-rating scale that a psychologist named William Zung had developed in the 1960s to diagnose depression.[153] The questions, which rephrased diagnostic criteria as questions and asked respondents to indicate how often they experienced them, are familiar to anyone whose routine physical examination includes questions about mental health; for example, patients are asked whether they "feel down and blue" a "good part of the time."[154] The Army's PDHA added seventeen questions on PTSD and four about alcohol use to Zung's original twenty about depression.[155] The survey was intended to easily allow medical officers to determine which troops needed a referral for further care and to gather information that might prove useful should that soldier subsequently complain of the Gulf War Syndrome–like symptoms.[156]

In truth, though, there weren't much data on the forms beyond the soldier's responses to the questions and basic demographic information about age, rank, and marital status. Among the questions, however, was two that asked about "number of months on the deployment" and "how many previous times the participant had been deployed to the Balkans."[157] That information, Adler and Castro quickly realized, could reveal whether deployment length and the pace of deployments affected mental health.

Unsurprisingly, longer deployments correlated with high scores for depression, traumatic stress, and alcohol use, though this was more likely among male soldiers than females and male soldiers not on their first deployment seemed to do better than those on an initial deployment.[158] Only 11 percent of soldiers who had been deployed for four months required further screening; for soldiers who had deployed nine months or more, it was nearly 21 percent. The research also illustrated that even soldiers who had not been in combat experienced psychological distress that worsened with longer deployments. In points that would prove prescient, they noted that "experience in a peacekeeping mission may aid soldiers, yet it may not help them cope in a future combat mission" and that, "although second deployments appear to be beneficial to wellbeing, it may well be that incessant repeated deployments on the same operation begin to take a toll on a soldier's wellbeing."[159]

This research, which didn't appear in print until 2006 but was doubtless circulating internally before then, didn't prove a cautionary tale for the Army, which adopted twelve-month deployments quite soon after the Iraq War began and by 2007 was sending troops on fifteen-month deployments. But it did reveal some important things about the Army's capacity to study mental health issues. On the one hand, it showed that the Army had the capacity for large-scale mental health screening, however blunt an instrument the PDHA turned out to be.[160] On the other hand, however, the initial research with the PDHA data highlighted not only what the Army could learn from these surveys but also their limitations. As a result, when the Iraq and Afghanistan Wars began, Adler and Castro knew that they wanted to rethink how the PDHA would be scored.[161] Since individual medics and unit medical officers scored them, how could consistency be assured? Was surveying troops for traumatic stress, depression, and alcohol use sufficient, or did other experiences also needed to be included? These questions, and others, would guide the revision of the PDHA in the wars' early years, and the results of that revision would prove central to understanding the wars' impacts on the troops.

* * *

In the 1980s and 1990s, the Army devoted considerable effort to mental and behavioral health and entered the Iraq and Afghanistan Wars with a set of assumptions, attitudes, and capabilities that defined its approach to those conflicts. Imagining a potential war with the Soviet Union as the most likely conflict the Army would face, Army researchers devoted considerable efforts to creating healthy units and developing a doctrine for the treatment of battle fatigue, which they assumed was inevitable but largely treatable. Their assumptions proved correct—good leadership and sleep matter considerably, and most soldiers do recover relatively quickly from the psychological shock of combat. Throughout these decades, however, they struggled to have that research published as doctrine, and more problematically it was never implemented effectively. Mental health units were often poorly prepared, understaffed, or not respected by the larger Army. These were issues that the Army would struggle with throughout the Iraq and Afghanistan Wars. Other research, however, proved more promising. As Castro shivered in Kosovo and as Adler hunched over the pencil-and-paper surveys that had been mailed to her from Bosnia, they became confident that troops' psychological well-being could be measured, understood, and treated and that they had the capacity to do all of this while those troops were in the field.

If the Army entered the twenty-first century with knowledge, capacities, and an awareness of the methodological, theoretical, and institutional challenges that affected mental health care, it did not imagine the wars that it would fight after the September 11 attacks.[162] By the time the Iraq War entered its second year, those assumptions were being sorely tested under the strain of wars that lasted longer than the Army expected and that placed unprecedented stress on the force.

2

"The Psychiatric Cost of Sending Young Men and Women to War"

Mental Health as Crisis and Enigma amid Growing Opposition to the Iraq War

A week before President George W. Bush announced that the United States had begun bombing Iraq, the *Washington Post* warned that, while "in Washington, there might be talk of swift victory in a war against Iraqi President Saddam Hussein and discussion of the shape of postwar Iraq . . . there are powerful reminders that jumping past war is impossible for those who actually fight it."[1] Four days after the invasion, the *New York Times* reminded readers that that "history also makes it clear that for some who fight, the exposure to the terror, carnage, chaos, exhaustion, grief and guilt of combat can prove as damaging as any bomb or bullet, inflicting serious mental injuries."[2] Despite the military's assertions that it had been, as a June 2003 *New York Times* article reported, "schooled by the experiences of Vietnam and the first gulf war [and] grown far more sophisticated in their approach to the psychological pressures of battle" and that, as the *Atlanta Journal-Constitution* explained, there was "an aggressive Pentagon program designed to help soldiers learn to come to terms with such issues as loss of buddies, deaths of civilians on the battlefield and lengthy deployment to a combat zone" that "help[s] a new generation of soldiers avoid the problems . . . [that] Vietnam veterans experienced," the Army's view of battle fatigue as acute but temporary had evidently not permeated Americans' consciousness; the specter of traumatized Vietnam veterans loomed over the war.[3]

As the initial success in toppling the Hussein regime descended into a chaotic, violent occupation, those early warnings seemed increasingly prescient. Between 2003 and 2007, Army researchers and leaders, members of Congress, and the general public increasingly encountered reports of psychologically struggling service members and veterans, sought to comprehend why these struggles were occurring, and drew upon these encounters and impressions to frame their impressions of the war and the policies that defined it. For Army researchers and providers, mental health issues were inevitable medical problems that could be successfully addressed. Their efforts thus took the form of improved surveillance through in-theater Mental Health Advisory Teams and a Post-Deployment Health Reassessment survey that sought to better understand how soldiers were faring during and after deployments, improved education through the BATTLEMIND training program, and an increased array of treatment options aimed at recuperating soldiers' pre-war mental health.

Many members of congress and a significant segment of the American public, however, viewed increasing reports of mental health issues with alarm. Recalling images of permanently traumatized Vietnam veterans and increasingly critical of the Iraq War, these groups seized upon reports that veterans were struggling to get treatment or were being returned to Iraq for another deployment to cast the Iraq War as a misbegotten debacle forged by a callous, cavalier, and duplicitous administration.

The debate over mental health during the first four years of the Iraq War thus pitted scientific and medical understandings of post-traumatic stress against popular ones. As Army researchers and leaders sought to reassure their soldiers and the nation that mental health issues were expected, that most soldiers would relatively easily recuperate and return to duty, and that they were effectively delivering care, reports of suffering soldiers increasingly provided ammunition for growing anti-war sentiment. The troops' mental health, then, was both a medical and cultural issue, one that fundamentally shaped Americans' perceptions of an increasingly violent, intractable, and unpopular war.

* * *

The Army's first indication that mental health might be a significant issue in the Iraq War came in the summer of 2004, when commanders in Iraq became increasingly troubled by a cluster of nineteen suicides among soldiers stationed in Iraq.[4] These figures were disconcerting in part because the deaths had occurred despite the Army's increased attention to mental health. The *Washington Post* pointed out that there were "nine combat stress company detachments [deployed] to Iraq" and that "each Army division in Iraq . . . has a psychologist, a psychiatrist and a social worker, with an emphasis on trying to treat soldiers' stress problems as close to the front lines as possible"—a commitment of resources that reflected the Army's continued embrace of the principles of proximity, immediacy, and expectancy.[5] James Stokes had, in fact, updated the pocket cards that he had created in the 1980s and 1990s for the current wars. These "hip pocket training guides" on subjects like "Helping a Soldier in Distress" and "How to Face the Injured and the Dead" no longer featured his sketches, but they reiterated the principles that had guided combat stress control since the 1980s.[6]

Ever image conscious, the Army, which, as Erin Finley explains, had been put "on the defensive against charges of negative mental health-care services," offered reassurances that it had redoubled its efforts in the wake of the suicides.[7] That action included an effort to ascertain the general psychological well-being of soldiers and the accessibility and quality of mental health care in Iraq. That question made its way to Deputy Surgeon General Maj. Gen. Kenneth T. Farmer, who created a Mental Health Advisory Team (MHAT) modeled on the Human Dimensions Research Teams of the 1990s and that would "assess and provide recommendations to address potential organizational and resource-limitation factors which may be related to the recent spike in OIF[Operation Iraqi Freedom]-related suicides."[8] Farmer's order eventually landed on Carl Castro's desk at the Walter Reed Army Institute of Research (WRAIR). It was not an assignment that Castro relished. Having just become the Surgeon General's research psychology consultant after four years in which he had deployed four times to the Balkans and twice to the Middle East on Human Dimensions Research Teams, the thought of another deployment wasn't all that appealing.[9]

Nonetheless, Castro and the rest of the team arrived in Iraq in August 2003. The deployments of the 1990s provided a template for how they approached the latest war. This included knowing both what orders had to be written so that commanders would have their troops fill out the surveys and that the researchers would have to wait because troops in the field were rarely "sitting there waiting to complete a survey when we show[ed] up with one."[10] Within a month, they had delivered 750 surveys, tabulated the data, and briefed in-theater commanders.[11] The results were not uniformly encouraging. On the one hand, the Army's Combat Stress Control doctrine seemed to be working, and "the forward elements of the OIF behavioral healthcare system demonstrated great effectiveness."[12] Moreover, many providers were deeply committed.[13] These clinicians represented exactly what the Army had been imagining throughout the 1980s and 1990s—forward deployed, embedded specialists effectively delivering quality mental health services.

Nearly one in five soldiers, however, had symptoms of post-traumatic stress, but only a quarter of those had been treated, perhaps because, as Finley notes, there was significant stigma and a lack of resources.[14] Perhaps more important, though, was that many mental health professionals were not sufficiently trained. Indeed, "more than half of the behavioral health providers interviewed reported either they did not know what [Combat Operational Stress Control (COSC)] doctrine was, or did not support it," and the same percentage "had not received adequate training in combat stress prior to deployment."[15] More troubling, "almost half of the Soldiers surveyed reported not knowing how to obtain [mental health] services."[16] Castro's team called on providers to be more visible within units, for personnel to be better trained to follow COSC doctrine, and for soldiers to be given more preventative treatment prior to deployments.[17]

The Army's first effort to evaluate soldiers' needs in the Iraq War and whether the Medical Department (AMEDD) was meeting them thus offered a disconcerting vision as the war became increasingly violent and seemed poised to last a long time. Although one general mistakenly insisted that troops shouldn't be complaining about a lack of air conditioning because every unit had climate-controlled sleeping quarters—Castro had just spent fourteen nights in an un-air-conditioned camouflage tent—most commanders were receptive to Castro's conclusions.[18] This included some who were initially skeptical of the significance of mental health issues in the force. Army Chief of Staff Gen. Peter Schoomaker, a Special Operations veteran who, in Castro's words, initially took the approach, that, "well, these things happen," accepted Castro's recommendation to commit additional resources to mental health.[19] Gen. George Casey, who was then the Vice Chief of Staff but who would soon lead the Iraq War, describes the MHAT as "my personal wake-up call, a realization that we had a whole generation of leaders who hadn't had to deal with PTSD or TBI at that level."[20]

When the Army made the report public in January 2004, it was accompanied by promises of improved care in Iraq.[21] These reassurances, however, competed with an increasing flow of bad news about soldiers' psychological well-being. In July 2004 the *New England Journal of Medicine*—the nation's premier medical journal—published the results of a study that revealed that many soldiers were likely returning from Iraq with mental health issues.[22] The article represented the first publication by WRAIR's Land Combat Study Team, which had been created in 2003 to run a five-year study of the mental health impact that the Iraq and Afghanistan Wars were having on soldiers.[23] The lead author, Col. Charles Hoge, had come to his position, and indeed the Army, somewhat untraditionally. Initially trained as an infectious disease doctor, he had begun his career in the Uniformed Public Health Service and transferred into the Army in 1991 and researched tropical diseases.[24] In 1997, he changed specialties and was sent to Walter Reed Army Medical Center for a psychiatric residency. His supervisors were so impressed with his ability to use the epidemiological skills that he had built in his first specialty to conduct psychiatric research that when his residency ended and he would normally have been assigned as a division psychiatrist, he was assigned to a grant-funded position at WRAIR.[25]

Hoge, along with Carl Castro, Dennis McGurk, and others on the Land Combat Study Team, had selected three demographically similar brigades—two from the 82nd Airborne Division and one from the Third Infantry Division—that had had similar experiences in Iraq. Using predeployment surveys to set a baseline, they administered surveys after redeployment—that is, return from Iraq—that asked whether soldiers were experiencing "current symptoms (i.e., those occurring in the past month) of a major depressive disorder, a generalized anxiety disorder, and PTSD," as well as whether they were drinking too much.[26]

The results were troubling. Compared to Afghanistan, Iraq was much more violent, and a deployment there had a more significant psychological impact.[27] "The percentage of study subjects whose responses met the screening criteria for major depression, PTSD, or alcohol misuse," Hoge and his colleagues wrote, "was significantly higher among soldiers after deployment than before deployment, particularly with regard to

PTSD."[28] Like the MHAT, the Land Combat Study research discovered that only about one-third of those who needed psychological help actually sought it, and those who needed it most were also most likely to fear that getting it would harm their reputations or careers.[29] Perhaps most troubling was the authors' caveat that "our estimates of the prevalence of mental disorders are conservative" and that "the magnitude of the differences between the responses before and after deployment is particularly striking, given the likelihood that the group responding before deployment was already experiencing levels of stress that were higher than normal."[30]

This sentiment was shared by Matthew J. Friedman, a VA psychologist who contributed an editorial that accompanied Hoge's piece in the *Journal*. "The data presented by Hoge and associates . . . force us to acknowledge the psychiatric cost of sending young men and women to war," he wrote, adding that "it is possible that these early findings underestimate the eventual magnitude of this clinical problem."[31] He made much the same point in the *New York Times*, cautioning that "the percentage of returning soldiers with post-traumatic symptoms could still go up" and raising concerns about how National Guard and Reserve troops who had not expected a combat deployment, as well as troops serving on extended tours, would fare.[32] In the *St. Petersburg Times*, he reflected, "'I'm not an alarmist, but I think this is a serious problem. It may be worse just because of the nature of the war.'"[33]

The notion that the particular nature of the Iraq War was creating mental health issues helped fuel Americans' increasing dissatisfaction with the conflict. Only *USA Today* and the *New York Times* mentioned that past research on Vietnam, Somalia, and Gulf War veterans formed the basis for Hoge's and Friedman's concerns. Other newspapers emphasized the Iraq War's supposedly unique aspects, which unsurprisingly aligned with anti-war critiques: a government indifferent to the wars' realities, the increasing use of National Guard and Reserve troops, the lengthening deployments, and the evolving mission. The *San Jose Mercury News*, which quoted Friedman, quoted one veteran's complaint that, "If they can't see your wound—like a broken arm—they think you're fine and want you to get back to work."[34] "Those costs, too, must be calculated when we assess the worthiness of war," wrote *Philadelphia Inquirer* columnist Jane Eisner, explaining that Hoge's piece "proved what common sense suggests: The more combat, the higher the risk of mental illness."[35]

For some Americans, then, Hoge's article illustrated why the war was disastrous. At WRAIR, however, it raised a clinical question. In Heidelberg, Amy Adler and Paul Bliese had been reviewing about fifteen hundred post-deployment health assessments from members of the 173rd Airborne Brigade. These soldiers had redeployed to Vicenza, Italy, in February 2004 after a ten-month stint in northern Iraq that had begun with a combat parachute jump on March 26, 2003.[36] The results were surprising. "The range of behavioral health was so incredibly low that we just couldn't believe it," Bliese remembers. "It was one or two percent depression, PTSD."[37] Why, they wondered, were Hoge's rates were so much higher?

There were two possibilities. The first had to do with stigma. Because the Land Combat Study sought to determine how widespread mental health issues were, the survey could be conducted anonymously, which meant that soldiers might be more forthcoming. The Post-Deployment Health Assessment (PDHA), in contrast, was a diagnostic tool, meant to funnel soldiers into an appropriate continuum of care, and thus, Adler explains, "it had implications for your medical records."[38]

Given that stigma—the belief that that "my leadership might treat me differently" or "I would be seen as weak" for seeking help—was evident in Hoge et al.'s population, soldiers might well have been loathe to admit that they were struggling if their names were on the survey.[39] Some soldiers feared that an affirmative response would delay their leave.[40] Nor was the Army always discreet when soldiers reported issues. When Rebecca Porter was screening returning troops at Schofield Barracks in Hawaii in 2005, for example, she was told to sit at the top of the bleachers in a stadium; if a medical officer concluded that soldiers needed a consultation, they had to climb up to see her.[41]

But there was another possibility. Adler and Bliese had surveyed the members of the 173rd during their first week back from deployment, but Hoge et al. had conducted their research three months after troops redeployed. Those twelve weeks, Adler and Bliese hypothesized, might represent a period in which soldiers began feeling the effects of deployment as the euphoria of homecoming wore off. With the PDHA, Bliese remembers, "the problem is that soldiers were really feeling pretty elated to be back. . . . They'd survived, they'd seen their families. It almost felt silly asking them a question about their depression."[42] Moreover, issues that might become a problem, like fighting with a partner or alcohol abuse, weren't yet apparent. Soldiers hadn't really seen their partners or had a chance to drink.[43]

To test their hypotheses, Adler and Bliese returned to Vicenza a few weeks after the Hoge et al. *New England Journal of Medicine* article appeared, and they administered two surveys—one anonymous, the other not—to about five hundred members of the 173rd, who by this point had been back in garrison for three or four months.[44] They were shocked by the results. "Anonymity," Bliese remembers, "didn't really seem to matter at all."[45] Timing, however, mattered greatly. Even when they identified themselves, the number of soldiers screening positive for PTSD, depression, "general psychological distress," "anger problems," or desire for counseling more than doubled.[46] The incidence of PTSD had tripled, and the number of soldiers exhibiting problematic anger had increased nearly as much; incidence of depression and "general distress" had gone up fivefold.[47]

These results illustrated the legacies and shortcomings of WRAIR's work in the 1990s. Adler and others' early studies provided a model for understanding soldiers' deployment experiences. However, the Iraq War's high level of combat and the trauma that accompanied it were notably different from those of peacekeeping missions, and soldiers' symptoms were manifesting in unanticipated ways. In particular, it took longer than the Army had anticipated for issues to surface, and the PDHA thus missed a lot of people who needed help. Bliese and Adler's results thus validated post-deployment screenings, but they also revealed the need to change when they were delivered.[48]

Adler and Bliese published this research in *Psychological Services* three years later.[49] The army, however, acted on their recommendation within days. When the two psychologists returned to Heidelberg, they briefed Maj. Gen. Carla Hawley-Bowland, the commander of the Europe Regional Medical Command, who in turn presented it to visitors from the Department of Defense's Health Affairs office.[50] On March 10, 2005, William Winkenwerder, the assistant secretary of defense for Health Affairs, issued a memorandum to all three branches of the military directing that a new assessment, the Post-Deployment Health

Reassessment (PDHRA), would be administered three or four months after troops redeployed.[51] "The purpose of this reassessment," he wrote to each branch's undersecretary for Manpower and Reserve Affairs, "is to proactively identify health concerns that emerge over time following deployments, to help remove potential barriers, and to facilitate the opportunity for service members to have their health needs and concerns more fully addressed following deployments."[52] Back in Germany, Adler and Bliese were pleasantly stunned. Adler had never seen the Department of Defense move so quickly to change policy based on medical research.[53] Bliese was so worried that no one would believe that a small team of researchers at a tiny unit in Germany had had so affected the entire Armed Forces that he asked his commander, Maj. Gen. Lester Marinez-Lopez, to write a memo that formally recognized their efforts.[54]

The contrast between the media's and the military's response to Hoge et al.'s article illustrates the distance between popular and medical understandings of mental health in this moment. For the popular press, service members' declining mental health portended a crisis that illustrated why the Iraq War was a disaster; to WRAIR researchers, it was a normal, if more severe than expected, problem that demanded rigorous research. These competing narratives would define how Americans in and out of the Army thought about mental health as the war in Iraq became more violent and placed increased demands on the troops during 2005 and 2006.

* * *

On July 27, 2005, a few months after Winkenwerder issued the order instituting the PDHRA, the House Veterans' Affairs Committee held its first hearing about returning Iraq and Afghanistan veterans' mental health needs. The main witness was Stefanie Pelkey, whose husband, Michael Jon Pelkey, had returned from Iraq suffering from what he had experienced and, as his widow put in her opening remarks, "died in a battle of his heart and mind."[55] The story that she unfolded was indeed tragic, and its elements previewed how many families would narrate the experiences of veterans who took their own lives. Her husband had joined the Army because "being a soldier was a childhood dream of his." He served in the First Armored Division, a unit that saw heavy action in the war's early months.[56] Sidelined in the initial invasion—much to the dismay of many of its soldiers—the First Armored arrived in late March 2003 and was given what one journalist called "an arguably more important task: helping to nail down the U.S. win by reinforcing the occupation and bolstering efforts to restore order and rebuild the country."[57]

That mission was challenging. Transformed with little training from artillerymen to policemen, the First Armored's fifteen thousand troops patrolled Baghdad's narrow streets on foot, raiding houses, discouraging looters, and enforcing the order by Paul Bremer, leader of the Coalition Provisional Authority in Iraq, that Iraqis disarm while attempting to convince residents that the United States could provide safety.[58] For soldiers used to Germany's cloudy winters, Baghdad's 100° F days were debilitating.[59] Their quarters at Baghdad's international airport were hot, dirty, and infested with incontinent pigeons.[60] The Iraqis were, at best, indifferent to their presence. Many strongly opposed them, and First Armored soldiers died when Iraqis tossed grenades into their guard posts or raked their vehicles with gunfire, when soldiers' vehicles hit roadside bombs, and when snipers shot at them as they guarded buildings or bought sodas.[61] Soldiers were scared and demoralized. "The depressing thing is that there's not a whole lot we really can do about those guys who are determined to try to kill us," one reflected.[62]

The deployment left Pelkey exhibiting exactly the symptoms that Adler and Bliese had discovered in their research with the 173rd. When he returned to the U.S. garrison in Baumholder, Germany, in late July 2003, Stefanie testified, "he seemed so happy to be home."[63] He completed his PDHA and saw his doctor, reporting that "he was worried about having serious conflicts with loved ones" and complaining of "feeling down, depressed and sometimes hopeless" and of being "constantly on guard and easily startled."[64] He followed his doctor's advice to seek counseling, but Baumholder's psychiatrists were overbooked.[65]

After a transfer to Fort Sill, Oklahoma, Pelkey's mood seemed to improve. He had a new house, a young son, and a challenging position.[66] But within a few months, he was struggling. He forgot simple tasks. His chest hurt. He couldn't perform sexually.[67] When he slept, he thrashed about.[68] Stephanie and he eventually sought marriage counseling, which was the only option readily available at Fort Sill, where soldiers often waited a month to see a psychologist.[69] When the civilian counselor diagnosed Michael with PTSD, it was the first time they had heard of the condition.[70] Both when the First Armored Division had deployed and when Michael was at Fort Sill, Stefanie recalled, "No doctor ever asked him about depression or linked his symptoms to the war."[71]

Like many people, Michael was relieved to have a diagnosis. It seemed to clear a path toward recovery.[72] Naming your demons, however, does not immediately slay them. The Sunday after Michael had been diagnosed was Halloween, and he took his two-year old son Benji, dressed as a brown-and-white floppy-eared dog, to a party. In a photograph taken that day, Michael's sadness is unmistakable.[73] That Friday, when his visiting father-in-law came home, Michael was in bed. "He looked as if he were sleeping peacefully," Stefanie remembered, "except for the wet spot on his chest."[74] Army investigators determined that Michael had killed himself. Stefanie disagreed, calling it a "horrible accident" that wouldn't have happened had his PTSD not compelled him to sleep with a loaded pistol.[75] Before Congress, she condemned the Army for not having enough mental health providers and not preparing families.[76] "My husband," she concluded, "died of wounds sustained in battle. That is the bottom line."[77]

Committee members responded to Pelkey's testimony with sympathy and outrage. For Bob Filner (D-CA), her story illustrated the hypocrisy of a system and a country that purported to care for its veterans. An unabashed anti-war progressive, the former Freedom Rider had joined the Veterans Affairs committee largely because then Sen. Alan Cranston (D-CA) had advised him that a commitment to veterans would help the liberal keep his seat in a conservative, military-friendly area like San Diego.[78] When he had arrived on the committee, he told then chair Sonny Montgomery (D-MS) that he had once toured Mississippi for several months, a reference to his time in the state's jails; Montgomery, it turned out, had commanded the National Guard unit that had arrested him.[79]

Filner situated Pelkey's story within a larger narrative of governmental failure to care for ailing veterans. "We've seen it with our atomic veterans, our Agent Orange from Vietnam, PTSD," he complained. "At first, no one wanted to recognize the truth. . . . And it looks to me, there's an institutional dynamic to deny illnesses, maybe because its going to cost money, or they don't want to admit mental problems on behalf of our brave young men and women."[80] The problem could be solved, he insisted, if Americans stopped ignoring veterans' needs and committed the resources necessary to address them.[81] Even as Stefanie Pelkey acknowledged that awareness of mental health issues had grown and more support had become available in the months since her husband's death, Filner remained skeptical. "Half of the people on the streets today are Vietnam vets," he explained, although that figure is incorrect.[82] "And that's partially because we didn't take [mental health] very seriously. . . . And we see it happening again."[83]

Others on the committee also raised Vietnam's specter. Shelley Berkley (D-NV) reflected that she "grew up during the Vietnam War" and that "so many of the kids I went to school with that went to Vietnam came back dramatically changed. . . . I suspect it's very similar to Iraq right now."[84] As Filner had, Berkley understood events like Pelkey's death as evidence that the country had learned nothing from that earlier war: "What astounds me now as a member of Congress is that knowing what transpired in Vietnam . . . I would have hoped by now that we would not . . . avert our eyes or don't put the necessary resources in."[85] Grace Napolitano (D-CA) echoed these remarks, asserting that "a lot of our soldiers from previous wars have not been able to deal with the issue of PTSD and end up being homeless"; she described conversation that she had had with a Vietnam veteran who had been unable to hold down a job for two decades, adding, "Something is wrong, that we are not helping our soldiers be able to cope with it."[86]

Each representative's comments drew upon a popular but inaccurate remembrance of the vast majority of Vietnam veterans as permanently and debilitatingly traumatized. In fact, only about 15 percent of Vietnam veterans screened positive for PTSD, and just over 5 percent had ever had a "major depressive episode."[87] The vast majority of Vietnam veterans, that is, went on to lead lives relatively unhindered by their wartime experiences. These statistics had little purchase, however, against the dominant remembrance of the troubled Vietnam veteran and the anxiety that Iraq veterans would face similar struggles, an anxiety stoked by numerous comparisons in the six months prior to Stefanie Pelkey's testimony. In March 2005, *USA Today* cited a *New England Journal of Medicine* article to argue that "as many as one out of four veterans of

Afghanistan and Iraq treated at Veterans Affairs hospitals in the past 16 months were diagnosed with mental disorders, a number that has been steadily rising" before quoting a VA psychiatrist who worried that "the soldiers didn't come right away after Vietnam, either. If they come in the numbers predicted . . . we could be overwhelmed."[88] In May, Congressman Tom Udall (D-NM) published an op-ed in the *Santa Fe New Mexican* that asserted that "recent studies show that troops who served in Iraq are suffering from post-traumatic stress disorder and other problems brought on by their experiences on a scale not seen since Vietnam."[89] Later that month, a Florida newspaper quoted a Vietnam veteran turned therapist who predicted that "the post-traumatic stress is going to be unbelievable" and asserted that "my mission is to alert the mental health field that they need to pay attention. They didn't pay attention when we came back."[90]

Nearly all of these articles also stressed that the military and the VA were, in fact, paying close attention, had tried to improve mental health care, and that, in fact, there was ample evidence that there had been improvements.[91] In January 2005, the second MHAT report found that, despite continuing concerns about long tours, morale had improved by nearly twenty points.[92] Fewer soldiers were "seeing dead or seriously injured Americans, handling human remains, or killing an enemy combatant," the sorts of things that led to significant mental health issues.[93] Positive screenings for Acute Stress Disorder and PTSD were about 5 percent lower than they had been in 2003, and more soldiers were accessing care.[94] The Army had cut the ratio of soldiers to behavioral health providers in half, and those providers were more visible and felt better prepared.[95]

Nonetheless, popular understandings of post-traumatic stress disorder posed a challenge for the Army and VA mental health experts who took their seats in front of the Veterans Affairs Committee after Stefanie Pelkey's testimony. Together, they sought to reassure Congress of their progress and complicated the notion that post-traumatic stress disorder was most often severe and permanent. Charles Hoge, for example, did not shy away from the reality that about a fifth of U.S. troops in Iraq had diagnosable mental health symptoms, but he also touted the Land Combat Study's research, the PDHRA, and the increased availability of services.[96] More important, he argued that "mental health symptoms are common and expected reactions to combat" and that "helping to normalize these reactions is a key to stigma reduction and early intervention."[97] The next witness after Hoge, WRAIR researcher Charles Engel, argued that "about half of those with initial PTSD and depression quickly improve, but overall, rates rise two- and three-fold in the next few months" and that "many are distressed after war, but most distress is not severe."[98] Engel was followed by Mathew J. Friedman, the director of the VA's National Center for Post-Traumatic Stress Disorder, who reiterated that "most people who return from a combat zone do not have psychiatric or psychological disorder. A large number of them do have transient problems, adjustment disorders, from which they recover quite quickly."[99] On the next panel, VA Undersecretary Michael J. Kussman testified that "almost everybody who serves has some kind of readjustment or reintegration problem. Most of the time it's not illness. It is normal reactions to abnormal conditions."[100]

This line of argument, which illustrates what Finley has called the military's "emphasis on normalizing the experience of combat stress," consistently framed the Army's and VA's public narrative about wartime mental health.[101] A year later, Hoge and Col. Elspeth Cameron Ritchie, the Army Surgeon General's psychiatry consultant, testified at another House hearing that figures showing "that 10 to 15 percent of Army soldiers develop post-traumatic stress disorder after deployment to Iraq," that a comparable number had other issues that "may benefit from care," and that about a third of the PDHRAs that Hoge and his team had studied noted "some sort of mental health concern" were not alarming because "it is normal to experience symptoms related to combat and deployment, and many individuals who express concerns do not have a mental disorder or need referral for further care."[102] Ritchie further explained that "PTSD is not a debilitating disease and can be managed effectively if diagnosed and treated early."[103]

These hearings and the media coverage of returning veterans reveal the distance between medical and popular understandings of psychological stress. If the dominant image of the traumatized veteran is one who, as the Associated Press put it, has been "forever changed," medical understandings of the disorder are more nuanced.[104] Walter Reed researchers and VA officials understood post-traumatic stress as one of many potential responses to a wartime deployment and that those reactions would vary in duration. Indeed, as the historian Ben Shephard explains, as trauma became popularized as a cultural and medical category in the 1990s, research indicated that a singular, one-size-fits-all definition of the symptomology, progression, and treatability was insufficient.[105] In 1994, the *Diagnostic and Statistical Manual of Mental Disorders* (*DSM-IV*) added the category of "Acute Stress Reaction," in which "symptoms manifested during the period from 2 days to 4 weeks posttrauma, whereas PTSD can only be diagnosed from 4 weeks" and PTSD can only be diagnosed as chronic when symptoms persist for three months.[106] Most people who are traumatized, researchers found, had acute stress reactions, not post-traumatic stress disorder. By the mid-1990s, research had found that nearly half of the population that had been diagnosed with PTSD had recovered within three years.[107] In layman's terms, this meant that most individuals who had been in an automobile accident might have a few nightmares or be anxious the next time they got behind the wheel; very few suffered indefinitely from intrusive thoughts and anxiety.[108] These results held true for combat veterans as well, most of whom also recuperated more quickly when they availed themselves of treatment.[109]

Alongside this evolving understanding of trauma had come innovations in treatment. In addition to traditional cognitive-behavioral therapy and group therapy, Cognitive Processing Therapy, in which patients write about and discuss their traumatic experiences, and Eye Movement Desensitization and Reprocessing, in which the patient thinks first about the traumatic event and, subsequently, about "positive beliefs about the event" while the therapist makes hand gestures that the patient follows visually, showed promise for veterans.[110] The peer-reviewed literature on efficacy of Cognitive Processing Therapy and Eye Movement Desensitization and Reprocessing were likely on Matthew Friedman's mind when he told Congress in 2005 that the "seat-of-your-pants type of an operation" through which Vietnam veterans were diagnosed and treated—likely a reference to the "rap groups" popularized by psychiatrists Robert Jay Lifton and Chaim Shatan and the storefront readjustment counseling that the VA had offered throughout the 1980s—had been replaced by "psychosocial treatments that are very, very effective."[111]

Moreover, the notion that mental health conditions could be treated and cured, or that people with such a diagnosis could still live productive lives, should not have been alien in a culture in which 11 percent of the population was taking prescription anti-depressants.[112] Nonetheless, the Army's and the VA's assertion that war-related stress was normal, temporary, and treatable faced resistance in the media, in Congress, and within the force. While some stories focused on veterans who had run afoul of the law, most portrayed veterans as chronically unable to readjust because of their combat experiences, a sharp contrast with the military's reassurances that stress disorders were normal and treatable and the persistent.[113] Readers found accounts of a soldier who "has dreamed over and over of attacks that killed a friend and other soldiers"; of soldiers who "report problems with 'nightmares, anxiety, anger, antisocial behavior, adjusting to family, [and] overindulging in drugs'"; and who were "a little more jittery and confused when he sees people on overpasses or dead animals or trash in the road."[114] Rarely did these articles suggest that those responses were typical but temporary.

These anecdotes undermined the military's assertions that it was aggressively addressing the problem. In January 2006, for example, the *New York Times* profiled Greg Papadatos, a forty-two-year-old medic in the New York Army National Guard who complained that he was struggling to readjust. Repeatedly, the article contrasted statistics about returning veterans' mental health and Papadatos's efforts to get help with the Army's reassurances. "Military studies already indicate that nearly one in five returning soldiers struggle with depression, anxiety, or post-traumatic stress disorder," the article explained before undercutting the Army's research with the observation that "many veterans suspect the numbers are much higher."[115] Similarly, the article's assertion that "military officials say they are determined not to repeat the aftermath of Vietnam, when thousands of traumatized veterans ended up dysfunctional, homeless, and suicidal" and the enumeration of their efforts seemed an ineffectual rebuttal to Papadatos's claim that "they won't give me the slightest bit of thought unless I kill somebody . . . and maybe not even then."[116] Other stories made similar points. In Rhode Island, a soldier complained that "there are guys out there with PTSD and they don't even know it. . . . The military needs to step up to the plate and admit this is an issue."[117] An Akron newspaper contrasted the National Guards' "proactive approach" with a veteran's complaint that he was "frustrated trying to get the care he felt he needed through the VA" and his complaint that "I am talking and nobody is listening."[118]

Media coverage also explicitly questioned whether the military was adequately addressing service members' mental health issues. In March 2006, Charles Hoge and his colleagues published an article in the *Journal of the American Medical Association* that reviewed 300,000 PDHA forms and found that about one in five "met the risk criteria for a mental health concern" but that only a fifth of them had been referred for additional treatment.[119] This did not, however, mean that Iraq veterans were not utilizing mental health services; Hoge and his colleagues found that nearly a third of veterans "were documented to have had at least 1 outpatient mental health care visit within the first year postdeployment."[120] For the authors, these figures affirmed that the Iraq War was producing a substantial number of mental health issues and raised the question of whether the healthcare system was up to the challenge.[121]

Media coverage of this study viewed these results as evidence of the military's inattention to mental health issues. The *New York Times* was openly critical, complaining that "the study . . . calls into question the military's effort to estimate the mental health costs of Iraq."[122] The *Washington Post* suggested that even these numbers were low, quoting a veterans' advocate who argued that "the military would have found larger numbers of troubled soldiers and Marines if it had done a better job of reaching out."[123] Paul Rieckhoff, the executive director of Iraq and Afghanistan Veterans of America, interpreted the study as revealing that "mental health issues and PTSD are one of if not the most critical issue facing Iraq Veterans" and "could very well be our generation's Agent Orange."[124] Against this coverage, Michael Kussman's familiar comment to the *New York Times* that "a large proportion of people coming back are dealing with normal reactions to abnormal conditions," although accurate, seemed dismissive.[125]

Convincing service members that their conditions were normal and transient, however, remained at the heart of the military's efforts. During the Iraq War's first two years, returning soldiers received a briefing on stress management and reintegration. "It was just something that they plugged in when soldiers were going through all of these reentry briefs," Amy Adler recalls. "You know, learning how to drive again, figuring out how to get stuff out of storage, dealing with finance, here's the stress education brief. . . . It was a day of PowerPoint briefings."[126]

For psychologists, there were real questions about the efficacy of large-group presentations. During Desert Storm, David Marlowe's staff had urged the Army to "undertake squad level group debriefings, with particular attention to units that suffered combat losses" and argued that units should be kept together "for as long as possible" in order "accomplish the (psychosocial) work required to achieve a recovery context for the restoration of soldiers."[127] However, what were clinically known as "Critical Incident Stress Debriefings" had become controversial among clinicians.[128]

In the summer of 2005, Amy Adler was the principal investigator of a study that sought to determine whether the post-deployment stress debriefings were meeting returning combat veterans' needs, and she sat in on such a session. As the soldiers described their emotions upon returning from Iraq, she suddenly realized the insufficiency of the Army's stress education program. "These soldiers were so raw and so angry," she remembers, "[that] it felt like [the PowerPoint presentation] was just going to be a drop in the bucket."[129]

Adler realized that the Army's approach had been too narrow, that focusing on post-traumatic stress ignored the range of issues that soldiers faced. Returning soldiers, she recalls, were "feeling disconnected from others around them, feeling really angry, very pissed off and edgy, feeling like it was very hard for them to get reconnected to those near them," but their symptoms often "were not necessarily classic PTSD."[130] At the same time, veterans were reporting that there had been some positive outcomes from their deployment, like their camaraderie with fellow soldiers. The presentation that soldiers were receiving didn't capture all of those responses.

This problem led to the creation of BATTLEMIND, the Army's psychoeducation program. Carl Castro had initially come up with the idea. Based on his work on the MHAT teams and with the Land Combat Study Team, he had learned that returning soldiers faced serious issues but that "reassurance is very, very powerful. Normalizing is very, very powerful."[131] His idea, then, was to create a program in which soldiers learned that the adaptations that were necessary to survive in a combat environment would take time to dissipate when they redeployed and that that process could be helped, if necessary, by treatment.[132] As Castro explains, "You've really got to know [in] what context you are seeing these symptoms and reactions. In one context they could be highly adaptive—in fact, your survival depends on it—and in another context you could be seen as abnormal and it could actually interfere with your functioning."[133] Hoge, who also worked on the program's development, agrees, citing what he calls "the paradox of PTSD"—that many symptoms of PTSD are, in fact, behaviors that enable survival in a stressful moment. "Hypervigilance," he explains, "is what soldiers call situational awareness in the combat environment, and [it] is often necessary and beneficial."[134] He makes the same point about "emotional numbing," explaining that while "you have to be able to control your emotions in the combat environment," at home "it seems like they are detached."[135] According to Adler, "Carl's genius" was recognizing that "the very thing that got you through combat is going to screw you up" at home.[136]

Castro's idea was a program that would explain these adaptations and their potential pitfalls to soldiers.[137] In 2005, Adler and Castro went to Fort Carson, Colorado, and, in addition to conducting additional research on the effectiveness of the reintegration briefings, studied whether the psychoeducation model worked. The soldiers they were working with, Adler remembers, were in "units that got really beat up"; nearly all of them had "received artillery, rocket, or mortar fire," and 92 percent reported "knowing someone seriously injured or killed."[138] The nearly nine hundred soldiers were divided into platoon-sized groups; some received the Army's standard debriefing, while others received what would become the BATTLEMIND psychoeducation program.[139] Each group completed a series of surveys designed to measure their mental health before the training and three months later.[140]

Soldiers liked the new training. More important, the results demonstrated its effectiveness. More than half of the soldiers reported that they had "learned a lot from [their] deployment experiences," and nearly 60 percent "realized [their] reactions to the deployment were normal."[141] More important, on almost every measure of post-deployment reintegration, from post-traumatic stress to drinking, soldiers who had been given the psychoeducation briefing performed better than their peers.[142] All of this came as a surprise to Adler and Castro. "We were thrilled,"

Adler recalls. "It was never expected that a one-hour training was going to have a measurable effect three to four months later."[143] In fact, the goal had been far more modest: "We just wanted to make sure it didn't hurt anybody, because of all of the arguing [that] debriefing . . . was toxic and bad. We wanted to show that for service members, it wasn't."[144]

The WRAIR psychologists quickly packaged the program, which was rooted in the idea that "*all* soldiers have the necessary skills to successfully transition home," for wider use.[145] This language reflects one of BATTLEMIND's goals, which was not to pathologize the soldiers.[146] As Adler explains, "We are dealing with a population of service members who are successful. They have been selected, they have survived training, they have been prepared for deployment, they've gone and come back. . . . These are not patients who are all sick."[147]

The name BATTLEMIND was similarly intended. Castro had heard a general use the term and subsequently defined it as "the Soldier's inner strength to face fear and adversity in combat with courage" because of their "existing skills and inner mental strengths."[148] Focusing on strengths, Adler explains, made it more appealing to soldiers, unlike "suicide prevention, which is difficult, because you are already putting the onus on the negative medical problem."[149]

And, because it was the Army, "Battle Mind" became an acronym. Castro, Adler and Dennis McGurk worked over the phone and by email to develop a deployment-related behavior for each letter and a corresponding risk after redeployment.[150] Thus, that soldiers had been with their "Buddies" was contrasted with the "withdrawal" that they would experience; holding people "Accountable" was juxtaposed with being "controlling"; and the "Targeted Aggression" that was appropriate in Iraq could manifest as "inappropriate aggression" at home. The acronym also reflected the researchers' recognition that, however much the program was an improvement over the earlier briefing, it was still one more thing a returning soldier had to do. "The closer you got to the D in BATTLEMIND," Adler recalls, "soldiers could see that it was ending."[151]

The Army quickly embraced the program as its primary means of helping soldiers prepare for and readjust after deployments, with Army Secretary Pete Geren touting it in 2007 as "the only scientifically validated resilience training program in the military" and "an effective method of reducing mental health concerns compared to standard stress education briefings soldiers typically received."[152] By April 2006, Castro, Hoge, and another researcher, Anthony Cox, had presented a conference paper touting the success of BATTLEMIND I and BATTLMIND II, two interventions timed to coincide with the PDHA and the PDHRA.[153] In September 2006, the Army rolled out PreDeployment BATTLEMIND, which prepared soldiers for a tour in Iraq or Afghanistan by warning them that "combat impacts every soldier both physically and mentally" and encouraging them to "expect success" because "other Soldiers have made it through year-long deployments successfully."[154] And yet these presentations also offered an unvarnished assessment of what war entails. The BATTLEMIND brochure, for example, made clear that "it is the Soldier's job to kill the enemy" and that "innocent women and children are often killed in combat."[155] If the last of these statements elides the reality that innocent men are also often killed in combat, it acknowledges a fact that, as the theorist Elaine Scarry has famously pointed out, is often obscured in accounts of war.[156] Similarly, BATTLEMIND training for leaders includes pictures of dead children, tells leaders that they "may hear . . . pleas for help or mercy [and the] wailing of mourners," and "may smell . . . burning flesh and hair."[157] These passages reveal a practical appreciation of the wars' realities. However unwise or unpopular the war was, WRAIR researchers did not critique it or suggest that the Army should stop sending soldiers; their task was doing their best to prepare soldiers to go and helping them readjust afterward; by naming and confronting these realities, they believed that war's psychological impacts could be blunted.

BATTLEMIND's theory, development, and implementation encapsulates WRAIR researchers' views that mental health issues were expected challenges that could be prevented and resolved through rigorous research. Yet neither WRAIR's openness about the psychological realities of war nor its evidence-based approach to treating them stemmed the increasingly frequent critique that the military was not doing enough to address service members' declining mental health. The perception that the government was not doing enough to identify and treat veterans deepened in May 2006 when the Government Accountability Office (GAO), Congress's investigative arm, reported that there were no consistent criteria according to which service members who responded affirmatively to the mental health question on the PDHA were referred for further assessment and that there was no "reasonable assurance that OEF [Operation Enduring Freedom]/OIF [Operation Iraqi Freedom] servicemembers who need referrals receive them."[158] In particular, the authors worried about the lack of "information on why some OEF/OIF servicemembers with three or more positive responses to the [PDHA's] PTSD screening questions received referrals while others did not" and complained that providers use their "clinical judgment to identify those servicemembers whose mental health needs further evaluation."[159] Demanding to know "the factors upon which DOD health care providers based their clinical judgments in issuing referrals," the GAO wanted objective criteria for treating conditions that the medical community approached with a great deal of subjectivity.[160]

Maj. Gen. Stephen Jones's response that that soldiers who had recently completed a PDHA might well be experiencing "symptoms associated with PTSD" but that "to label those symptoms as denoting a disorder may not be appropriate at this time" and that, "frequently, individuals are provided the advice that they should pay attention to the symptoms and return if they do not dissipate or they get worse" was, if predictable, also predictably frustrating to the wars' critics.[161] Ignoring this somewhat nuanced debate on the criteria for referring troops for counseling, a *USA Today* article with the damning headline "Many War Vets' Stress Disorders Go Untreated" explained that "only about one in five Iraq and Afghanistan war veterans who screen positive for combat-related stress disorders are referred by the Pentagon for mental health treatment."[162] *Washington Post* readers found stories with the headlines "Study Shows Combat Stress Aid Falls Short" and "Few Troops Are Treated for Disorder," while CBS posted a story entitled "Stressed Vets Not Getting Help."[163] Clinically grounded assertions by officials that, "in some cases, a medical referral or treatment may prolong symptoms that could disappear naturally," or claims that "many people are referred and were referred in this process not to a mental health professional but to a primary care professional, their doctor," so that "the fact that 22 percent were referred to a psychiatrist or psychologist is not an indication the [others] did not get the support or help they needed" did little to assuage fears among a public already poised to imagine PTSD as an epidemic-in-waiting and the government as uncaring, if not malicious, and that the military was, as the *Washington Post* put it, "short-changing returning service members who are in distress."[164] Many Americans likely agreed with the sentiments of a St. Paul Pioneer Press editorial that reflected, "If officials can't say how doctors determine whether at-risk troops need treatment . . . it's not clear how they can know whether troops are getting the help they need" and that "the Pentagon has a lot of work to do in this area."[165]

Congressional opponents of the war also seized on this study as evidence of military inaction. Mike Michaud (D-ME), a member of the House Veterans' Affairs Committee, released a statement arguing that, "when 78% of the service members who are at risk of developing PTSD do not get a referral for further evaluation, then it's clear the assessment system is not working."[166] In the Senate, Barbara Boxer (D-CA) responded with a scathing letter to Kevin Kiley, the Army's Surgeon General. Arguing that the report "illustrates the emptiness of [the] promise . . . that troops would get the help they need," she argued that "it is clear that we have an emergency on our hands."[167] Like the reporters covering the story, these legislators imagined few distinctions between screening positive for a PTSD symptom and actually having PTSD, assumed that everyone with such symptoms required mental health treatment, and imagined that the military willfully ignored veterans' issues. The failure to refer every symptomatic service member thus affirmed for many Americans that the Bush administration and the Department of Defense were cavalier at best and callous at worst in their treatment of those who had fought in Iraq and Afghanistan.

Outrage over this perception heightened in response to a *Hartford Courant* series that began the same day that the GAO report went public. "The U.S. military is sending troops with serious psychological problems into Iraq and is keeping soldiers in combat even after superiors have been alerted to suicide warnings and other signs of mental illness," the *Courant* wrote in the first article, even though "military superiors were aware that soldiers were self-destructing."[168] The article, which was referenced in newspapers around the country, explained that these deployments occurred as the lengthening wars demanded repeated deployments of a small and stretched force.[169]

Amid devastating anecdotes of "a 20-year-old soldier who had written a suicide note to his mother [and] was relieved of his gun" but who was dismissed as a malingerer before taking his life, and of a soldier who, despite telling a psychologist at Fort Polk, Louisiana, that he was "wracked by nightmares and depression . . . [but] was given a higher dose of medication and the sleeping pill Ambien and told that he was going back to Iraq," the military once again hewed to its clinical argument and came off as callously dismissive. "'The challenge for us . . . is that the Army has a mission to fight. And, as you know, recruiting has been a challenge," Elspeth Ritchie told reporters. "And so we have to weigh the needs of the Army, the needs of the mission, with the soldiers' personal needs."[170] Asked whether the Army should have known better than to send troubled soldiers on additional deployments, she responded with a comment that reflected the Army's complex understanding of the nature of mental health concerns: "When you go back, in retrospect, there may be some warning signs. . . . What you don't see from that are the other cases that perhaps had the same warning signs and were kept in [the combat] theater and went on to do OK in their job."[171]

In fact, determining the circumstances under which soldiers with prescriptions could be deployed or how medications could be managed in the combat environment deeply concerned the AMEDD. Partly this was because Army medicine has two sometimes contradictory missions. "All doctors in the military wear two hats," Charles Hoge explains. "We are the advocate for the patient and the provider, and we also are occupational medicine, physicians or clinicians evaluating whether that individual is capable of doing their job in a safe and effective manner. And sometimes, those two roles conflict with one another."[172] Thus Army clinicians were faced with both helping soldiers recover from the psychological effects of a deployment and returning them to health so that they could deploy again. As a result, doing what the Army needed, the soldier wanted, and what was medically most advisable was sometimes challenging. On the one hand, the Army needs deployable soldiers, and, as Army psychiatrist Col. Chris Ivany explains, "most of the time soldiers want to deploy. They sign up and they want to do this and they don't want to be singled out and kept at home to be treated."[173] On the other hand, there's the question of what "their mental health condition is and how severe it is, whether or not it can be treated safely in theater, and whether or not the act of deploying them may make their mental health condition worse."[174]

Early on in the wars, there were no official guidelines on which mental health conditions or medications would render a soldier non-deployable, and providers made decisions essentially on a case-by-case basis. What that meant in practice, as Ritchie explained, was that, "if we can keep somebody in the fight who happens to be on Prozac, then we ought to continue to treat him on Prozac."[175] This was a view that Ritchie had long held; one of her first scholarly articles, based on her experiences in Somalia in 1992, had argued that such deployments were acceptable.[176] And yet she understood that the demands of the Iraq War were causing the Army to sometimes make medically indefensible decisions. "Back in 2005, 2006, we were really looking for soldiers," she concedes, "and there were probably soldiers that we were returning to the fight that shouldn't have been."[177] And yet, in her view, the *Courant* articles tended to oversimplify the issue to "people shouldn't deploy on medications."[178] Like Ivany, she believed that most soldiers wanted to deploy and, in fact, had joined the Army with the expectation of going to war, and the prescriptions allowed many soldiers the chance to do that. "This is an all-volunteer army," she explains, "so people are joining because they want to. If the medication helps them do their job, if they can get better, then why not?"[179] To not allow a soldier that opportunity, she feared, would further stigmatize mental health issues.[180]

On the issue of medications, Army Surgeon General Kevin Kiley remembers, "I think we kind of defaulted to 'If you are doing well, you can take those back with you.'"[181] In fact, though, there were some implicit guidelines, although they were largely commonsensical. "Typically, you could deploy on SSRIs [selective serotonin reuptake inhibitors] as long as you were stable," Ritchie explains. "You shouldn't deploy on anti-psychotics for pretty obvious reasons. If you're treating somebody with an antipsychotic they shouldn't be in Iraq with an assault rifle."[182] Beyond that, there were questions of specialty, diagnosis, and access to treatment in theater: "What is someone's job? If they are an 11-Bravo, an infantryman, well, we don't want them deploying on medication . . . because we can't guarantee them the medication" because medications that need to be monitored and refrigerated or that might require regular blood tests can't be managed at a remote forward operating base.[183] "On the other hand," Ritchie explains, "if we have a neurosurgeon and we are short on neurosurgeons and if he needs a little bit of Ritalin . . . and he's going to be in a hospital environment where he can be monitored," that was an easier decision.[184]

A similarly flexible set of guidelines surrounded the deployability of soldiers with PTSD. Clinicians sought to take into account how acute the PTSD was and how well it was being managed. On the one hand, if a soldier is "barely functioning despite medications," Kiley asserts, "he shouldn't be deployed."[185] On the other hand, there are soldiers at the other end of the spectrum, for example, a "guy on medication for four or five months, and he's no longer incapacitated by the PTSD."[186] A deployment for that soldier, Kiley asserts, may even be beneficial and aid in his recovery. "If he wants to stay with his buds—maybe he's a squad leader and maybe they've been good to him—[maybe] he got a heroism award over there [or] a Purple Heart, and the Army is looking pretty good," the former surgeon general reflects, "should we not send him over there?"[187] In Charles Hoge's view—one that echoes BATTLEMIND's ethos—deploying that soldier might also be good for the mission. "The reality was that a lot of those guys who had PTSD [and] who had learned to cope with their PTSD symptoms were exactly the guys you want on your team down range because they are highly effective," he explains, adding that "they have that emotional control. They have that ability to channel anger. They have that hypervigilance that's necessary for that environment." As a result, he explains, "the way the military approached those concerns was really very simple and straightforward. If the soldier is able to do his job . . . they go back to doing their job."[188]

This decision making was controversial, Kiley asserted, because people outside of the military medical community rarely had a nuanced awareness of the range of severity with which PTSD can manifest. "This is a level of discussion," he argues, "that generally doesn't happen in the *Congressional Record*."[189]

Clinicians thus made decisions about deployment based on a number of factors—their understandings of particular medical conditions and their Hippocratic responsibility to do what is best for individual soldiers, the Army's demands and their obligations as officers to meet them, and their understanding of the force as a group of volunteers who in many cases had joined wanting to go to Iraq or Afghanistan. In doing so, Army medicine had to overlook the reality that racial and economic factors have historically driven marginalized populations to enlist, and it had to abandon the exceptionalism often accorded, and asserted by, the military to claim that military service is a job that, like any other, a person should not be prohibited from performing simply because he or she has a mental health issue.[190] Decisions about deploying troubled troops, then, intersected with, challenged, and required the dismissal of significant historical realities and cultural narratives that, had they been embraced, might well have led providers to less liberal interpretations about who could be deployed.

* * *

If May 2006 was a month of bad publicity, it was also a month in which the Army's internal studies began to show that its efforts were struggling in the face of a prolonged war. The third MHAT report, released on May 26, betrayed the Army's increasing concern over its ability to provide high-quality care as the wars went on and required more deployments. The Army was able to point to continued improvements in behavioral health care, with a corresponding reduction in stigma and more accessible care.[191] These improvements, however, could not keep pace with declining mental health in the force. Soldiers reported that they were exposed to greater levels of violence than soldiers had been a year earlier, levels of depression and anxiety had returned to 2003 levels, as had the suicide rate, and soldiers were dissatisfied with the Combat Operational Stress Control training that they had received.[192]

More significant, the MHAT-III report was the first to devote considerable attention to the effects of multiple deployments. Paul Bliese, who first deployed to Iraq on MHAT-III, remembers this question as the most significant one that he and his WRAIR colleagues sought to study. One hypothesis was that soldiers on their second deployments would have few mental health issues, Bliese explains, because "if [they]'ve been over in a combat zone before, and [they]'ve been successful at it, then coming back again will make [them] more resilient."[193] As it turned out, however, the differences between soldiers on their first tour and those who had returned for another were substantially worse by almost every measure. "Soldiers who had at least one prior deployment to Iraq reported significantly higher levels of acute stress (18.4%) than those on their first deployment (12.5%)," the report concluded, not because they were performing more dangerous tasks but because they were "returning to OIF with unresolved acute stress problems . . . related to their previous deployments."[194] For the reports' authors, long and frequent deployments loomed as the most significant source of stress.[195]

Six months later, the fourth MHAT report's conclusions were equally grim. The MHAT-IV team, which Carl Castro led, was in country between August 28 and October 6, almost three years to the date after MHAT-I.[196] Unlike the previous three MHAT reports, this one was written entirely in theater so Castro, McGurk, and a specialist could present a finished report to the commanders before they returned to the states.[197] That report reflected the worsening conditions in the country and the corresponding decline in soldiers' psychological well-being. Most soldiers and Marines found themselves in life-threatening situations and "described an event which occurred during the current deployment that caused them 'intense fear, helplessness, or horror.'"[198] This language, drawn directly from the *DSM* criteria for post-traumatic stress disorder, implied that a significant percentage of U.S. troops were at risk of a stress reaction. And indeed, soldier reports bore this out. Although rates for anxiety, depression, and acute stress had remained static, the rate of "severe stress, emotional, alcohol, or family problems" had nearly doubled since 2003, and soldiers who had experienced significant combat screened positive for acute stress reactions with much greater frequency.[199]

So, too, were conditions worse for those on their second or third deployment. Research that Adler, Bliese, Castro, and another psychologist, Ann Huffman, had done on troops deployed to peacekeeping missions in the Balkans had suggested that while long deployments were detrimental, multiple deployments produced fewer mental health issues.[200] By 2006, however, research—including some that Castro had conducted—was suggesting that that was not the case for troops deployed to an active combat zone.[201] The MHAT-IV team's research revealed that mental health issues were 10 percent higher in those who had deployed more than once compared to those on their first deployment.[202] Perhaps more important, the study concluded that "factors generally found to be protective or restorative" had little bearing on mental health outcomes and that "previous deployment experience does not 'inoculate' Soldiers against further increases in mental health issues."[203]

On top of increased stress, soldiers had few chances for relaxation. Much has been written about the availability of fast food, first-run movies, and other niceties for soldiers serving in Iraq, but the MHAT-IV report offered a withering assessment of those services' accessibility to soldiers who had been in combat and most needed them.[204] "The Soldiers or Marines who execute the 'outside the wire' missions who are at greater risk of being seriously injured or killed," Castro wrote, "can rarely take the afternoon off to attend MWR [Morale, Wellness, and Recreation] events or concerts."[205] Soldiers were thus facing increased stress with fewer opportunities to rest, a revelation that rendered statistics showing that the psychological impact of multiple deployments was worse than it had been six months earlier even more troubling.[206]

Passages like this were part of an explicit effort on Castro's part to highlight both the reality that the war was as traumatizing as any other war had been, if not more so, and that the psychological danger was not parceled out equally. In Iraq, Castro found that some senior officers didn't take into account that "a lot of people never left the base camp once we got there."[207] When one general reflected that he was on his sixth deployment with no mental health effects, so perhaps the issue was soldier resiliency, Castro was forced to point out that generals might experience a deployment somewhat differently than enlisted infantrymen.[208] The data bore this out. Troops who had experienced high combat were nearly three times as likely to have some sort of mental health issue as those who had not, and in the report Castro emphasized that "not everybody in the brigade is equally at risk" and that resources had to be apportioned corresponding to that risk.[209]

Castro had another bone to pick as well. Anecdotally, he had encountered veterans and others who described serving in Iraq as somehow less strenuous than previous conflicts, and that was a view that he "want[ed] to try to dispel."[210] In his view, Iraq was more psychologically stressful because, in "World War II and in Vietnam, you got pulled off the line . . . and you could be in the rear." In the twenty-first century, by contrast, "It was really like being on the line . . . for a year straight with very little break." He wanted his readers to grasp that "we've never fought a war like that before, ever."[211]

For all of the comparisons to Vietnam that the Iraq War engendered—and by late 2006 they were legion— Castro is correct that there were crucial differences between those who fought in each. When Army physicians attempted to discern why so few men in Vietnam were suffering psychiatric disorders, they concluded that "no one served in the theatre of war for longer than a year; there was plenty of rest and recreation during the tour of duty; battles were short; soldiers had to endure few major artillery barrages; and their morale was high."[212] By 2007, though, the opposite was true on nearly every count. Most personnel were on their second or third tour, and the Army extended tours to fifteen months in April 2007. Rest and recreation options were not widely available, and there was no discernable rear where soldiers were safe from improvised explosive devices (IEDs) and mortar attacks. Thus it is not that military mental health professionals took lessons from the wrong war, as military leaders in general were often accused of doing. It was, rather, that they had not planned on, and had no model for, fighting a prolonged, violent insurgency with a small volunteer force, and they did not have the resources necessary to ensure troops' mental health.

The desire to dispel the myth that Iraq was somehow easier on soldiers than the Second World War or Vietnam had been led Castro and McGurk to vigorously critique the situation in Iraq as unsustainable. "Although it has long been recognized that mental health breakdown occurs after prolonged combat exposure . . . a considerable number of Soldiers and marines are conducting combat operations everyday of the week, 10–12 hours per day seven days a week for months on end," they railed. "At no time in our military history have Soldiers and marines been required to serve on the front line in any way for a period of 6–7 months, let alone a year, without a significant break in order to recover from the physical, psychological, and emotional demands that ensue from combat."[213] They then unfavorably compared the Iraq War to Vietnam, pointing out that, "even during Vietnam, week-long combat patrols in the field were followed by several days of rest and recuperation at basecamp."[214] In a blistering assessment, they condemned as foolish anyone who doubted how much the war was damaging the troops: "Arguing that the intensity of combat operations in Iraq is not comparable to those of previous wars . . . demonstrates a lack of appreciation of what constitutes combat in general, and an ignorance as to the level of combat Soldiers and Marines are experiencing in Iraq."[215]

The report's recommendations, and Castro's strident tone, illustrated the Army's growing concern with the psychological impact of long tours and multiple deployments to a brutal war. The Army had poured significant resources into providing troops with mental health education and care. By the time of MHAT-IV, the ratio of behavioral health providers to soldiers was lower than it had been at any time, and those providers were better trained and more committed than any in the Army's history.[216] None of that was sufficient, however, in the face of the Iraq War's realities. It was increasingly violent. The soldiers and marines who fought it were perpetually at risk, and they were required to fight longer, with fewer breaks, and on more deployments than any previous fighting force that the United States had fielded. The very nature of the Iraq War, and the failure of the Bush administration to plan for it and commit the resources necessary to sustain the force during it, had left the AMEDD overwhelmed and soldiers struggling.

* * *

For the war's critics in the House and Senate, Hoge's *JAMA* article and the media response that it generated, the GAO report, the *Harford Courant* articles, and the MHAT-IV report provided yet more evidence that the military—and, more important, the Bush administration—was sacrificing soldiers' mental health to a futile war.[217] While congressional Democrats critical of the Iraq War often mentioned the death toll, the number of injuries, or the cost of the war, they also pointed to the psychological damage that the war was causing as a primary indicator that it was misbegotten. In the month that followed these publications, Rep. Jim McDermott (D-WA) announced that "the American people have heard enough to know the trust they placed in the President over his justification to invade Iraq was misplaced" and denounced Bush's declarations of progress as "a Presidential declaration that more American soldiers will . . . return home traumatized by post-traumatic stress disorder."[218] In February 2007, Sen. Bernie Sanders (VT-I) called the war "a disaster" that "we were misled into and a war many of us believed we never should have gotten into in the first place" and one "which the administration was unprepared to fight." The cost, he explained, was not just the dead and wounded but "tens of thousands [who] will come home with post-traumatic stress disorder."[219] Sen. Robert Byrd (D-WV), who had eloquently opposed the war in 2003, responded to Republicans—who had opposed a resolution that cut off funding to the war[220]—by invoking PTSD: "How can some express unwavering support for the troops if they quake in the face of a debate about their safety?" he asked, explaining, "Our troops are stretched thin. They are weary after deployment and redeployment. Post-traumatic stress disorder and mental problems—yes—are rife in the troops."[221]

Perhaps the most biting critique came on March 28, 2007, a week after the Iraq War passed the four-year mark and as U.S. troop numbers increased with the surge. Debating another supplemental funding package, Sen. Barbara Boxer (D-CA) took the Bush administration to task for its failures in Iraq, linking an unachievable mission to the continued deployment of troops who had been diagnosed with PTSD. "If you love the troops," Boxer concluded—condemning with an equal mixture of sarcasm and contempt the pro-war maxim that the country must support the troops by not critiquing the war—"you don't send them back to fight with a post-traumatic stress disorder and a bottle of antidepressants."[222] Although these comments problematically made Americans the war's primary victims, they also highlighted how much discourse of mental health had come to define Democratic opposition to the war.

Mental health, however, turned out to be more than a rhetorical cudgel with which the war's opponents could chastise the Bush administration. In October, the Senate passed the 2006 National Defense Reauthorization Act, the major Department of Defense (DOD) funding bill. Senate Democrats inserted several provisions aimed at addressing the troops' mental health, including funding for a fifteen-year longitudinal study on traumatic brain injury and a revision of the mission of DOD's commission to study mental health needs and access to care that stipulated that the DOD would particularly "consider the specific needs with respect to mental health of members who were deployed in Operation Iraqi Freedom or Operation Enduring Freedom upon their return." Demanding "an identification of mental health conditions and disorders (including Post-Traumatic Stress Disorder, suicide attempts, and suicide) occurring among members who have undergone multiple deployments" and "recommendations on mechanisms for improving the mental health services available to members . . . who have undergone multiple deployments," the amendment was the first legislation that asserted that the particular conditions of the twenty-first-century wars were producing a specific set of mental health issues.[223] In a second amendment introduced within a month of the *Hartford Courant*'s series, Boxer stipulated that the Department of Defense specify the criteria under which service members with mental health conditions or who were prescribed drugs could be deployed.[224]

These amendments indicate the extent to which the events of 2006 had fundamentally reshaped the Senate's understanding of the centrality of mental health issues to the Iraq War. For the war's critics, a focus on mental health allowed them to place the troops' well-being at the center of their critiques and thus to condemn the war without being accused of not supporting the troops. These demands also illustrated, however, that, even as the military was coming to terms internally with the degree to which the war had overwhelmed its initial efforts to address service members' mental health needs, addressing these issues required an active and engaged Congress that both mandated action and provided resources. In making these demands, the war's opponents in Congress—almost universally Democrats—began the process of countering the Bush administration's refusal to acknowledge the psychological costs of the war and to help the military reverse its decades-long marginalization of these issues.

Nonetheless, the military's response to the second of Boxer's amendments continued to rely on the long-standing medical discourse that had defined its approach throughout the wars. On November 7, a month after the National Defense Reauthorization Act passed, Undersecretary of Defense William Winkenwerder issued a memorandum that detailed "guidance on deployment and continued service in a deployed environment for military personnel who experience psychiatric disorders and/or who are prescribed psychotropic medication."[225] The document provided more precise guidelines for deploying troops with psychological diagnoses, but it also formalized the notion that diagnosis and treatment usually had to be individualized, that few precise prescriptions could be made, and that, therefore, decisions about deployment had to be based on the individual service member's capacity, the clinician's judgment, and the mission requirements. Service members could deploy if their condition was "in remission or whose residual symptoms do not impair duty performance," the memorandum declared, and "disorders not meeting the threshold for [unsuitability] should demonstrate a pattern of stability without significant symptoms for at least 3 months prior to deployment," while Service members can deploy on medication provided that the prescription could be easily managed in theater.[226] Those who had not achieved "restoration to full functioning within one year of onset of treatment," in contrast, were to be sent to a medical evaluation board for possible discharge.[227]

If on first glance these criteria seem like they resolve the problems raised by the *Hartford Courant* articles, they in fact leave much to the provider's clinical judgment and the commander's needs. If a condition developed in theater, for example, clinicians were instructed to take a variety of factors into account, although it offered no specific criteria for doing so. With the exception of the stipulation that "Psychotic or Bipolar disorders are considered disqualifying for deployment" and that soldiers diagnosed with those conditions in theater "will be recommended for return to their home station," the memo makes no mention of particular conditions, much less how they should be handled. The only advice was that decisions about medically redeploying soldiers should be made in consultation with their commander.[228] Similarly, the memo offers no blanket restrictions on medication, noting that "there are few medications that are inherently disqualifying for deployment for all military occupational specialties, to all potential operational locations, and at all times during the conduct of operations."[229] Moreover, the memo countered the GAO report's assertion that the military needed to establish more objective criteria for referring troops to care by asserting that mental health referrals were mandated only when "the healthcare provider determines that a concern or condition demonstrates a potential negative impact on performance in an occupational specialty or fitness for military service" and that "it is the responsibility of the Service member to report past or current physical or mental health conditions or concerns."[230] In effect, the memo codified the informal practice of allowing individual providers to rely on their judgment when considering who could be deployed and who should stay or be sent home, who needed treatment and who did not.

* * *

By the time George W. Bush declared that he was sending an additional thirty thousand troops to Iraq that January, the mental health of U.S. service members had emerged as a significant concern in the Army, in Congress, and in the public sphere. And yet these groups approached the issue very differently, and their visions often conflicted with one another. In Congress and in media coverage, the deteriorating mental health of U.S. troops was further evidence that the Iraq War was misbegotten and that the Bush administration was pursuing a failed policy. For the war's critics, an emphasis on the troops' mental health insulated them from criticism that in opposing the war they were failing to support those fighting it. These critics, however, maintained a relatively simple view of mental health that did not distinguish between ordinary and temporary readjustment challenges and long-term, debilitating psychological issues. This anxiety shaped their opposition to other policies, including the subsequent deployment of troops who had a mental health diagnosis or who were prescribed drugs to teat mental health conditions.

The Army Medical Department, in contrast, viewed these issues with more nuance, maintaining that most mental health issues were temporary and treatable and that better education and surveillance could restore most soldiers' mental health. For them, conditions like post-traumatic stress disorder existed on a Gaussian curve, and it was thus impossible to make precise specifications about which soldiers could deploy with which conditions, and which could not. To the war's critics, these assertions sometimes seemed like obfuscations and further evidence that the military, and by extension the administration, were cavalier in their commitment to those who served. And yet, even as Army researchers and clinicians sought to assert that they were using their best clinical judgment, the Army's own research revealed that the worsening wars were causing soldiers' mental health to deteriorate.

This tension between popular and clinical understandings of mental health, and the increasing internal anxiety about how the wars were affecting the force, would define the debates over troops' psychological well-being for the remainder of the war. In both the Army Medical Department and the Department of Veterans Affairs, questions concerning how to identify and treat mental and behavioral health issues always devolved to the question of what diagnostic measures and treatment research were most effective. For those with lay understandings of psychology and psychiatry, however, the notion that veterans were returning from the wars struggling with severe mental health issues came to define in significant measure their understandings of the war itself. And nowhere was this more true than when Americans began encountering stories of Iraq War veterans who had taken their own lives.

3

"Callous Disregard of Veterans' Rights Is of a Piece with the Administration's Entire Approach to War"

Veteran Suicide and Anti-war Sentiment

On March 19, 2005, the Iraq war's second anniversary, President Bush sought to reassure Americans that that U.S. casualties had "added to America's security and the freedom of the world."[1] Many Americans were not convinced. Near-daily news of sectarian violence, attacks against U.S troops, and American deaths had led many to agree with the *New York Times*'s assessment that "it's clear that the presence of American troops is poisoning the situation."[2] More than seven hundred protests in locales ranging from Chicago, Illinois, to Sarasota, Florida, marked the anniversary.[3]

In Galva, Illinois, an hour outside of the Quad Cities, however, Mike and Kim Bowman waited for the convoy of buses carrying their son Tim and his fellow soldiers in Foxtrot 202, a combined unit of the Illinois National Guard. March 19 marked their return from a sixteen-month deployment in Iraq, including nearly a year attached to the First Cavalry Division.[4] Their mission had been harrowing. After leaving for Iraq on March 4—Tim's twenty-second birthday—their duty was providing security on the "Route Irish, the road to the airport from the green zone. Or as it has been called so many times, 'The most dangerous Road in the World.'"[5] The road had a terrible reputation. After U.S. troops killed an Italian intelligence agent at a roadblock in March 2005, the *New York Times* wrote that the road was so dangerous that the military flew high-ranking officials from the airport to the Green Zone rather than risk the drive.[6]

Mike and Kim knew about some of their son's experiences. Communication between deployed troops and their families was more consistent in the Iraq War than it had been in any previous conflict, and they were able to video chat with their son regularly.[7] A week before his return, when his mother had asked if he was excited to come home, he replied, "I'm not sure. Everything's changed. I'm not the same person anymore."[8] And when Tim walked off the bus, Mike immediately realized that Tim had told the truth. "He was a different man," he recalled. "He had a glaze in his eyes and a 1,000-yard stare."[9]

No one from the Veterans Administration (VA) met the 202nd's buses, however, and Tim went home and started working in his father's electrical contracting business, which the family assumed he would inherit.[10] He soon grew irritable and began drinking. He missed his Guard buddies.[11] He avoided talking about what was bothering him. If someone suggested counseling, he told them that he could handle it on his own.[12] On the night before Thanksgiving, however, Tim seemed more like his old self, "joking and affectionate" with his family.[13] Nonetheless, his friend Dustin pulled Mike and Kim aside, telling them, "Tim's just not getting better."[14] They decided that on Sunday they would hold an intervention to force the issue.[15] On Thanksgiving, though, Tim didn't show up for dinner.[16] Mike found Tim at his electrical shop. He had shot himself twice. When the first bullet didn't kill him, he had put the gun in his mouth.[17]

The inner turmoil that led Tim Bowman to take his life remains unknowable. However, suicide, like any death, is both a biological fact and a cultural phenomenon.[18] "In suicide," Richard Bell writes of eighteenth-century Americans, "they found a tool of political argument equally capable of unifying, validating, and mobilizing as it was of dividing, delegitimizing, and enraging."[19] Such was the case for twenty-first-century Americans. As public opinion turned firmly against the Iraq War in 2006 and 2007, Americans mobilized around the suicides of veterans like Tim Bowman as further evidence of the Bush administration's disingenuousness in prosecuting the war and unwillingness to acknowledge its failed strategy.[20] In media coverage of veterans' suicides and in the meeting rooms and chambers of the Senate and House of Representatives, profiles of suicidal veterans explicitly countered the claims that military recruiters made about the value of military service, while accounts of their post-war trauma highlighted the most violent and controversial aspects of the war. Accounts of the VA's unpreparedness and seeming apathy became a microcosm of the Bush administration's lack of planning, duplicity, and unwillingness to acknowledge its failures, while the growing awareness that so many veterans suffered in silence became emblematic of a culture that purported to support the troops but was unwilling to apprehend their experiences.

* * *

That Tim Bowman had served in Iraq at all was itself an indicator that the war was not going as well as the Bush administration had promised. Although Bowman had enlisted after the September 11 attacks, he had likely not expected to deploy overseas, and certainly not to a combat zone.[21] By Bowman's 2005 return, however, a January 2002 article's speculation that "the strain on the military—including 1.3 million reservists, many of whom had been called to active duty several times in the past decade—could become overwhelming" now seemed prescient.[22] By 2005, the regular forces and the Guard and Reserve were struggling to fill the ranks.[23] About the time Bowman returned, the Army announced that it had about one-third as many newly trained infantry troops as it projected needing.[24] A year later, as anti-war sentiment increased and Americans increasingly turned their attention to Iraq veterans' mental health, the Army missed its recruiting goal by more than six thousand inductees, the worst rate since 1999; nearly 40 percent fewer high school graduates planned to enlist than in 2001.[25]

The Army met this challenge at the end of 2005 with a new advertising campaign that targeted not only potential volunteers but also adults who might shape their decisions.[26] With the tagline "You made them strong, we'll make them Army strong," the Army presented the "Help Them Find Their Strength" campaign as "an effort to help parents understand why they should support their sons' and daughters' decisions to join the Army," promising parents that it would help their children transition to adulthood.[27]

The campaign's central motif was that parents concerned about their children's decisions would ultimately be impressed by their transformation. As Carrie Anne Platt explains, "The Army Reserve is presented as a *nurturing* environment in which young men and women grow into well-educated and well-qualified adults."[28] One commercial begins with a rural, working class father—he wears a flannel shirt, and the ad features him and his uniformed son walking down a dirt road flanked by cornfields, the family sitting in their modest kitchen, and the parents leaning on a large piece of farming equipment—admitting that he "was pretty nervous, apprehensive." The son, however, intones that "it's given me a whole bunch of confidence," while the parents confess that, "now, I'm very proud of him" and that "he's just a stronger, more driven individual."[29] Another commercial featured a father whom one writer called "Mr. Middle America, with his denim jacket and his steely glint," talking to his son on a porch of what appears, again, to be a farmhouse in the rural Midwest.[30] As Carrie Anne Platt highlights, the ad clearly frames the conversation as occurring between two men.[31] The father tells his uniformed son, "You've changed" because "you shook my hand, and you looked me square in the eye," and he asks, "Where's that come from?"[32] As Platt argues, "the [Army Reserve] logo answers the question," and the intent of the advertisements is clear: In the midst of a war that increasingly drew service members from the rural Midwest, South, and Mountain West, parents could rest assured that their children would return better than they had been when they left.[33]

As Platt explains, in these commercials, "the military is understood as a masculine rite of passage," a place where young men "will learn . . . how to be a man."[34] The Army's head recruiter put it similarly: "What will happen if you come into the Army is what soldiers know happens: you become better."[35] And if the Madison Avenue firm McCann Erickson, which designed the campaign, claimed that they "didn't want to avoid" the reality that enlistment in 2006 almost certainly meant deployment to Iraq or Afghanistan, for some critics this was hardly a reasonable claim.[36] "Who wants to let their son enlist when soldiers are getting killed every day," Seth Stevenson asked in *Salon*, complaining that the campaign "just sidesteps the improvised explosive device in the middle of the room."[37]

This critique paralleled criticisms of the war. After the 2006 midterms, the *Augusta Chronicle* called the mandate "stay the course," which Bush had resolutely promised to do, "as little more than stubborn intransigence in the face of mounting casualties."[38] In December, the Iraq Study Group, which Rep. Frank Wolf (D-VA) had first proposed in the fall of 2005 and which would, the *Washington Post* explained, be "a bipartisan commission of well-respected policymakers to bore deeply into the Iraq dilemma and recommend solutions," released its report.[39] The findings were not encouraging. "The situation in Iraq is grave and deteriorating," it announced, before explaining that "if the situation continues to deteriorate the consequences could be severe. A slide towards chaos could trigger the collapse of Iraq's government and a humanitarian catastrophe. Neighboring countries could intervene. Sunni-Shia clashes could spread. Al Qaeda could win a propaganda victory and expand its base of operations. The global standing of the United States could be diminished."[40] In the wake of the report, the *Washington Post*'s Eugene Robinson declared it "unconscionable to think about dispatching more young men and women to Iraq without the realistic expectation that their presence will make a difference in a war that is no longer in our control."[41] John Kerry (D-MA), whom Bush had defeated in 2004, declared that "refusing to change course for fear of the political fallout is not only dangerous—it is immoral. I'd rather explain a change of position any day than look a parent in the eye and tell them that their son or daughter had to die so that a broken policy could live."[42]

In media coverage, in congressional testimony, and in the remarks of liberal legislators, veteran suicide became further evidence that the war had been poorly planned and that the promises made about it had either deeply underestimated the war's violence or been intentionally misleading. Despite the reality that the military was increasingly populated by African Americans and Latinx soldiers, accounts of veterans who had died by suicide almost universally featured white working-class young men of the sort frequently featured in advertisements who had been asked to perform violent actions beyond the scope of their training and who had returned home irrevocably damaged. Accounts of veteran suicide thus explicitly countered the "Help Them Find Their Strength" campaign's logic. Far from becoming men who were credits to their community, Tim Bowman, Joshua Omvig, Chuck Call, and Jeffrey Lucey had lived post-military lives marked by personal and professional struggles, failed masculinity, and psychological regression. The description of these men, their experiences in Iraq, and their ultimate suicides thus cast into specific relief broader criticisms of the Iraq War.

Newspaper profiles of suicide victims consistently present them as innocent and altruistic, precisely the sort of young men whom the military promised to improve. Jeffrey Lucey, a Marine veteran who hanged himself in his parents' Massachusetts basement in 2004, "was just an ordinary kid from small-town America," a widely reprinted Associated Press piece began, before describing a nearly idyllic childhood: "He grew up loving his parents, his high-school sweetheart and backyard Wiffle ball games in this quiet picturesque community."[43] His motivations for joining the military seemed plucked from the "Help Them Find Their Strength" campaign's narrative. "His parents," the *New York Times* reported in 2007 "were not happy. They had hoped their son would go to college."[44] But, as a 2004 account of his death made clear, he joined up in 1999 for a "run-of-the-mill [reason]"; as his mother put it, "'He just wanted to prove he could cut it."[45]

This tone echoed throughout other accounts of veteran suicides. Joshua Omvig, for example, was "a former Boy Scout with a newspaper route" who had "always wanted to be a policeman" and who "with his parents' guidance . . . became a soldier."[46] Here was another portrait of an ideal childhood leading to a self-improving military career. "The choice was logical for an aspiring policeman or sheriff's deputy," the article explained.[47] Jonathan Schulze, according to the *Minneapolis Star Tribune*'s description, was "a tough kid from rural Minnesota" who "visited aging veterans in the state homes, helped anyone in need [and] bragged of adoration for his young daughter, Kaley Marie." To his fellow Marines, "he was the life of the party."[48] Another article noted a devastating detail from Schulze's VA intake form; asked to list his strengths, he had written, "Big heart."[49]

Other descriptions of veterans who took their own lives were nearly identical. Chuck Call, a West Virginia Army reservist who took his life in February 2005, was a thirty-year-old man who "often spent time at the nursing home where his mom worked. On Halloween, he would sit with residents and hand out candy to kids. Before he left for Iraq, he went to the nursing home and shook each veteran's hand, personally thanking them for their service."[50] Call thus appeared as another man whose life seemed cut from an army recruiting commercial: a model citizen willing to do his part in the war. Indeed, "he chose to go to Iraq" even though "his unit wasn't going" because "he wanted to fight for his country."[51] Twenty-five-year old David Fickel, who killed himself with a shotgun blast on Memorial Day 2006, "was a level-headed, caring kid [who] made people laugh and loved to play baseball and golf" and was "an average student who liked history and hoped someday to teach school."[52] And then there was Tim Bowman, "a charming jokester, a small-town kid who played musical instruments in high school, attended some junior college, and then went to work in his family's electrical business."[53] At a 2007 hearing on veteran suicide, Bowman's father testified that "Tim was a life-of-the-party, happy-go-lucky young man who joined the National Guard in 2003 to earn money for college and to get a little structure in his life."[54]

The similarities among these descriptions defined the terms of the suicide debate. While there is no question whether they in fact evinced these qualities, and an analysis should not minimize the tragedy of each man's suicide, the parallels illustrate how attention to veteran suicide evoked but subverted narratives that validated military service as an avenue toward personal growth while minimizing the war's violence. Accounts of the war's impact on these youthful, innocent, family-focused, and community-oriented young men instead demanded attention to the war's impact on American service members and furthered critiques of the war and the Bush administration.

Indeed, each man was presented as having been destroyed by a war more gruesome than they had anticipated and for which they were ill prepared. Joshua Omvig "probably participated vigorously" in combat, an article in his hometown newspaper explained, and his parents "realized the person they got back . . . was not the young man they sent."[55] A similar sentiment surrounded Lucey's death. "By the time he came home," Bob Herbert wrote in the *New York Times*, "Jeffrey Lucey was a mess. . . . This once-healthy young man had been shattered by his experiences."[56] Fickel had "had changed from being a happy, fun-loving, really outgoing person into being . . . very angry, a clean freak, uptight. He was not the same kid."[57] Mike Bowman immediately followed the portrayal of his son as "happy-go-lucky" with the assertion that his son returned "a different man."[58] All of these portrayals, but particularly Bowman's, recall the "Two Things" commercial, in which the father declares that his son has "changed," but here those changes are emphatically negative. "A glaze in his eyes and a 1,000 yard stare" is the opposite, after all, of looking your father in the eye.

Accounts of suicidal soldiers' post-deployment lives also countered images of happy homecomings and parents sizing up their newly mature adult children. Omvig's mother told reporters that "He'd say, 'Mom, I don't want you to hate me,'" while he repeatedly told his father "Dad, I just want to be happy like you."[59] Images of happy reunions were likewise contradicted by Mike Bowman's story of his son committing suicide while the family gathered for Thanksgiving, Joshua Omvig's suicide taking place "four days before Christmas"—and his mother initially thinking that her son's suicide note "was a Christmas list"—and Jeffrey Lucey's suicide coming after "a moment of deep despair on the Christmas Eve after his return from Iraq, [when he] hurled his dog tags at his sister Debra and cried out, 'Don't you know your brother's a murderer?'"[60] For these veterans and their families, holidays were not times to celebrate accomplishments. They were moments at which they reckoned with the despair that the war had created.

More explicitly, these articles countered assertions that military service was a crucible that forged masculinity. A lengthy article about Bowman's death described a scene where, having been "home for a short leave, Tim and his father stopped for a beer after a softball game," referencing the American pastime and a father and son meeting as equals in one of the most time-honored homosocial rituals—going out for a beer.[61] In fact, it's not difficult to imagine an Army recruiting commercial that relocated the father and son from their porch to local bar.[62] Yet the Bowmans' conversation departed gruesomely from these comforting images, as they "talked about an episode in which Tim, as the last line of defense, was forced to shoot at a car—with a family inside—that had failed to stop at a checkpoint."[63] Here, the homosocial moment is not an opportunity to reflect on the son's growth, but rather one of confession.

Lucey's also spoke of atrocities and civilian casualties. "The most chilling story Jeff told his family," the *Boston Globe* reported, "was of being ordered to shoot two Iraqi prisoners and then keeping their dog tags" and "of running from his truck and scooping up a dead child."[64] Perpetrating such violence resulted not in Lucey's maturation but, rather, in his failure to achieve manhood. The most devastating account in media coverage of suicidal veterans' post-war lives may be Lucey's father explaining that on the night before he took his life, Jeff "asked me if he would be able to sit in my lap. And so for 45 minutes we rocked in silence."[65] This is not a story of an adolescent who became a man through his service; it is a story of regression that asserted that the war had profoundly unmade a man, returning him to infancy.

These narratives of broken young men confessing to atrocities aligned with broader critiques of the war as a poorly planned disaster that was bereft of morality. Chief among these were accounts of U.S troops killing civilians and perpetrating war crimes of the sort that Lucey and Bowman cited as having precipitated their traumatic memories. Stories of American atrocities began in 2004 with accounts of the torture of prisoners at Abu Ghraib and continued with revelations about the murder of civilians at Haditha, where U.S. marines killed fifteen civilians in November 2005.[66] By 2007, accounts of smaller-scale atrocities were commonplace in U.S. newspapers. And while George W. Bush's claim that "99.9 percent of our troops are honorable, decent people who are serving our country under difficult conditions" has merit, that frequent atrocities were evidence of a systematic failure to develop appropriate policies has been central to critiques of the war.[67] "It's not just 'a few bad apples' responsible for Abu Ghraib, Fallujah and Haditha," one editorial explained. "And what one U.S. official called 'a total breakdown in morality and leadership' is not just confined to one Marine platoon in Iraq. There is a total breakdown of morality and leadership in this country also."[68] Indeed, by 2007, one in ten Iraq veterans would admit to abusing a civilian.[69]

Americans increasingly understood the sort of violence that drove veterans to suicide as endemic to an immoral war. Tim Bowman's account of killing Iraqis at a roadblock, for example, intersected with the growing sense that such violence illustrated the United States's broader failure to develop a successful military strategy. In early 2005, the *New York Times* had editorialized that, while "no one can fault an American G.I. at a checkpoint who fires on a car that refuses to stop," it was nonetheless the case that, "with every additional civilian who is killed by American fire, the human cost rises—both in terms of the lives lost and the psychological damage suffered by the Americans in uniform."[70] The article placed blame on "those at the top of the chain of command" who had "told soldiers to shoot first and ask questions later" and concluded that "none of us want our soldiers killed by suicide bombers who get too close. But neither do we want these soldiers to have to live forever with the knowledge that they . . . mowed down the parents of four Iraqi children in front of their very eyes, by mistake."[71] Likewise, in July 2006 Andrew Bacevich, one of the war's most relentless critics, wrote that, while Haditha belonged "alongside Sand Creek, Samar and My Lai in the unhappy catalogue of atrocities committed by American troops," Americans should also pay attention to the much more frequent civilian casualties occurring at roadblocks.[72] Such violence, Bacevich argued, was endemic and illustrated the Iraq War's failure. "Plenty of evidence suggests that in Iraq such mistakes have occurred routinely," he explained, "with moral and political consequences that have been too long ignored. Indeed, conscious motivation is beside the point: Any action resulting in Iraqi civilian deaths, however inadvertent, undermines the Bush administration's narrative of liberation, and swells the ranks of those resisting the U.S. presence."[73] Bacevich's critique of Bush's optimistic rhetoric about Iraq was matched by his indictment of "the intellectually bankrupt policymakers who sent U.S. forces into Iraq in the first place and now see no choice but to press on."[74] Most important, though, is his sense that roadblock killings illustrated the war's immorality. "One at least ought to acknowledge," he concluded, "that in launching a war advertised as a high-minded expression of U.S. idealism, we have waded into a swamp of moral ambiguity."[75] For these critics, the everyday violence that occurred at U.S. roadblocks emblematized the war's violence and futility, and the consequences of this violence affected not only Iraqi bodies but also American minds.

Stories that cast Lucey and Bowman as traumatized by their actions in Iraq contributed to critiques of the Iraq War as not only disastrous but immoral by personalizing the Iraq War's moral impact. That is, if Americans might have struggled to empathize with Iraqis who were murdered in their cars, they could better appreciate the moral consequences of an American teenager accidentally killing an innocent person.[76] Those consequences amounted to what psychologists term "moral injury," a condition that "requires an act of transgression that severely and abruptly contradicts an individual's personal or shared expectation about the rules or the code of conduct. . . . The event can be an act of wrongdoing, failing to prevent serious unethical behavior."[77] According to the journalist Ann Jones, moral injury is an overlooked cause of veteran suicide; as she explains, Lucey's killing Iraqi prisoners was clearly such an act because he "had gone to war as a boy who liked to help people but had returned, in his own estimation, 'nothing but a murderer.'"[78] Certainly, the same could be said for Tim Bowman. In highlighting the moral damage that these men had suffered, accounts of veteran suicides thus encouraged Americans to criticize the manner in which the war was being fought and acknowledge the immorality of the entire enterprise.

* * *

Descriptions of suicidal veterans also highlighted another aspect of the war that its opponents increasingly critiqued—the overreliance on the National Guard and Reserves. The use of these units was not initially anticipated. In the immediate aftermath of the September 11 attacks, the National Guard had been mobilized primarily for security at airports, national monuments, and the 2002 Winter Olympics.[79] In fact, the Army National Guard's posture statement from 2003, the year Tim Bowman joined, makes no mention of Guard members fulfilling combat roles.[80] By 2004, however, National Guardsmen were dying at a higher rate than their regular-service counterparts, partly because they were often given the dangerous task of guarding convoys.[81] In 2005, nearly half of combat brigades in Iraq were composed of reservists and Guard members, "all of the National Guard's first-line brigades [had] been mobilized," and the Guard "account[ed] for about a quarter of those killed."[82] Two years later, the Army National Guard reported that "since July 2002, overall unit readiness has decreased by 49.25%."[83] In the ensuing years, Americans became increasingly concerned that the National Guard was overtaxed and unprepared.[84] In the months leading up to the 2006 midterm elections, newspapers routinely reported on Guard members' psychological and financial stress, the lack of training that they received for their missions, and the challenges posed by multiple deployments.[85] This heavy use of Guard members and reservists seemed further evidence that the wars had been poorly planned and were taking an undue toll on the nation.

The suicides of Omvig, Bowman, and others highlighted this toll by emphasizing that members of such units had generally not expected to deploy to a combat mission in Iraq and were not adequately prepared to do so. Omvig's father, for example, remarked that, "when he signed up," the reserve company to which he belonged "hadn't been activated in more than 30 years," a comment that made clear that Omvig, who enlisted with the goal of getting experience to become a police officer, had not expected to fight a foreign war during which he was "mortared daily" and "encountered close combat in urban conditions."[86] Similarly, Mike Bowman offered a prosaic explanation of Tim's reasons for joining—"to earn money for college and to get a little structure in his life"—and highlighted that National Guardsmen in Iraq were performing missions far outside of their normal training.[87] "Our National Guardsmen from the F202 were not filling sandbags," he told Congress. "Their tour took them directly into combat."[88] His juxtaposition of "filling sand bags" to collecting their peers' body parts contrasts the National Guard's expected mission—responding to natural disasters—with the gruesome details that they were assigned. Veterans' suicides thus resulted from the failure to appropriately plan for and manage the war that unexpectedly put undertrained troops in danger.

The failure to anticipate and provide for veterans mental health needs likewise became evidence of the Bush administration's disingenuousness in beginning the Iraq War and continued willingness to fight it without recognizing its consequences. A January 2007 *Minneapolis Star-Tribune* editorial about Schulze's suicide acerbically took this line of attack, asserting that "mental wounds are a given of any war, which is why Americans should be absolutely sure war is necessary before they ever agree to put the lives of U.S. troops on the line."[89] The editorial became more explicit as it layered on critiques of how the war had been fought: "When you compound the ordinary mental health risks of any war with the confusion of purpose, repeat deployments and guerilla nature of the war in Iraq, you have a situation guaranteed to twist the emotions of many soldiers in ways so painful and hopeless that some choose death instead."[90] Veteran suicide was thus not simply endemic to war; it was the product of an unnecessary and poorly managed war, "and the honest solution is to quit sending Americans into the maw of Iraq."[91]

In the wake of this editorial, letters amplified its attack. "How is it that we can afford a war in Iraq and yet not be prepared on our own shores to take care of the courageous soldiers who return?" asked a writer who complained that Schulze's "country failed him."[92] Another wrote that "I blame elected officials with puffed-up promises about supporting our troops; I blame a health care system that puts the almighty dollar first; I blame an American public that prefers to focus on gay marriage rather than veterans' benefits."[93] Amid these wide-ranging critiques of a culture that had failed veterans, the author concluded by condemning the profits that defense contractors had reaped in Iraq, suggesting that, "perhaps in the meantime, Halliburton can make a generous contribution to the VA to help free up the waiting list."[94]

Three weeks later, an editorial in Lucey's hometown newspaper made nearly the same point. "When it comes to our nation's fighting men and women," the *Pittsfield Berkshire Eagle* wrote, "President Bush sees them as either fodder for his unending war in Iraq or as a cudgel to use against anyone in Congress who dares suggest a change in the White House's failed policy. The utter lack of compassion the Bush White House has for the soldiers it has put in harm's way is revealed by the national disgrace that is America's treatment of veterans returning from Iraq and Afghanistan."[95] Like the *Star-Tribune* and its readers, the *Eagle* presented the Bush administration as solely responsible for a misbegotten war and ignorant of its consequences: "Assuming the Iraq invasion would be a 'cakewalk,' the Bush administration did nothing to prepare for this onslaught of traumatized veterans and, even now, the Bush budget cuts veterans' health care two years from now just as the president is escalating the war."[96] Here again, the tragedy of Schulze's suicide was not simply that young men and women were psychologically damaged by combat; it was that this war had been unnecessary and poorly planned, and neither the president nor the public had made good on their promises to honor those who had served.

Indeed, it was not only the Bush administration that came in for critique. The American public, which had remained largely disengaged from the war, was also a target. Critics have noted that, when most Americans sacrifice little for an ongoing war and rely on a few Americans to sacrifice much for that war instead, this poses problems for a democratic society. This contention has become central to academic critiques of the war. Mary Dudziak, for example, argues that the era of the War on Terror has produced a "culture of irresponsibility" in which "we are routinely asked to support our troops, but otherwise war requires no sacrifices of most Americans, and as conflict goes on, Americans pay increasingly less attention to it."[97] Andrew Bacevich, somewhat more cuttingly, writes that "no longer seeing war as an endeavor requiring collective effort on a national scale . . . Americans swallowed hard, averted their gaze from the consequences of actions undertaken in their name, and did as President Bush bid them to do."[98]

Political and media discourse about veterans' suicides anticipated this line of critique, framing the epidemic as one for which the public should also be held accountable. In March, 2007 the *Star-Tribune* published yet another editorial about Schulze, this time condemning a disinterested public unwilling to make any sacrifices for those who had fought it. Referring to "the outrageous and sad story of Jonathan Schulze," the editorial concluded that

> it is obvious that from coast to coast, this nation is not keeping its bargain with its military men and women. The bargain is quite simple: You risk your life for us, and we will see to it that if you are wounded you will be cared for, and if you die your family will be cared for. That's what should happen, but not always what does happen. . . . Here's the question that will soon be answered: Does this nation have what it takes to make this right, or are the wounds of war a part of the price of war it would just as soon not pay?[99]

Other papers followed suit. On the same day, a column in the *Sarasota Herald-Tribune* began a critique of the nation's treatment of wounded veterans with a graphic account of Schulze's suicide before suggesting that "maybe all of us bear some of the blame."[100] The author, Larry Evans, extensively quoted an *Air Force Times* editorial that railed against the VA and concluded that "what's needed as well, is more public outrage of the depth contained in an editorial in a newspaper that serves people in the military."[101]

Similar ideas reverberated through Congress. Representative Bob Filner (D-CA) marked the war's fourth anniversary with a gesture towards bipartisanship that also condemned a disinterested culture. Arguing that "no matter where you are on this war in Iraq," both Democrats and Republicans understood the necessity of mental health care, the California democrat nonetheless explained that "We cannot say, support our troops, support our troops, support our troops, and then forget them when they come home."[102] Filner's critique spoke directly to the distance between the military and the nation in the twenty-first century. The problem was not simply that a callous bureaucracy had failed a suicidal veteran; it was that the nation as a whole seemed rather disinterested. Stories of veterans taking their own lives were thus more than personal tragedies. Discussion of suicides by Omvig, Bowman, Lucey, and others became moments for indicting of both the administration's poorly planned war and the broader public's apathy. These representations also raised questions about why the nation had been so ill prepared, and the answers facilitated broader critiques of the war, the administration that fought it, and the culture that enabled it.

* * *

In February 2007, Sen. Tom Harkin (D-IA) introduced the Joshua Omvig Veterans Suicide Prevention Act, which increased training for VA mental health providers, provided more funding for research, and improved the mental health assessment of veterans and the training of family members.[103] Harkin anticipated Bowman's statements, calling Omvig's suicide the result of his family having been given insufficient resources to acknowledge and address their son's mental health issues: "This was a preventable death. If Josh and his family had had better access to mental health services; if they had been trained to recognize the symptoms of PTSD; and if they had known where to turn for help; then the tragedy of his death might well have been avoided."[104] Omvig's parents agreed, disparaging the training that they had received as woefully inadequate. They were told, "Give them space. Don't push them to talk. Give them time to acclimate," Randy Omvig explained, echoing some of the principles that family members were taught in the Army's BATTLEMIND program, but "it didn't work."[105]

Eight months later, the Bowmans made nearly the same point, asserting that National Guard troops had received neither the respect nor the treatment that the country claims to accord its veterans: "Family members of F202 were given a 10-minute briefing on post traumatic stress disorder (PTSD) before the soldiers returned, and the soldiers were given even less."[106] Bowman insisted, though, that he and his wife had been provided no tools to help men and women like their son who "were hardened combat veterans now, but were being treated like they had been at an extended training mission."[107] "He had shown us small signs but not enough to trigger anything," Bowman explained, "because we did not know what we were looking for."[108]

In the estimation of both the Omvigs and the Bowmans, the VA and the military had adopted a cavalier attitude about identifying and treating Iraq veterans' mental health issues. The VA, Bowman explained, had been distant, and training had been nearly nonexistent. Mental health "gets shoved to the side" among the many issues that veterans and their families had to address after a deployment. "Then they hand you a magnet with a bunch of phone numbers on it that says, 'Here is where you call for help,'" he explained.[109] Families needed better information, Bowman insisted, because they were the ones most likely to encounter a troubled veteran. National Guardsmen, he explained, "can suck it up for weekend drill and they will look like the most normal human being you will every find," but it is the families who "are going to see the breakdowns."[110]

Senate Democrats embraced this line of attack. In March, Sen. Amy Klobuchar (D-MN) took to the Senate floor to critique both the president's lack of foresight and his unwillingness to support the VA, as well as the nation's seeming indifference. She cited Schulze's death to argue that the VA was underresourced, pointing out that in the previous decade the number of psychiatric beds at the hospital where he had sought help had been cut by 93 percent and that the Department of Defense underestimated the number of veterans who would require treatment by a factor of four.[111] "It is as if nobody even realized that we have been at war for the past 4 years," she complained.[112]

Klobuchar explicitly took aim at those who had thought that the war could be fought easily and on the cheap. "Just as this administration sent our soldiers into battle without a plan for victory," she complained, "it also failed to develop a plan to address their needs once they got home."[113] She argued that "for the past several years, this administration has submitted a budget request for the VA that significantly underfunded the needs of America's 25 million veterans. This is from the same administration that each year asks Congress to authorize tens of billions of dollars for projects in Iraq."[114] Declaring that these were "the wrong priorities for our country," Klobuchar identified the failure to provide sufficient mental health care as ample piece of evidence that the Bush administration had woefully misunderstood the toll that the Iraq War would take on the nation and then cavalierly ignored the war's impacts.

Klobuchar followed these statistics with an indictment of the Bush administration's efforts to fight the war with minimal resources. In particular, she condemned the administration's reliance on the National Guard and Reserves. Reminding her colleagues that "the National Guard was not built to serve as an Active-Duty force for prolonged periods of time. Yet that is exactly what we are requiring them to do," she condemned the administration for raiding National Guard armories to send equipment to Iraq and critiqued the administration and the nation for asking so much of an all-volunteer force and then not supporting them upon their return.[115] "Our veterans didn't stand in long waiting lines when they were called up or volunteered to serve our nation," Klobuchar argued. "So why are we asking them to stand in line now for medical care?"[116] In a closing that referenced men like her father who, during the Second World War "could count on the fact that their Government would stand by them" and Lincoln's Second Inaugural Address's commitment to "bind up our Nation's wounds and to care for those who have borne the battle," Klobuchar cast the failure to provide mental health care to Iraq veterans as a abrogation of the nation' the best traditions.[117]

Klobuchar's Democratic colleagues echoed her in a hearing a month later. Sen. Patty Murray (D-WA) complained that the VA hadn't even managed to spend the funds that it had been allotted the previous year for mental health care, and Sen. Sherrod Brown (D-OH) remarked that the epidemic was further evidence that "we cheer them on as they go to war, and do so little in so many cases when they come back."[118] West Virginia Democrat Sen. Jay Rockefeller picked up this theme, chiding the administration for being willing to add to the national debt to fight the war. "This Nation sort of has a tendency . . . of honoring the warfighter while the warfighter is fighting," he explained, "and then the warfighter gets hurt, visibly or invisibly or both, and gets subjected to a budget which . . . is entirely inadequate because it is within the budget of the United States, as opposed to the warfighters war, which is subject to the loans of China, Japan, and South Korea."[119] For Democrats, the reckless spending—reckless because it wasn't formally budgeted and because it took money away from priorities that they deemed more important—was another failure of the Bush administration's Iraq War policy, and the debate about veterans' suicides became another place where those failures became visible.

These claims, however, were more than political theater. In fact, the twenty-first century had not been a good one for VA mental health care.[120] In 2004, the VA had unveiled its Mental Health Strategic Plan, which promised to expand mental health services, incorporate them better into primary care settings, and increase focus on evidence-based treatments for mental health.[121] The strategic plan, which was entitled "Achieving the Promise: Transforming Mental Health Care in the VA," took its subtitle and mission from the final report of President Bush's New Freedom Commission on Mental Health.[122] However, the VA received relatively little mention in the Commission's final report—it was cited only as a model for electronic record keeping, a recognition that must have seemed ironic later, when controversies over the VA's scheduling erupted—but "Achieving the Promise"'s goals were clearly structured along its lines.[123] The VA's plan called for more focus on rehabilitation, mental health treatment in the primary care setting, and increased attention to suicide and twenty-first-century veterans.[124]

Improving mental health care in the VA was a tall order, though. "The VA was not planning for a flood of new veterans when September 11th suddenly changed the stakes," the anthropologist Erin Finley writes, and the comments of VA officials bear this out.[125] As Antoinette Zeiss and Bradley Karlin, both leaders of the VHA's mental health program during the 2000s, explain, there was a "decline in VA psychology staffing numbers from fiscal year 1995, a historic high point, to a low in 2000, with only minor increases up to the end of fiscal year 2003."[126] Beginning in 2004, however, the VA put some weight behind this plan, as well, as "over $850 million cumulatively was committed specifically to implement the strategic plan and enhance VA mental health services" by 2008.[127]

How that money was spent—or not spent—became controversial as the wars dragged on. The VA budget did not fare well under the Bush administration. Rather than increasing the budget in anticipation of the increased stresses that the Iraq and Afghanistan Wars would bring, the administration proposed few increases, refused those requested by Bush's first VA secretary, and endeavored to cut spending and programs.[128] Veterans Administration mental health, however, was not even spending all of the money that it had been given. An October 2006 GAO report found that in 2005 the VA had spent only $88 million of the $100 million that it had intended to spend on mental health, claiming that "there was not enough time during the fiscal year" to spend the money.[129] To make matters worse, $35 million of the money that the VA did spend appeared to have been misdirected; VA officials, having made decisions about how to use that money four months after they determined their other FY2006 allocations, "did not notify network and medical center officials that these funds were to be used for [Mental Health Strategic] plan initiatives."[130] Some hospital officials explained that "the length of time it took to hire new staff" made spending the money more difficult.[131] The next year, the VA did even worse, failing to allocate more than 20 percent of the money that it had set aside for strategic plan implementation.[132] Officials explained that they had done so because that process was taking longer than they had expected.[133] Perhaps more troubling, the VA's record keeping on these allocations was terrible; in fact, it was non-existent in 2005.[134] By the summer of 2006, the VA, anxious not to repeat the failure to spend down its allocations, was telling its facilities that any spending on mental health "would constitute spending on mental health strategic plan activities."[135]

Calls for a more robust VA mental health and suicide prevention program endeavored not simply to avert a public health crisis but to meet a moral obligation as well, and they often juxtaposed the promises made to soldiers with the cost of the war and the realities of veterans' post-war struggles. Tom Harkin, for example, ended his speech introducing the Omvig bill by asserting that the United States had reneged on its responsibilities: "The Federal Government has a moral contract with those who have fought for our country and sacrificed so much," he argued, concluding that "our service men and women deserve to know that we will not forget about their service."[136] Indeed, when the House debated the bill on March 21, 2007—not incidentally the fourth anniversary of the Iraq invasion—and again on October 23, nearly every legislator who spoke in favor of it construed its passage as meeting an unpaid debt.[137] The Omvig bill's passage, congressional Democrats insisted, would move the nation past disinterested, disengaged, celebration-of-service members who demanded no sacrifices on their behalf. As Rep. Jerry McNerney (D-CA) told his colleagues in March, "We owe them more than a debt of gratitude. We must also provide them with the support and care they need."[138] Seven months later, Filner echoed this point, arguing that "we cannot ask them to fight and then forget them when they return from battle."[139] Suicide prevention thus became a means of actually acknowledging the war's impact on those who had fought it.

* * *

The Omvig Bill passed unanimously on October 23, 2007, and became law on November 5. The bill's passage, however, did not quell debate over veterans' suicides. On November 13, *CBS Evening News* aired a story about "a troubling and tragic epidemic among our military veterans, an epidemic of suicide" and announced that as many as 120 veterans were committing suicides each week.[140] The story profiled the families of several veterans who had died by suicide, including the Kevin Lucey, Jeffrey's father, who complained, "There's the crisis going on, and people are just turning the other way. . . . They don't want the true number of casualties to be known."[141] Ira Katz, the VA's top mental health official, appeared dismissive, telling CBS that "There is no epidemic of suicide in VA, but suicide is a major problem."[142] Katz's argument, CBS implied, was that the VA didn't imagine veteran suicides as more significant than any other national health trend.

Katz hardly meant that, but he quickly became the public face of the VA's perceived intransigence. Over the next several months, Katz, a geriatric psychiatrist who had also earned a PhD in organic chemistry from Columbia, was repeatedly thrust into the national spotlight.[143] It was not a role to which he was well suited. He knew a lot about suicide, but he was not, at heart, a bureaucrat. In 1996, he had appeared before the Senate's Special Committee on Aging to talk about suicide among elderly people.[144] Even then, his message had been optimistic. "When older people kill themselves, it is because they are experiencing or anticipating intolerable suffering," he explained, but "by ensuring that late life depressions are recognized and treated, we should be able to prevent suicide."[145] Moreover, he was an advocate for more governmental resources; programs to address elderly suicide, he insisted, "demand public action" in the form of increased funding.[146] Katz's research had increasingly brought him into the orbit of the Philadelphia VA, and in 2004, Katz retired from the University of Pennsylvania and was recruited by Jonathan Perlin, an acting undersecretary of the VA whose father, Seymour Perlin, had been one of the earliest suicidologists, to become Deputy Chief Patient Care Services Officer for Mental Health.[147]

Katz's academic credentials were beyond critique, but his professorial demeanor did not translate well to television news. When he explained that research on veteran suicide rates "is on-going," CBS cut to an explanation of the study that they had done. Asked whether the suicide rate was “an epidemic,” Katz sought to offer a nuanced answer but came off as fumbling. “You know, veterans, I had mentioned,” he began, only to be cut off by correspondent Armen Keteyian, who went in for the kill, asking “Yes or no, Dr. Katz?” Katz could only reply that “suicide in America is an epidemic, and that includes veterans.”[148]

Katz’s responses were partly constrained by what the VA knew about suicides at that moment. As he would tell Congress a few weeks later, the suicide rate for Iraq and Afghanistan veterans was, in fact, on par with the suicide rate of the general population.[149] Moreover, veteran suicide was not a topic that the Bush administration was eager to acknowledge or have as part of the public debate about the war, and while there was no direct pressure from the White House to avoid the subject, VA officials sensed that it was too politically sensitive to address directly.[150] Katz thus had the unenviable job of defending VA suicide prevention without saying much about veteran suicide.

* * *

The morning after Katz’s interview, Washington Democrat Patty Murray gave a blistering Senate floor speech on veterans’ suicide. As it had been for Klobuchar and in editorials, talking about suicide was a means of talking about the war’s larger failures, and Murray’s claim that “no matter how any of us feel about this current conflict, we know our troops are serving us honorably [and] we must do better” reflected critics’ use of mental health to subvert conservatives’ claims that condemning the war amounted to not supporting the troops.[151]

Murray’s speech mobilized every element of the discourse that had surrounded suicidal veterans, offering a blistering critique of the Bush administration and closing with an appeal to a public that seemed willing to ignore the war. The suicide statistics, she inveighed, “reflect an administration that has failed to plan, failed to own up to its responsibilities, and . . . dropped the ball.”[152] It was a stinging indictment, and Murray was just getting warmed up. The real problem, she argued, stretched beyond VA funding; it was the Bush administration’s larger failure to acknowledge the Iraq War’s realities and its willingness to continue deploying troops in the service of a failed strategy. “Five years ago, when the President asked us to go to war in Iraq, he talked about weapons of mass destruction, he talked about al-Qaida, he talked about the mission to fight the war on terror," she reminded her audience, "but he never talked about policing a civil war. He never talked about the stress of living months without a break and constantly waiting for the next attack. He has never talked about, in my opinion, taking care of those men and women who have served us honorably when they finally come home."[153] She went on to critique repeated deployments and the lack of dwell time between them, explaining that "some men and women are now serving their second, third, fourth, and now even fifth tour of duty. They are stretched to the breaking point."[154] Veterans' mental health issues, she asserted, resulted from the Bush administration having misled the nation into an unnecessary war that took an unconscionable toll on the relatively small number of men and women sent to fight it.

Those service members once again appeared as childlike, innocent, altruistic, and abandoned. "These are young men and women," Murray explained, defining them as family members rather than soldiers, "They are someone's son, brother, sister, wife, best friend."[155] Her primary examples were, unsurprisingly, Joshua Omvig and Justin Bailey, a marine veteran who had overdosed at a Los Angeles VA hospital. Both men appeared once again as dedicated volunteers forsaken by their nation and who left behind flummoxed parents, and embedded in her descriptions of them were further critiques of the administration's policies. Bailey had been "about to separate from the Marines in 2003 when his service was involuntarily extended because of the war in Iraq," a comment that reminded her audience of the controversial stop-loss policies through which the Bush administration had maintained troop levels in Iraq. For Murray, Bailey's death was made more tragic because of the VA had betrayed his family's expectations. "Justin's parents assumed that he would get proper supervision in the VA program," Murray explained after describing the slew of drugs that he had been prescribed, sometimes improperly and without the advice of a psychiatrist. Instead, "Justin took too many pills and overdosed."[156] Here again, Murray cast the VA's failure as a preventable breach of the nation's fiduciary responsibility to its soldiers.

The proposed solution was unsurprising—more practitioners at the VA and better funding from Congress. But Murray also appealed to a nation apparently oblivious to veterans' psychological suffering. She made this point twice, first arguing that "it is time for all of us to wake up to the reality and consequences of this war. It is time to wake up our neighbors and our communities," and then closing her speech with a much longer and more emotional appeal in which she argued that "we owe them so much more than we have given them so far. We can do better. We must do better" and that it was incumbent upon all citizens to "to reach out . . . and [say]: I am here for you if you need me."[157] Murray's speech thus rivaled Klobuchar's in its condemnation of both an administration portrayed as exploiting idealistic young service members in its misguided, underfunded war and a culture that supported the troops in word but not in deed. It modeled the extent to which the issue of veterans' suicide was intertwined with broader critiques of the war.

That evening, CBS continued its reporting, once again casting Katz as hapless and defensive. Confronted with the complaint made by Paul Rieckhoff, the head of the organization Iraq and Afghanistan Veterans of America, that "the VA system is not at all prepared. This country has not ramped up resources to meet this flood of people coming home," Katz said only that "we're deeply sorry to hear about any death."[158] When CBS interviewer Keteyian said, "I can tell you honestly, Dr. Katz, a lot of the parents that I talked to harbor enormous anger towards the VA," Katz replied that "one of the factors that's led us to really develop suicide prevention programs is precisely these tragedies."[159] What seemed like Katz's indifference outraged the families. In December, Mike Bowman would tell Congress that, "when I first saw Dr. Katz's reaction on the TV, the first thing I wanted was to reach through the screen and choke him."[160]

Katz appeared at the same hearing. His testimony didn't get off to a good start. Moments into his opening statement, Filner interrupted Katz, telling him, "I've read your whole thing, and I couldn't figure out what you're doing to stop suicides, frankly."[161] It went downhill from there. When Katz touted that the VA's suicide prevention hotline had saved three hundred lives, Filner pointed out that that was a modest success among six thousand veteran suicides in a year, accused Katz of "acting as if everything is goodness and light" and "using activity as a substitute for effectiveness," and he entreated him to respond directly to the Bowmans about why the VA had failed their son.[162]

Katz was determined, however to present the VA as successfully meeting the challenge. He told the committee of the new counselors the VA had hired, the outreach efforts it had made to returning units, including National Guard units like Bowman's, and that they had implemented the Omvig Law's provisions even before the law had been signed.[163] He even told the committee that VA suicide prevention was sufficiently funded.[164] Filner, unsurprisingly, had none of this. Accusing Katz of offering statistics that wouldn't arouse public outrage and pointing to the contradictions between the Bowmans' testimony and his accounts of improved VA care, Filner raged, "How would you have helped Timothy Bowman or the Timothy Bowman that is coming in tomorrow? . . . How are you responding to their cry for help?"[165] Katz, however, stuck to his talking points: "Sir, I really think we want to emphasize the message that treatment is available and treatment works."[166]

This hearing turned out to be only the beginning of Katz's problems. If in December 2007 he had seemed an out-of-touch bureaucrat, in April he became the face of VA intransigence. On March 20, CBS marked the fifth anniversary of the Iraq War by reporting that "even vets who get help from the VA are still at risk" and that the suicide rate among young veterans receiving VA treatment had quadrupled during the twenty-first century.[167] This story, like many others, cast the Iraq War as having irrevocably damaged the country's most promising youth. "All of our kids were pretty much in the same boat," Mike Bowman explained as the story opened:

> They were happy go lucky, life of the party people who got along with everybody, who went over there and had to most life-changing experience that they could have, and came home not able to deal with it because their personality did not allow them to process the hate, rage, anger, and death that they saw.[168]

Later, Bowman explained that when those soldiers come home, "That's when they become abandoned," placing the blame squarely at the feet of a VA that hadn't bothered to meet his son when he returned from Iraq.[169]

A month later, CBS dropped a bombshell. In a story quoting Katz's claim that "There is no epidemic of suicide in VA," Paul Sullivan reported that Katz had apparently tried to prevent CBS from learning how many veterans had attempted suicide in the previous year, exposing an email in which he instructed readers to "SHH!, as in 'Keep it quiet,'" and asked whether statistics citing twelve thousand [suicide attempts] per year—despite public proclamations that they numbered fewer than one thousand—were "something we should carefully address in some sort of release before someone stumbles on it?"[170] The story had legs. CBS covered it each night, revealing on Friday that Katz had written "I don't want to give CBS any more numbers on veteran suicides or attempts than they already have—it will only lead to more questions."[171]

Critics of the VA responded with outrage that was again linked to wider critiques of the war itself. Both Sen. Patty Murray and Sen. Daniel Akaka (D-HI) called for Katz's resignation.[172] "When you bring a 3-year-old up and tell them they have to stop lying, they understand the consequences," Murray told reporters. "The VA doesn't. They need to stop hiding the fact this war is costing us in so many ways."[173] On May 1, Rep. Tom Udall (D-NM) and other House Democrats wrote President Bush to demand that Katz resign.[174] The same day, the *Spokane Spokesman-Review* wrote that "Dr. Ira Katz and others in the top VA echelons deserve the scolding they got for concealing the severity of the problem," arguing that "the job of a suicide prevention counselor is demanding enough without top agency brass glossing over uncomfortable realities to avoid political embarrassment."[175] Newspapers equated the failure to effectively treat suicidal veterans with sending troops to Iraq in unarmored Humvees and with demanding multiple deployments, and they alleged a cover-up.[176] Once again, the suicide debate was evidence of the Bush administration's fundamental dishonesty about Iraq.

This coverage had to be painful for Katz, who had devoted his life to suicide prevention and had taken a job with the VA because he thought he could save more lives there than he could writing academic journal articles in his office at Penn. Within days, he was back before Filner's committee, and the chairman was outraged. Wondering why Katz hadn't owned up to the figures on CBS, he noted with withering sarcasm that "either Dr. Katz knew that the *CBS News* figures were indeed an accurate reflection of the rates of suicides at that hearing or he had a sudden epiphany just 3 days later" and pointed to emails between Katz and Michael Kussman, the undersecretary for health, that confirmed the figures.[177] Arguing that "the pattern is deny, deny, deny. Then, when facts seemingly come to disagree with a denial, you cover up, cover up, cover up. When the cover-up falls apart, you admit a little bit of a problem and underplay it. . . . Then, finally, maybe you admit it's a problem and then, way after the fact, try to come to grips with it," he accused the VA of "criminal negligence" and asked whether either Katz or Kussman would be forced out.[178]

Katz was preceded by James Peake, the secretary of the VA, who did his best to tamp down the outrage by pointing to areas in which the VA had made progress—by hiring more therapists, conducting more outreach, and establishing a suicide prevention hotline. He also explained that the high attempt rate that CBS had reported resulted from suicide prevention coordinators taking too expansive a view of what actually constituted a suicide attempt. To get a sense of the suicide rate among veterans receiving VA care, Katz had in October 2007 asked each facility to track suicide attempts.[179] The problem was that there was no clear definition of "suicide attempt" among the 153 facilities, and when Katz received a thousand reports, he determined that they needed more analysis before they could be released. Peake explained by asking, "Does a cry for help with an overdose of a non-threatening medication ingestion or a cut on the wrist so slight as to not really risk serious injury [amount to] an attempt?"[180] For Peake, the answer was no, and "the data was clearly not accurate"; he even speculated that given how many veterans receive VA care, it was surprising that the suicide rate wasn't higher.[181] Here again, scientific and popular understandings of mental health proved irreconcilable.

Filner wasn't impressed. Calling the response a paragon of bureaucratic indifference, he parodied Peake's testimony as "Everything is fine, we have it under control, we are going to study our data."[182] If these comments implicitly recalled the Bush administration's dismissal of critiques of the Iraq War, Filner's withering questioning of Katz more closely associated the VA's handing of veteran suicides to the its refusal to acknowledge bad news about Iraq. Filner began by asking Katz, "Why shouldn't you go to court for perjury or resign because you didn't tell us the truth?" and when Katz explained that he had only been critiquing a subset of CBS's data, and not the larger assertion that eighteen veterans took their lives each day, Filner exploded. Referring to George W. Bush's May 1, 2003, declaration aboard the USS *Abraham Lincoln* that the United States had "prevailed" in Iraq, he remarked, "So the 'Mission accomplished' should have said, 'Mission accomplished only by these sailors who were aboard this ship on those 2 days.' We didn't see the fine print."[183]

The impression that the VA was negligent, secretive, and defensive was heightened, as one representative pointed out, by the RAND Corporation's *Invisible Wounds of War* report, issued just days before CBS broke the story about Katz's emails. The report was not entirely critical of the VA, noting, for example, that the quality of care was high and that the VA had made great strides in treating major depression in the primary care setting.[184] Nonetheless, the report also found that "significant challenges remain, including maintaining the quality of care with the increasing demand for services resulting from benefit enhancements and with the influx of veterans who have served in [Operation Enduring Freedom/Operation Iraqi Freedom]."[185] And the need was clearly high; about one-fifth of Iraq and Afghanistan veterans who had come to the VA were being treated for PTSD.[186]

More damning was a lawsuit filed by Veterans For Common Sense, a non-profit critical of the Iraq War. The suit, filed in San Francisco in July 2007, leveled damning accusations against the VA's "shameful failures."[187] The 257-paragraph complaint went on to link PTSD to suicide, complain that "the VA's health care system has also collapsed with the drastic increase in demand for services, particularly in the area of mental health," and argue that "the VA has not only shortchanged the wounded veterans for whom it is supposed to provide care and benefits, but also has consistently presented misleading statistics to the American public" regarding the screening and care of veterans.[188] In particular, the plaintiffs cited a high rate of suicide among veterans, noting that "a number of veterans have committed suicide shortly after having been turned away from VA medical facilities either because they were told they were ineligible for treatment or because the wait was too long."[189] Arguing that veterans had a "fundamental right to hospital care and medical services" and that there was a congressional mandate to furnish effective care, the complaint asked not for financial compensation but merely that the VA be forced to deliver that care.[190]

Although the suit had been wending its way through the courts for nearly a year, the trial itself coincided with CBS's story about Katz's email—which had become known during the trial's discovery phase—

and the subsequent hearing. Coverage of the testimony portrayed it as following the by-now familiar arguments that had been made before congress. "Opening arguments painted sharply different pictures of the department's success," the *New York Times* reported. "The veterans groups said the department was ignoring a mental health crisis and was so swamped that former soldiers were dying needlessly. The defense countered that the country's largest medical care system was adding the personnel needed to cope."[191]

Katz's "SHH!" email figured prominently in the trial's opening statement, as did other emails from Katz discussing the suicide rate, and the former was quoted in the *Times*. These messages formed the basis of the plaintiff's critique. "We certainly think there was a cover-up in some sense," one attorney explained.[192] Likewise, the *Times*'s portrayal of the closing remarks did little to alter the dominant perception of the VA as coldly bureaucratic. The VA's attorney, Daniel Bensing, pointed to an overwhelmed system. As the *Times* explained, "While acknowledging suicide as a serious problem, Mr. Bensing also emphasized that change takes time given that the V.A. runs the largest health care system in the country. 'It cannot all be done immediately like plaintiffs seem to think.'"[193]

The RAND report, the lawsuit, and the May 6 hearing contributed to the continued thrashing of the VA and, particularly, Ira Katz in the popular press. In Palm Beach, Florida, "the long-term costs of the war—both in dollars and in human terms" were called "a picture that the VA and the Bush administration have distorted for too long."[194] The Sunday following Katz's testimony to Filner's committee, the *New York Times* editorial page called the VA "tragically unready" for the wars' aftermath and again compared the VA's handling of suicide to the Bush administration's management of the Iraq War: "The last thing [veterans] need is the toxic blend of secrecy, arrogance and heedlessness that helped to send many of them into harm's way."[195] Twelve days later, a *USA Today* editorial derided Katz for seeking "a smart public relations strategy" and called his email "just one indication that the agency . . . is more interested in minimizing the extent of mental health problems that today's veterans face than it is in tackling them" and cited the Bowmans as evidence that the VA was failing returning troops.[196]

Katz's was not the only email to garner critical attention that week. On May 28, Gordon P. Erspamer, the lead attorney for the plaintiffs in the *Veterans For Common Sense* case, forwarded to Judge Samuel Conti an email from Norma Perez, a psychologist at the Temple, Texas, Veterans Administration. In March, Perez had written to her staff, "Given that we are having more and more compensation seeking veterans, I'd like to suggest that you refrain from giving a diagnosis of PTSD straight out. Consider a diagnosis of Adjustment Disorder, R/O [Rule Out] PTSD." She further explained that "we really don't have time to do the extensive testing that should be done to determine PTSD."[197]

As Finley points out, "her e-mail was read by some veterans' organizations as evidence of a larger VA effort to deny veterans their rightful compensation."[198] Adjustment disorder, Erspamer wrote to Conti, was "a false medical diagnosis," and the email provided further evidence that the "VA actively discourages diagnoses of PTSD," of "systematic deficiencies in the implementation and enforcement of official policies formulated by VA's central office, and that "the overwhelming majority of VA facilities are not complying with VA official policies regarding suicide."[199] Two weeks earlier, on May 15, one of Erspamer's colleagues at the San Francisco law firm Morrison & Foster, Arturo J. Gonzalez, had written to the Justice Department lawyer handling the VA's defense that Perez's "instruction is directly linked to the VA's inability to provide appropriate PTSD screening" and "the e-mail demonstrates VA's current policies that encourage the callous mishandling of veterans' mental health care and claims for benefits."[200]

Newspapers agreed, and the email blossomed into a scandal. In *The Washington Post*, the leader of the anti-war group VoteVets.org explained that "many veterans believe that the government just doesn't want to pay out the disability that comes along with a PTSD diagnosis, and this revelation will not allay their concerns."[201] On the floor of Congress, Rep. Bobby Rush (D-IL) declared his "astonishment and indignation," calling himself "appalled" and "incensed" over Perez's email, asking, "At what point do we determine that certain medical diagnoses would be too expensive to treat?"[202] Another Illinois politician had similar feelings. "Simply put, Ms. Perez's email is outrageous," wrote Sen. Barack Obama (D-IL), who was in the midst of a presidential campaign that was based largely on his opposition to the war. "To hear that a VA official is promoting misdiagnoses of soldiers to save money," he wrote, "is unacceptable and is tantamount to fraud."[203]

As with the broader debate over veteran suicide, Perez's email came to illustrate the folly of the entire war. "Callous disregard of veterans' rights is of a piece with the administration's entire approach to war," the *St. Louis Post-Dispatch* wrote. "It sent troops into combat with inadequate body and vehicle armor, issued so-called stop-loss orders that forced them to serve beyond the expiration dates of their contracts and repeatedly sent them into combat zones. They came home to discover that the benefits and care they were promised either were not available or required extended and expensive waits."[204] In West Virginia, the *Charleston Gazette* quoted these lines and added, "President Bush's Iraq war will be remembered as a needless U.S. tragedy. All the alleged reasons for it turned out to be imaginary. . . . The burden of combat is borne by a tiny segment of American families. It's a shame for the administration to treat these volunteers shabbily."[205]

Initially, the Bush administration seemed perfectly willing to throw Perez under the bus. Responding to Gonzalez's letter on May 22, Justice Department lawyer James J. Schwartz wrote that "the contents of the psychologist's email do not reflect VA's policy regarding diagnosis of Post Traumatic Stress Disorder" and cited Secretary James Peake's statement that "a single staff member, out VA's 230,000 employees, in a single medical facility sent a single email with suggestions that are inappropriate and have been repudiated at the highest level of our health care organization."[206] On June 4, Schwartz wrote directly to Conti, assuring the judge that the "VA acted quickly to correct any misperceptions and clarify the facility's actual policy" and that the email was "the action of a single individual that in no way represented the policies of VA, that, once discovered, was dealt with quickly and appropriately."[207] Norma Perez, the Justice Department was eager to assert, was an individual bad actor.

Schwartz closed his letter by noting that the Senate Veterans Affairs Committee had called Perez and Michael Kussman to testify about the email that day. The psychologist, understandably, was terrified. She was, as she put it, "real new to the VA." In fact, she been there for less than a year, having completed her doctoral work at the University of Rhode Island in 2004 and begun at the Temple VA on June 10, 2007.[208] Getting publicly reprimanded by six U.S. senators did not seem a promising step in her career. For those senators, in contrast, reprimanding a VA staffer had already proven an excellent avenue for critiquing the war, especially in an election year. This was particularly so for a relatively inexperienced senator running for the presidency on a platform in which opposition to the war was a significant plank. That the hearing was being held at all was primarily at the behest of Obama, who "wrote a letter to committee chairman Sen. Daniel Akaka (D-Hawaii) asking him to look into the matter."[209]

As the hearing opened, Chairman Akaka made a point of tying Perez's email to Katz's and cited another case in which the VA's failure to deliver care had caused a veteran to take his own life.[210] Patty Murray made that connection as well, calling the email "a sad reminder that this Administration's attempt to play down the cost of war . . . has begun to actually affect the way that VA employees view their own work."[211] Here again, failures in mental health care were evidence of the Bush administration's duplicity.

Perez was able, however, to defuse much of the criticism. Her approach was not, however, to simply apologize, as the excerpt of Peake's statement that Schwartz had sent to both Gonzalez and Conti assured she would.[212] Rather, the psychologist attempted to explain her email in a manner that once again highlighted the difference between lay and clinical understandings of how mental health issues are diagnosed.[213] Central to her explanation was a differentiation between the temporary condition of "combat stress"—which the VA, like the Army, saw as "a normal reaction to abnormal events" and which would warrant a diagnosis of adjustment disorder—and longer lasting post-traumatic stress disorder.[214]

The term "rule out," she explained, had a different connotation in the medical field than in lay usage. It meant to begin a treatment for the most likely ailment while continuing an examination to look into other causes. "Many conditions look very similar to one another and sometimes it is important to identify the likely diagnosis while noting in the patient's record to test for possible alternatives," she explained. "For example, a patient with chest pains could have indigestion or could be experiencing the early signs of a heart attack. Based on initial information, a clinician would determine the most likely diagnosis, heartburn, but note in the record the need to rule out heart attack."[215] Post-traumatic stress disorder, she likewise maintained, can often not be recognized or diagnosed in a half-hour intake session, so it was important to begin treating people while still working to ascertain whether they had PTSD.[216] Directly pushing back on Erspamer's contention that adjustment disorder was "a false medical diagnosis," she asserted that "I believed that it was important to remind the team clinicians of the diagnosis of Adjustment Disorder, which is a clinically sound diagnosis and will result in the appropriate treatment while continuing the assessment process for a possible PTSD diagnosis.[217]

This was a point that both Kussman and Katz seconded, and one that further highlights the gap between popular and medical understandings of mental health care. Katz explained that, "after one-half hour or an hour and one-half with a patient, you do not know enough to make a diagnosis" and that it was thus important "not to commit ourselves prematurely to the presence or absence of any diagnosis."[218] This may make sense on its face, but for many lay people, the notion that a diagnosis of PTSD cannot be quickly and easily made is confounding. Moreover, when that diagnosis is tied to the provision or denial of benefits, that a clinician may make one diagnosis over another contributes to an already extant narrative of an agency choosing not to meet veterans' needs.

In the last few minutes of the hearing, Sen. Jon Tester (D-MT) asked, if there was a disagreement between two clinicians as to whether a person had PTSD or adjustment disorder, whether "reasonable doubt goes to the veteran" or "the tie goes to the runner," meaning that the veteran will receive the PTSD diagnosis and, with it, greater benefits.[219] Retired Rear Admiral Patrick Dunne, the VA's undersecretary for benefits, explained that that was so, but Katz made a different point.[220] "When we are talking about treatment rather than benefits, the whole issue of the tie going to one side or the other does not count," he explained. "The patient needs the most accurate diagnosis to allow the most precise and predictive treatment planning. Sometimes you do not get it right the first time."[221] His comments point to a fundamental challenge that the VA faced in suicide prevention in particular and in mental health treatment more generally: Developing appropriate care models ran headlong into questions of benefits and funding, and in the highly politicized environment of the wars, diagnoses that may have been medically correct but did not convey benefits furthered a narrative of the VA health system malevolently conspiring to deny veterans care.

* * *

In the end, both Ira Katz and Norma Perez managed to keep their jobs. But their tribulations reveal much about the extent to which veterans' suicide became deeply politicized as the Iraq War reached its nadir. Stories of promising young men who returned from war so damaged that they took their own lives challenged the arguments that the Army had made about the value of military service, and the war's opponents increasingly tied veteran suicide and the VA's failures to resolve it to the Bush administration's duplicity in beginning and prosecuting the war. In the media and in the halls of Congress, veteran suicides became not simply the inevitable result of combat but the particular result of a misguided, mismanaged, and immoral war.

That narrative at times relied upon a general mistrust of the VA, one that was sometimes warranted. The VA did enter the wars ill prepared to address mental health issues, and VA employees felt unable to speak openly about suicide rates. At the same time, however, the linkage of the VA to the Bush administration's failed and deceitful policies sometimes revealed a misunderstanding of the practices of medical surveillance and diagnosis and cast them as further evidence of the administration's indifference to veterans' suffering. At times, the political potential that those critiques held led to some VA employees being unfairly maligned. Beginning in 2008, the VA became more ambitious in its suicide prevention efforts. The political controversy that surrounded those endeavors, however, hardly dissipated, and the political uses to which veterans' suffering were put would become more pronounced.

4

"The Culture of the Army Wasn't Ready"

Stigma, Access, and the Politics of Organizational Change

In January, 2007, George W. Bush made two important decisions that reshaped the Army. The first was to adopt a counterinsurgency strategy that sent thirty thousand additional U.S. troops to Iraq and required that the Army extend deployments to fifteen months. The second was to nominate Gen. George Casey to replace Peter Schoomaker as Chief of Staff. Casey, whose father had been the highest-ranking American killed in the Vietnam War, had been Vice Chief of Staff in the early twenty-first century before presiding over the Iraq War's bloodiest years.[1] For this, he earned significant critique. During his confirmation hearing, Sen. John McCain (R-AZ) bitingly remarked that, "while I do not in any way question your honor, your patriotism, or your service to our country, I do question some of the decisions and judgments you have made."[2]

In preparation for his confirmation hearing, Casey had written responses to dozens of questions from the committee's chairman, Sen. Carl Levin (D-MI). Only four of them dealt with mental health, and in answering them Casey was largely optimistic, highlighting that "the Army has implemented most of the recommendations of the MHAT reports" while noting that "several challenges remain," particularly regarding "access to care, and reduc[ing] stigma."[3]

Such was the sort of answer that one becomes very good at giving over a three-and-a-half-decade career as a military officer. Casey, however, was genuine in his concern. "I can't say that I went into Iraq with the notion that I needed to do something about PTSD and TBI," Casey reflects.[4] But once he got there, the troops' psychological struggles were evident. He saw senior leaders struggling after attacks that left their men dead. He saw junior soldiers stoically trying to hold it together, "coming out of a memorial service and getting into a Humvee and going out again."[5] "I'm in Iraq almost three years," he remembers. "I'm watching the impact of combat on soldiers and leaders . . . and I know that we need to do something. And I knew that I was going to take on stigma."[6] Tucked into one of the many briefing books was a survey that he recalls "said that ninety percent of men and women in the Army would not [choose to] get behavioral health care . . . because they thought it would affect their career," Casey remembers. "Ninety percent. That's a cultural issue."[7] In his view, "The docs were doing their job, but the Army didn't want to hear it. . . . The culture of the Army wasn't ready."[8]

However unready it may have been, the Army faced a clear need. By the summer of 2008, Army researchers estimated "that the total number of newly identified PTSD cases for CY08 will be around 12,000. . . . The vast majority [of whom] will remain on active duty."[9] This represented a significant increase from the previous year's total of 10,523 diagnoses, which in turn dwarfed figures from the first three years of the war. In 2006, the figure had been 6,845.[10] More soldiers were on their second, third, or fourth deployments, and the tours themselves had gotten longer, with less time off in between. Commanders were increasingly aware that soldiers were suffering, and public and legislative pressure demanded action.[11]

Changing the culture around mental health thus became a primary goal after 2007. In particular, the Army faced four challenges: ensuring that soldiers who suffered from post-traumatic stress disorder or traumatic brain injury weren't misdiagnosed with personality disorders that left them ineligible for VA care; reducing the stigma that prevented many soldiers from seeking care and made some commanders dismissive of their troops' needs; making quality care more accessible; and addressing the reality that lengthy and multiple deployments were substantially damaging the force's psychological well-being.

Some of these challenges turned out to be simpler to solve than others. Erin Finley has rightly argued that, as the result of "mounting public pressure at mid-decade, military leadership began harnessing the massive resources of the armed forces to push for top-down efforts aimed at improving prevention and treatment of acute and post-deployment stress reactions."[12] Some effective solutions thus did could come from Army leadership. Addressing personality disorder discharges, for example, was relatively easily addressed through a new set of guidelines that corrected a long-standing, flawed, but unquestioned policy.

Other issues, however, proved more challenging. Leaders like Casey and his vice chief, Peter Chiarelli, were deeply invested in improving access to, and the quality of, mental health care, and under their leadership the Army introduced several programs aimed at reducing stigma and improving access to care. Yet cultural change of this sort could not be accomplished solely by issuing new orders. Instead, the Army's often halting, still incomplete, and never universally successful culture shift on matters of mental health occurred through a more organic, grassroots process of organizational change that was supported by senior leaders but driven by clinicians and researchers. It happened when U.S. Army Medical Department (AMEDD) psychiatrists convinced soldiers that they understood their experiences, commanders that they could help make their units more effective, and primary care doctors that they should take mental health seriously. It also happened when senior leaders began openly discussing their mental health issues and their successful treatments. These individual encounters grew into new programs that were increasingly—though never quickly, rarely smoothly, not always without opposition—embraced by unit commanders and senior leaders, spread throughout the Army, and came to define the Army's behavioral health efforts.

The fourth issue that the Army faced—the continued strain on the force brought on by multiple deployments—fit in yet another category. It was not a problem that leaders could solve internally. If leaders like Casey and Chiarelli were used to giving orders to their subordinates, they also had to follow their civilian superiors' directives. By 2008, however, several senior Army leaders became increasingly critical of the demands that the country had placed on the force, implicitly chastising a culture that had decided to fight two protracted wars with a small, all-volunteer force as they argued that the short rest periods between repeated deployments denied soldiers the opportunity to restore their mental health.

Addressing personality disorder discharges, stigma, access, and dwell time—in short, getting soldiers into care, making sure that their care was effective, and creating the conditions under which they could recover—required an increased budget, better policies, and more clinicians. But it also required cultural change. The history of each of these four efforts reveals where and how leaders, researchers, and clinicians succeeded at changing the Army's internal culture, where they struggled as they sought to change attitudes in a large, complex organization of nearly half a million people, and where they confronted a broader culture that remained largely disengaged from the wars.

* * *

A week after the Iraq War's fourth anniversary, the *Nation* published a scathing indictment of the Army's handling of troops' mental health issues. The article by Jonathan Kors focused on Jon Town, a specialist left struggling with deafness, memory failure, and depression after an insurgent rocket attack. Upon his return to Fort Carson, however, doctors "did something strange: They claimed Town's wounds were actually caused by a 'personality disorder'" and "told [him] that under a personality disorder discharge, he would never receive disability or medical benefits."[13]

Kors's article cast a national spotlight on an issue that had been simmering in local newspapers since the day after Christmas, when the *Denver Post* published an article on the use of these discharges at Fort Carson. Over the next year, personality disorder became an oft-discussed issue.[14] For the public, and particularly for the war's opponents in Congress, it was further evidence of the Bush administration's malfeasance. For the Army, still reeling from the Walter Reed scandal, it was another black eye. But it was also evidence of how the Army's established protocols were insufficient to the challenges that the wars presented, which meant that even well-meaning Army psychologists and psychiatrists were hurting soldiers with inappropriate discharges.

The American Psychiatric Association's *Diagnostic and Statistical Manual* specifies a variety of personality disorders, some of which, like obsessive-compulsive disorder, are familiar to many Americans. In general, they are conditions that impede an individual's functioning, and they typically manifest "in adolescence or early adulthood."[15] Perhaps more so than in civilian life, these conditions can impede an individual's ability to succeed in the military. As Navy psychiatrists pointed out in two 1975 studies, "Adjustment problems generally appear to be incompatible with effective military service where standards of conduct tend to be uniform and well-defined, reliability of performance is emphasized, and close interpersonal relationships cannot be avoided" and the condition "account[ed] 50% of the 300,000 sick days and 75% of the more than 4,500 medical discharges from the naval services per year."[16]

For the military, diagnosing a service member with this condition had historically been an expedient way to discharge someone who was problematic or not adapting to military life. Many providers, however, worried about this use of the diagnosis. One study feared that it had become a means of medicalizing what were fundamentally disciplinary issues and that sailors ended up in treatment largely because their commanders had "become exasperated by the behavior of the men and want psychiatry to assume the responsibility of coping with their problems."[17] The authors of another study, also from 1975, worried that, because the category was vaguely defined, it could be used indiscriminately to get rid of problematic people and that the "potential dangers to the individual patient of inappropriate clinical or administrative actions are quite real."[18]

In the Army, the diagnosis was similarly vague and served a similar purpose. "Personality Disorder was a very easy way to get somebody out of the service, and there was minimal documentation to go with it," Elspeth Ritchie explains.[19] As the Army's chief psychiatric consultant in 2007, she led the investigation of personality disorder discharges that Kors's article prompted. As an Army psychiatrist with nearly two decades of experience, she had referred many soldiers for them. Doing so, she explains, amounted to "saying you don't think it is a more major illness. It's not depression. It's not something somebody should be medically boarded for."[20] Prior to the Iraq and Afghanistan Wars, that a soldier could be discharged because of a personality disorder and not be put through a medical evaluation board that would determine a level of disability and, in turn, determine a level of benefits seemed immaterial; the Army expected that most soldiers discharged into a civilian economy would receive private health care.[21]

By 2007, however, the fears that Navy doctors had expressed in the 1970s seemed to be playing out for soldiers like Jon Town. The story began on the day after Christmas, when the *Denver Post* ran a story on personality disorder discharges at Fort Carson.[22] Within a few weeks, concern over them was appearing in newspapers across the country.[23] By summer, the issue had gained congressional attention, and on July 25, Bob Filner (D-CA) gaveled the House Veterans' Affairs Committee to order. The discharges, he explained in his opening remarks, were "an incredible disservice to our young men and women who are serving this nation" and the result of "policy made at higher levels."[24] As with veteran suicide, personality disorder discharges provided the wars' opponents with further evidence that the Iraq War was unethical.

Joshua Kors and Jonathan Town were the central witnesses. The latter testified that an Army psychiatrist had told him that he was "going to push this [personality disorder discharge] because it will take care of both the needs of the Army and the needs of you," a narrative that accords with Ritchie's description of the discharges' historical application.[25] However, he called that assertion "all lies," noting that by accepting the discharge he became ineligible for VA benefits.[26] Kors was even more disparaging, calling the diagnosis "a contradiction in terms" because "recruits who have a severe pre-existing condition like a personality disorder do not pass the rigorous screening process and are not accepted into the Army."[27] This assertion is debatable—not all personality disorders are apparent at the time of enlistment—but some of the cases that Kors cited suggested a more nefarious use of the diagnosis and strained credulity. Some soldiers' hernias or eye injuries were attributed to personality disorder, for example, and he cast the diagnosis an example of medical malpractice and malfeasance that illustrated the disingenuous of those responsible for the war. It was not, Kors argued, only that the Army wanted to save money; these discharges also obscured the war's impact on the troops.[28]

The committee's response was overwhelming but partisan. Coming as it did in the aftermath of the Walter Reed scandal, Kors's article and the broader attention to personality disorder was easily widely accepted as proof that the military and administration had little interest in caring for the soldiers that it had put into harm's way. Rep. Phil Hare (D-IL) declared himself "beyond even angry," Rep. Stephanie Herseth Sandlin (D-ND) called it "additional evidence that we were not prepared to take care of another new generation of veterans," and Chairman Rep. Bob Filner (D-CA) called Town's treatment "absolutely disgraceful" and "chilling."[29] In contrast, the committee's ranking Republican, Indiana's Steve Buyer, impugned Kors's credibility because he had not served in the military and attacked his assertion that a recruit with personality disorder would not have passed the Army's intake screening.[30]

Buyer was clearly partisan, and his evident intent was to discredit Kors, but his point was not entirely incorrect. Recruit screening is, at best, imprecise. Recruits are not required to be screened by a mental health professional, and the military expects them to disclose pertinent diagnoses.[31] Thus there might be evidence that a recruit had had serious mental health issues if he had been hospitalized, had undergone prolonged treatment, or had been prescribed a selective serotonin reuptake inhibitor (SSRI). But it was also true that a person eager to enter the Army could hide that information, might never have received treatment, or might even have never been diagnosed.[32] As well, psychological issues often do not emerge until later, either as a response to the stresses of Army life or, more generally, because mental health conditions can emerge at any time and often evolve.[33] And on top of that, the presence of an adjustment disorder discovered during training did not automatically indicate a recruit's unsuitability.[34] As with civilians, a mental health diagnosis does not automatically disqualify a soldier from pursuing his or her chosen profession admirably. Alternately, a previously manageable condition can worsen and become a barrier to service.

All of this meant that it was in fact quite possible that a soldier could either have made it into the Army with a personality disorder or be diagnosed with one after some time in the Army—especially, given the nature of twenty-first-century military service, after a deployment. At the July hearing, VA psychiatrist Sally Satel pointed this out, explaining that "clearly, some soldiers are going to fall into that personality disorder category; there is a chance, though, that he or she has gotten too far into his tour of duty by the time that diagnosis is made" and that "those who have a personality disorder that hasn't manifested in gross ways like for example, as an arrest record, at enlistment, may well show maladaptive behavior as they progress through active duty."[35] Satel was also clear, however, that those symptoms would appear relatively early in a soldier's career and that personality disorder and PTSD can be relatively easily differentiated.[36] Kors's reporting and the congressional hearing indicated that some soldiers were clearly being misdiagnosed.

Throughout the summer and fall of 2007, reports that personality disorder diagnoses were being used to quickly discharge service members continued to appear in a variety of places. *An Achievable Vision: Report of the Department of Defense Task Force on Mental Health*—the report that Sen. Barbara Boxer's amendment to the 2007 National Defense Authorization Act had requested and that provided a scathing assessment of the military's approach to mental health—noted "instances in which returning service members were pressured by commanders and peers to accept an administrative discharge so they could be expeditiously cleared from the unit and replaced with a fully functional person."[37] At his confirmation hearing to become Secretary of the Army, Pete Geren declared himself "concerned" when Sen. Susan Collins (R-ME) cited Kors's article to ask whether "personality disorder discharges are being used as a tool to quickly discharge servicemembers who have service-connected mental health conditions."[38] On National Public Radio, Daniel Zwerdling reported "that the personality disorder diagnosis has been used to discharge problem soldiers quickly" and that "some commanders, in fact, have kicked troubled troops out of the military instead of trying to help them" and cited a 40 percent increase in the discharge's use.[39]

In the Army, in fact, the number of personality disorder discharges did not significantly increase during the wars and continued to average about a thousand a year.[40] The real question was how frequently Army doctors were misdiagnosing soldiers. The answer was not immediately clear to Maj. Gen. Gale Pollock, a nurse anesthetist who became Acting Surgeon General when the Walter Reed scandal forced Kevin Kiley's retirement. A critical question for her was whether the Army, which had lowered its personnel standards as recruiting became more difficult amid the unpopular wars, had accepted soldiers who were more likely to have mental health issues and was now living with the consequences.[41] On the one hand, "If they'd been good troops before they deployed, and now they come back and they are demonstrating issues and behavioral changes," she explained, it was obvious that "something happened while they were gone."[42] On the other hand, she wondered, "Were we talking about soldiers who hadn't been good soldiers, we deployed them anyhow, and now they are still bad soldiers?"[43]

Pollock assigned Ritchie, the psychiatry consultant, to study the issue. With a team of ten staffers, the colonel set about reviewing about seven thousand personality disorder discharges, dating back to 2001.[44] There was no smoking gun that showed traumatized soldiers being railroaded out of the Army. However, there was also little evidence that psychiatrists had arrived at the diagnosis after careful deliberation. "[In] some cases," Ritchie recalls, "there [were] literally a couple of lines saying 'Soldier has personality disorder not otherwise specified' with no background description."[45] The ease with which the discharge could be achieved had led to sloppy record keeping and left the clinical question—how many soldiers had been misdiagnosed?—and the political questions—had there been an intentional effort to deny soldiers benefits?—unanswerable.[46]

It is possible, of course, that some Army psychiatrists were bad actors. However, neither Ritchie's own internal investigation nor Zwerdling's reporting turned up clear evidence of this. What seems more likely is that the misuse of personality disorder discharges occurred because of three intersecting factors. The first, as *An Achievable Vision* and Zwerdling speculated, is that the pressure to maintain units prepared to meet a heavy deployment schedule led psychiatrists to use the diagnosis as they always had: as an easy way of dismissing a soldier. As Pollock put it, "Commanders . . . need a certain percentage of their unit there and functioning and training." As a result, "if they have what they call a 'broken soldier,' . . . they want them off their books, so they can put in a request and get a new person."[47] Indeed, commanders knew that mental health providers could be useful if they needed to get rid of someone who was damaging the rest of the unit. "In the Navy they call the mental health people down range 'the wizard[s],'" explains Col. Rebecca Porter, who succeeded Ritchie as the Army's top behavioral health consultant, "because we can make you disappear from the battlefield. . . . If you need a soldier to disappear, [behavioral health is] where you send them."[48]

The second factor that likely contributed to a misuse of personality disorder discharges is that many Army psychiatrists, at least in the war's early years, may not have been attuned to the realities of post-traumatic stress disorder and thus may not have recognized and diagnosed it. As the first chapter explained, in the early twenty-first century there were very few soldiers in the Army who had that diagnosis, and very few psychiatrists had experience treating it. The reality of Army psychiatry in the first three years of the Iraq War was that many providers may not have been thinking very much about PTSD and may not have had the training to recognize it, and in the climate of increased pressure to keep units fully functioning, psychiatrists likely defaulted to the process that had served them so well in peacetime.

Third, most members of the AMEDD had a limited understanding of the relationship between diagnosis and benefits, particularly with regard to what Veterans Health Administration benefit accrued to which diagnoses. "I don't think there was awareness that if you processed them out it affected their veteran status and their ability to get support through the VA," Pollock explains; Jonathan Jaffin, who had been hospital commander at Walter Reed when the war began and by the time of these hearings was interim commander of the Medical Research and Materiel Command, agrees. "Whether I discharged somebody on a personality disorder or I do a MEB [medical evaluation board] on him as a commander, it makes no difference to me. . . . Those decisions are made locally, and the benefit is ascribed nationally."[49]

Pollock went as far as maintaining that some commanders and clinicians might have mistakenly thought that they were doing a soldier a favor by giving them the diagnosis. Often, soldiers with personality disorders have disciplinary issues: They drink too much, they get into fights, they show up late or are insubordinate—the sort of things that can lead to a dishonorable discharge. "I think they thought they were helping people by giving them a diagnosis and discharging them instead of . . . giving them a dishonorable," Pollock explains, adding "I just don't think people think about the second- and third-order effects of what they are doing.[50] Indeed, many providers may have thought that everyone came out on top: The unit commander got rid of a troublesome soldier, and the soldier got out of the Army before their trouble deepened. No one, apparently, considered that a protocol that had gone uncriticized in peacetime might not work in wartime.

If discerning the nature of these dismissals remained impossible, the AMEDD had a somewhat easier time issuing new policies. Two weeks after the House's hearing on the matter, Pollock issued a memorandum ordering each facility's chief of behavioral health to personally review any personality disorder discharges "to ensure accurate diagnosis" as well as better record keeping. At the bottom of the memo, Pollock added a hand-written message: "We must insure [*sic*] the best outcomes for our soldiers! Please get personally involved in this on your installations!"[51] If the fact that the Surgeon General's office had to proffer this explicit instruction reveals how little oversight this process had historically received, the memo also reveals how alarmed Army leaders had become over the issue and the negative publicity that came with it. In nearly seven hundred pages of documents that the Army released detailing the treatment of mental health issues from 2007 to 2012, this is the only memorandum containing a handwritten note.

Asking the behavioral health staff to become more involved, however, was not enough. By 2008, Ritchie had developed specific guidelines aimed at ensuring that soldiers suffering from post-traumatic stress or traumatic brain injury were not inappropriately diagnosed with personality disorder. On May 1, Lt. Gen. Eric B. Schoomaker, who had replaced Pollock as Surgeon General, issued a memorandum stipulating that any soldier being considered for a personality disorder discharge must also be screened for PTSD and mild TBI.[52] The same month, the Army extended that requirement to all soldiers being discharged "for any reason related to misconduct."[53] By the spring of 2009, the requirements had become even stricter, explicitly stating that "a Soldier will not be processed for administrative separation . . . if PTSD, mTBI, and/or other comorbid mental illness are significant contributing factors to a diagnosis of personality disorder."[54] The screening became more substantial as well. First, the diagnosis had to be made by a doctoral-level psychologist or psychiatrist, and then it had to be confirmed "by a peer or higher-level mental health practitioner" before going to the Surgeon General's office.[55] What had once been a relatively quick and easy way to discharge a soldier now required three layers of review.

More important, however, was that in February 2009 the Army limited which soldiers were even eligible for a personality disorder diagnosis: Only those who had been in the Army for less than two years could be diagnosed with a personality disorder and given a Chapter 5-13 discharge, the Army designation for that condition.[56] The reason for this, Ritchie explains, is that, "by the definition of a personality disorder, it should have emerged early enough in a soldier's career. If you are that impaired by personality disorder, you shouldn't make it for the first two years."[57] Instead, soldiers would be discharged under Chapter 5-17, "Other Medical and Mental Conditions," a category that left them eligible for VA benefits.[58] In May 2009, the requirements became even more stringent, requiring any potential diagnosis to include significant documentation "to establish that the behavior is persistent, interferes with assignment to or performance of duty, and has continued after the Service member was counseled and afforded an opportunity to overcome the deficiencies" as well as requiring "a specific statement that the disorder is of sufficient severity to interfere with the soldier's ability to function in the military" and "clinical documentation that the symptoms or behavioral problems existed prior to enlistment, and do not simply represent maladjustment to the military."[59] The memo also took the 2008 requirement to screen troublesome soldiers for PTSD a step further, explicitly denying providers the ability to use the diagnosis to dismiss troublesome soldiers. "Separation for personality disorder is not appropriate, nor should it be pursued," it explained, "when separation is warranted on the basis of unsatisfactory performance or misconduct."[60] Clearly, a few lines in a soldier's file were no longer sufficient.

As a result of these efforts, the number of personality discharges dropped precipitously. In 2003, the Army had discharged 731 soldiers under the diagnosis, 55 of whom had deployed. The discharges reached a high point in 2007, when 1,066 soldiers, 336 of whom had deployed, were discharged. The growing public concern, and the new policies that the Army instituted in response to that scrutiny, quickly drove the number below the 2003 figures. In 2008, only 641 soldiers were discharged with personality disorder; two years later, the number was down to 365.[61] When the Army reviewed these diagnoses in 2010, it found that every one met the standards that Ritchie had set forth.[62] Moreover, by 2010 the Army was also reaching out to soldiers who had deployed to Iraq or Afghanistan and had been discharged with a personality disorder between 2003 and 2008 to alert them to the potential to have their diagnosis changed and be evaluated by the VA.[63]

In the media and before Congress, the Army's use of personality disorder discharges became another example of the Bush administration's malfeasance and its desire to obscure the damage that the war was causing at the expense of the troops. Certainly, it's likely that in isolated cases soldiers were intentionally misdiagnosed. However, to view the Army as a bad actor oversimplifies how an entrenched institutional culture and an unquestioned set of practices collided with the realities of two protracted wars. Prodded by unpleasant media attention and vigorous congressional oversight, the Army worked to effectively address this controversy, and it did so in ways that likely improved outcomes for soldiers by issuing and enforcing new regulations. Other challenges, however, proved more challenging.

* * *

Making sure that a soldier got the appropriate diagnosis was paramount, of course, but that couldn't happen if a soldier wouldn't come in for treatment. As in the civilian world, many soldiers look askance at mental health care, thinking that whatever issues they are facing do not warrant it, that those who seek it are weaker than those who do not, or that there is risk in being public about their struggles. In the civilian world, it's one thing to tell your employer that you will be coming in to work an hour late every Tuesday for a few months because you hurt your knee running and need to see a physical therapist or that you can't attend a friend's party because you have the flu; it's another to say that you'll be late because that's when your psychologist can treat your depression or that you're staying home because crowds make you anxious. As a result, nearly half of all Americans who might benefit from treatment thus never seek it.[64]

These same issues exist, and are often amplified, in the military, a culture that has not readily embraced mental health treatment. Sometimes that stigma is self-imposed, a refusal to acknowledge that one might need help. "When your ethos is 'I'll never quit, I'll never accept defeat,'" Casey reflects, referring to the Soldier's Creed, "You have a hard enough time getting these people to go in for a broken bone, let alone something you can't see."[65] But stigma is also the result of a culture that looks askance at mental health. In the military's masculinized culture, soldiers fear—often with good reason—that admitting a psychological problem will be perceived as a sign of weakness.[66] Research conducted by military psychologists, including those at Walter Reed Army Institute of Research (WRAIR), confirmed that many service members, and especially those with significant need for care, did indeed feel the pressure of stigma, as did the *Achievable Vision* report, which noted that "many active duty members fear loss of security clearances, assignment to non-combat positions, damage to their promotion potential, and ridicule by peers if they seek help under the program's current implementation and extant policies."[67] Another study found that "only 35% believed that they or someone else in their platoon might experience combat stress and 40% would not trust a returning stress casualty to be an effective soldier."[68] By the spring of 2007, newspapers were regularly reporting that soldiers were reluctant seek care and that even when they did so they were sometimes ill-treated. Again, Fort Carson was the epicenter of these stories.[69] In the *Denver Post* article that broke the personality disorder story, Private Tyler Jennings complained that he "was ostracized in his unit for his inability to cope," and a few months later the *New York Times* profiled Alex Lotero, whose "superiors treated his diagnosis disdainfully, showering him with obscenities and accusing him of insubordination when he missed training for doctor's appointments."[70]

Other soldiers worried that they might not remain eligible for promotion or the security clearances necessary to take a lucrative private-sector defense-contracting job if they sought care.[71] As Stephen Hunt, a VA official in Washington State, explained before a field hearing that Sen. Patty Murray (D-WA) had organized in Seattle in the summer of 2005, "There are real reasons sometimes that people are reluctant to be labeled as having PTSD, either because it may impact a security clearance, or working in law enforcement, or may result in a military career being curtailed."[72] Jimmie Keenan—who would serve as Chief of Staff of the Warrior Transition program after the Walter Reed scandal, command the hospital at Fort Carson, and retire as a two-star general in charge of the Army Nurse Corps—witnessed that impact firsthand in the summer of 2005 as a student at the Army War College in Carlisle, Pennsylvania. Many of her classmates, she noticed, had just returned from battalion commands in Iraq and "were really struggling with what they'd seen and with the soldiers they had lost."[73] Keenan, whose father's October 2001 suicide had made her an outspoken advocate of mental health care, encouraged them to seek help, but many demurred, fearing that doing so would imperil their careers.[74]

Surgeon General Kevin Kiley told a Colorado Springs newspaper in 2007 that "the Army is trying to change a culture in which admitting to mental or emotional problems after combat sometimes is seen as a weakness that could hurt a soldier's career," but those changes were slower to emerge than ones that could be mandated from the top down.[75] Ensuring that soldiers who sought care didn't lose their security clearances, for example, only required that the Army operationalize *An Achievable Vision*'s recommendation that the form be revised to specify only those conditions that "are indicative of defects in judgment, reliability, or emotional stability that are potentially disqualifying or raise significant security concerns."[76] The report's other recommendations—for example, "reinforcing in the minds of both service members and commanders that needing and receiving mental health services is normal" and that "just as commanders and others understand today that physical illnesses or injuries can be treated and in most cases cured or repaired, in a world-class system everyone understands that the same is true for mental illnesses"—proved more of a challenge.[77]

Walter Reed psychologists had, however, been working on improving mental health education programs for two years by the time *An Achievable Vision* appeared, particularly through BATTLEMIND, which the task force report embraced as a model.[78] Following *An Achievable Vision*'s release, however, the Army mandated additional training. On July 15, 2007, Pete Geren, the former Texas congressman who had become Secretary of the Army after the Walter Reed scandal had forced his predecessor's ouster, added another education program as a result of the Walter Reed scandal and ordered that every soldier in the Army—an organization of over one million people—be taught about post-traumatic stress by October 18, 2007, as part of "The Army MTBI/PTSD Awareness Program."[79]

Central to the effort was "chain teaching," in which leaders taught material to their subordinates. In an hour-long session that featured "a standardized script and supporting audiovisual products," including the ubiquitous PowerPoint deck that is the foundation of nearly every Army briefing—and thus a subject of some derision—it aimed to provide "an understanding by every soldier and leader in the Army of the causes, symptoms, and treatments available for MTBI and PTSD" and "ensur[e] soldiers and leaders are aware of the risks, symptoms, and response procedures."[80] Commanders would learn not to be skeptical of soldiers who sought treatment, soldiers would learn not to worry that their buddies would judge them, and those who didn't think that mental health issues really required medical care would learn the error of their ways.[81] In particular, the effort specifically targeted stigma, mandating that "the Army will care for soldiers and families of soldiers diagnosed with MTBI/PTSD with dignity, fairness, and respect" and that "commanders will address every MTBI/PTSD incident seriously and provide soldiers access to medical care as soon as possible."[82]

In 2009 the Army would claim that the chain-teaching program "will reach more than 1 million soldiers, a measure that will ensure early intervention."[83] The program likely benefited many soldiers. But presenting material to soldiers and changing the Army's culture are two different things, and the latter is probably too ambitious a goal for an hour-long session. As Erin Finley explains, "There is top-down pressure to improve access to mental health care and to decrease stigma, yes, but there are also other pressures, from other directions."[84] Efforts to address stigma thus had to come from other directions as well.

One significant shift in this regard occurred when senior leaders began speak out about their own mental health issues. By 2007, almost every colonel or general officer in a position of significant responsibility had shared the experience of Casey, who had spent almost three years in Iraq "watching the impact of combat on soldiers and leaders."[85] Having faced these issues themselves and seen them in their troops, Casey suspected, these commanders had become more willing to take action. "What I was banking on," he recalls, "was that there were enough people with that kind of experience that they would recognize that someone needed help."[86] He and his first Vice Chief of Staff, Gen. Richard Cody, thus encouraged leaders to be open about mental health issues and to encourage their subordinates to seek help. As Gale Pollock explains, shifting the culture of mental health in the force meant that "the senior leaders . . . had to have the experiences themselves in order to understand that, 'Oh shit, all of my troops are having these experiences too, and if I'm feeling this bad, I know they are feeling that bad.' And it became Army leaders wanting to take care of their own."[87] In her view, "It took Cody and his team, the four-stars, telling people that it was okay. It took senior leadership saying 'we need to do the right thing.' . . . If we lead into the behavioral health center and say 'You know, all this combat stuff is hard,' then they are going to have permission to talk about it, too."[88]

Given entrenched Army culture, speaking out about one's own suffering amounted to taking a considerable risk. In his masterful book about Maj. Gen. Mark Graham's efforts, amid great personal tragedy, to force the military to become more responsive to mental health issues, Yochi Dreazen argues that, even late in the war, "virtually no generals or admirals were willing to admit that they'd suffered the same kinds of PTSD and depression as their troops" and that the Army continued to "sen[d] a simple and dangerous message: If you want to get promoted, keep your mouth shut."[89] More troubling, he asserts that at least two generals—Graham and Maj. Gen. David Blackledge—saw their careers effectively ended because of their willingness to speak about mental health issues.[90]

In the wars' later years, however, some senior leaders did become outspoken about their mental health issues. The most notable example was Carter Ham, a four-star general who in 2004 had witnessed the aftermath of a suicide bombing that killed twenty troops in Mosul.[91] Casey, who flew to Mosul after the bombing, recalls how deeply effected Ham had been. "To go in and see parts of their soldiers blown all over the place," he remembered, "left courageous men shaken."[92] Indeed, that carnage, along with the daily stress of sending soldiers on missions in which they were injured or killed, haunted Ham.[93] In 2008, however, he began speaking out about his struggles as a means of encouraging lower-ranking soldiers to seek care. "'You need somebody to assure you that it's not abnormal,'" Ham told *USA Today*, embracing the Army's rhetoric about combat stress. "It's not abnormal to have difficulty sleeping. It's not abnormal to be jumpy at loud sounds. It's not abnormal to find yourself with mood swings at seemingly trivial matters."[94]

The open acknowledgment of mental health issues happened further down the chain of command as well. When Col. Rebecca Porter returned from Iraq in 2008 feeling as though the mundane details of her everyday life were intolerably frustrating, she and her husband decided to seek counseling.[95] Her primary care doctor referred her to Walter Reed's psychology department, which was problematic because Porter was the chief of that department and thus would have become the patient of one of her subordinates, a situation that raises a number of ethical issues. Yet when her chief resident approached her with the referral, she used the opportunity to teach her staff that "it doesn't matter what rank a person is or whether they are a psychologist or an infantryman. They can have issues. . . . Behavioral health issues are not dependent on rank or job."[96]

Porter's story reveals that perhaps more important than leaders acknowledging their mental health struggles was their visibility as they sought help at post clinics. This helped in two ways. First, when leaders benefited from mental health treatment, they became greater advocates of their soldiers getting it. "If you help the NCO, or the company commander, or the platoon sergeant with their mental health issue," Army psychiatrist Col. Chris Ivany explains, they become "much more likely to recommend that their guys go over to mental health."[97] The inverse was also true. When leaders were seen getting treatment, lower-ranking soldiers became more willing to go. As Gale Pollock asserts, when "people could say that 'Hey, did you know that Command Sergeant Major was over there? Did you know the Old Man was there?' it started to give them feelings of assurance that it was okay to go and say 'I need to talk about this.'"[98]

Nonetheless, many soldiers remained reluctant to acknowledge their issues, much less to seek help for them. For these soldiers, what some four-star general said mattered a lot less than what the people they worked with every day thought. Making mental health visible and available in the units where soldiers spent their time thus became critical. Ivany discovered this when he served as the 4th Infantry Division's division psychologist, his first assignment after his residency. In early 2008 he was in Baghdad, where the 4th Infantry Division had deployed as part of the surge. It was a brutal deployment. In the first few months of 2008 newspapers regularly reported news that the division's soldiers had been killed, and much of Ivany's clinical time was spent doing combat and operational stress control with units that had suffered a casualty.[99]

Because so many leaders had deployed before, by 2008 many of them were amenable to Ivany's presence and recommendations.[100] His division commander, Maj. General Jeffrey Hammond, was particularly supportive, telling his senior staff that Ivany spoke for him on matters of behavioral health.[101] Where Ivany had more trouble was in convincing soldiers further down the chain of command. As in the rest of the Army, the issue was less what Hammond said and more what the other members of the squad thought when the division psychiatrist came calling. "One of the big variables was whether or not they felt like their squad leader or their platoon leader [or] their platoon sergeant . . . were genuinely open and accepting of behavioral health."[102]

More than this, though, he struggled to ensure that soldiers actually had access to care. The embrace of counterinsurgency doctrine meant that troops were rarely on large bases, which meant that they had less access to mental health resources. In fact, between Medical Health Advisory Team IV (MHAT-IV) at the end of 2006 and Medical Health Advisory Team V (MHAT-V), which went to Iraq in late 2007, there was a nearly 50 percent decline in the number of mental health providers who "reported that there were sufficient behavioral health assets to cover the mission across the area of operations." One provider whom the team interviewed complained that "we have one psychologist, and two 68Xs [enlisted mental health specialists] per 4000 Soldiers spread out across one FOB [Forward Operating Base] and five outposts."[103]

Ivany thus had the paired tasks of making care available and making it something that soldiers were willing to take advantage of. To accomplish these goals, the Army began "sen[ding] those officers and those techs out to the smaller locations so they lived there."[104] In part, this was simply an extension of the Army's long-standing practice of attempting to treat soldiers as close to the front as possible. But it was also an effort to make behavioral health providers more visible, integral parts of the unit who were credible because they had had the same experiences as the soldiers whom they were treating.

The most visible example of an embedded mental health provider in U.S. popular culture is likely Lt. Col. John Cambridge, the hapless and pitiful reservist in Katherine Bigelow's 2008 film *The Hurt Locker* who is rebuffed after he dispenses pabulum to the clearly traumatized Specialist Eldridge before being killed because he lacks the situational awareness to be suspicious of Iraqis planting a roadside bomb—hardly a confidence-inspiring example of the AMEDD's credibility among the force.[105] Ivany found, however, that living among soldiers did lead more of them to seek care. "Being there, wearing the same uniform as they are, sleeping in the same kind of locations as they are, sharing some of the hardships, speaking their language a little bit—quite a few guys came in," he recalled.[106] The success of putting behavioral health providers in individual units was one of the most significant lessons that he would take from his deployment.

Around the same time that Ivany was figuring this out in Iraq, the Army was learning a more gruesome lesson about the potential consequences of junior soldiers fearing that their peers and immediate superiors would deride them for seeking help for behavioral health issues. Once again, Fort Carson proved the epicenter of the Army's mental health challenges. In the fall of 2007, Sen. Ken Salazar (D-CO) had written to Army Secretary Pete Geren about an "apparent clustering of violent behavior" at the Colorado post.[107] In particular, Army epidemiologists would later explain in their report that, "allegedly, 8 homicides in the previous 12 months were perpetrated by 6 Soldiers from units at Fort Carson," and one platoon of the Fourth Infantry Division were particularly overrepresented among the alleged perpetrators.[108]

The request to investigate this violence wound its way from Salazar to Geren to Fort Carson's commander, Maj. Gen. Mark Graham, who requested an epidemiological consultation from the Surgeon General's office. In Army parlance, epidemiological consultations are investigations, usually of violent events, conducted by the AMEDD Public Health Command. Already during the twenty-first century there had been several, including one at Fort Bragg after a spate of murder-suicides, and another at Fort Carson to understand a suicide cluster.

Graham's request eventually reached Lt. Col. Mike Bell. Bell had taken a somewhat circuitous path into the AMEDD. He had begun his military career in artillery and military intelligence. After serving in Operation Desert Storm, he earned a degree in biology at Syracuse University and attended medical school at the Uniformed Services University, specializing in environmental and occupational medicine.[109] Bell was medical director at the Army's chemical weapons stockpile until the Walter Reed scandal broke. Bell believed that his training in occupational medicine was ideally suited to "help[ing] individuals receive the highest level of function that they can within any given illness or injury," so he volunteered to work on the first Warrior Transition Units.[110]

Bell's next assignment was creating the Army's Behavioral and Social Health Outcomes Program, an initiative designed to apply epidemiological principles to mental health issues. He had barely gotten the program off the ground when the Surgeon General's Chief of Staff told him to put aside the work he had been doing to focus on Fort Carson. "It was framed as, 'There's a concern that a unit has come back from the war and is kind of running amok,'" Bell recalls, and "there were a lot of theories about why. . . . They wanted to understand from a scientific basis whether this was actually truly an unusual circumstance, or whether . . . there was a real problem at Fort Carson."[111] In particular, the Army worried that lowered recruiting standards meant that more troubled and trouble-prone soldiers were in the force.

Bell put together a twenty-seven person team that included, among others, epidemiologists, a chaplain, and a representative of the Army's

Criminal Investigation Division. Together, they framed out seven questions aimed at understanding what factors might have contributed to the spike in violence. Four issues were of particular interest. First, the Fort Carson epidemiological consultation (EPICON) team was concerned that the brigade combat team's experiences in Iraq had created a population with an increased propensity for post-deployment violence. Second, they wondered whether violence had increased because the Army had accepted soldiers with criminal and substance abuse histories. Third, the EPICON wanted to sort out whether soldiers had adequate access to behavioral health treatment. Perhaps most important, they wanted to know whether these issues were affecting only this unit, Fort Carson as a whole, or the entire Army.[112]

Over the course of November and December 2008, Bell and his team conducted interviews with several of the soldiers who had been accused or convicted of homicide, held fifty-nine focus groups with more than five hundred soldiers, studied Fort Carson's culture, and compared the brigade combat team that had seen so many arrests to another one "in order to assess the potential cumulative effect of operational tempo (OPTEMPO) and deployments."[113] Both the depth and the pace of the work are impressive. The team arrived at Fort Carson on November 3, presented the final report to Maj. Gen. Graham on March 24, and released it to the public in July.[114]

The speed with which the report was produced is all the more remarkable because of the climate in which the research was done. Graham was a fierce proponent of mental health—his own son had taken his life a few years earlier, and the issue was increasingly personal for him—and he sought to ensure that the EPICON would have the resources and cooperation that it needed.[115] The further down into individual units the team penetrated, however, the more reluctance there was to speak with them. Although people ended up being forthcoming, Bell remembers, "You could tell. The lower-ranking people were a little bit more reticent."[116] In Bell's view, this was typical because "units in the Army, or really any organization, like to keep their dirty laundry in house."[117]

When the report was finished, there was also considerable debate about whether to make it public. The Army had been battered by reports that it wasn't taking care of its soldiers, and there were fears that some in Congress and the media would view the report as further evidence of this failure.[118] After four months of debate, however, Eric Schoomaker and Mark Graham succeeded in convincing Geren to release the report. Their rationale, Bell recalled, was that the report was "an indication that the Army wanted to investigate a very serious situation, look at itself really hard, and then implement corrective action."[119]

Releasing the Fort Carson report was in keeping with the Army's transparency about behavioral health throughout the wars. All of the MHAT studies, for example, were also made public. Nonetheless, the skepticism of Army leaders had some merit, because the Fort Carson EPICON revealed some difficult truths about the Iraq War's impact on the troop and about attitudes toward behavioral health throughout the Army. The brigade combat team at the center of the investigation had had a particularly tough tour in Iraq. Originally part of the 2nd Infantry Division and stationed in Korea, it had deployed to Iraq in August 2004 and participated in heavy combat in Al Anbar Province. When its tour was up, however, it was reassigned to the 4th Infantry Division and sent to Fort Carson, where it quickly trained up and returned to Iraq.[120] When they returned to Fort Carson, some of their personal belongings that had been in storage were missing; other belongings had been ruined by water and mold.[121] On almost every measure, this unit had had it worse than the comparison brigade combat team: more fierce fighting, more soldiers killed in combat, and higher rates of mental health issues, traumatic brain injury, and substance abuse.[122] The report concluded that there was "a possible association between increasing levels of combat exposure and risk for negative behavioral health outcomes."[123] However, this deployment history was only one of the factors that the report concluded were likely linked to the high rate of alleged criminality. In fact, relatively few of the soldiers who had gotten into legal trouble had deployed on both tours.[124]

Access to care and willingness to seek it turned out to be a most significant issue. For starters, members of this brigade had significant difficulty getting treatment. On the first deployment, a mental health provider had not been attached to the unit. On the second, it did so with a provider who had been temporarily assigned.[125] At Fort Carson, the mental health staff was at about two-thirds strength, despite a 400 percent increase in the number of soldiers seeking care.[126] Most important, though, stigma had deeply penetrated the unit, and its leaders were disdainful of struggling soldiers.[127] "Psychological problems were dealt with in a very heavy-handed, unsympathetic way," Eric Schoomaker, the Surgeon General who had ordered the EPICON, explains. "Soldiers were humiliated if they had problems."[128] Some soldiers who had received treatment had subsequently been "viewed as 'weak' or labeled as 'bad' Soldiers" and were "ridiculed, treated differently, and referred to as a 'shitbag.'"[129] The jailed soldiers told Bell's team that leaders often made statements that paid lip service to behavioral health treatment but that "their subsequent actions contradicted their message." They complained of leaders violating soldiers' privacy through "public announcement of BH [behavioral health] appointments in formation [and] discussions about personal BH issues where other Soldiers can overhear," acts that made already reticent soldiers more reluctant to seek help.[130]

The events at Fort Carson were an extreme example of what happened when troubled soldiers were not encouraged to seek help, but Michael Bell and Amy Milliken's report spoke to the wider need to reduce stigma by fostering a better climate within individual units.

Recommending that leaders be better trained to be more sympathetic to their subordinates seeking behavioral health care in order to "ensure that there is no humiliation or belittling of Soldiers who seek or receive [behavioral health] care or [substance abuse] assistance," the report highlighted the reality that an hour-long PowerPoint was not going to be sufficient and that there were wider cultural issues that had to be addressed.[131]

* * *

Chris Ivany, who had learned similar lessons as explicitly, albeit less tragically, in Iraq, arrived at Fort Carson in the summer of 2009, just as the EPICON report was being made public. A year later, he became chief of the Department of Behavioral Health at the Evans Army Community Hospital. Under Mark Graham's aggressive leadership, Fort Carson had already become an innovative leader in behavioral health.[132] When Ivany arrived, he became deeply involved in one of these initiatives, the Embedded Behavioral Health program. In fact, according to Rebecca Porter, who succeeded Elspeth Ritchie as behavioral health consultant, Ivany "was the mastermind."[133]

Embedded Behavioral Health was intended to address the two realities that Ivany had seen firsthand in Iraq and that Bell and Milliken highlighted in the EPICON. First, stigma remained high, and many soldiers did not feel comfortable visiting a mental health provider. Second, many leaders were skeptical of behavioral health's benefits. This new strategy sought to meet these challenges by bringing into the garrison environment the strategies that Ivany and others had employed in Iraq.[134]

Doing so required making mental health more present in soldiers' and leaders' daily lives. When Ivany arrived at Fort Carson, the care model was the same as it had been before the wars began: A soldier who needed behavioral health care would visit the post hospital and see a clinician. This arrangement created problems for soldiers and their leaders that in turn led some soldiers to be reluctant to seek care and some leaders to mistreat those who did. One issue was that soldiers had to leave their units, and thus miss work, to see a provider. Their conspicuous absence raised privacy issues and heightened the likelihood that they might be seen as bad soldiers—malingerers who didn't pull their weight.[135] Another issue was that, to leaders, the hospital seemed disconnected from their units, which inherently made leaders skeptical of recommendations that sidelined a soldier.[136] As Ivany remembers, the prevailing attitude among leaders when he arrived was, "I just send our guys up to the hospital. I have no idea what happens to them when they go up there, who sees them, if they are getting better" and that it was hard to learn whether a soldier was deployable.[137] The officers and noncommissioned officers (NCOs) didn't know the providers, which made them skeptical of their understanding of the unit's needs.[138]

The solution was to implement the same solution that had worked for Ivany in Iraq: Place mental health providers in close proximity to soldiers and their leaders and build relationships that facilitated better care delivery. Although they weren't officially part of the units that they served, mental health providers were situated within the brigade combat teams' physical space, rather than at the post hospital.[139] The clinical care didn't change much, if at all. Soldiers still got counseling and prescriptions. What was different was that both leaders and soldiers had daily, face-to-face contact with the mental health provider.[140] Practically, this meant that psychologists or psychiatrists still wore the caducei of the Army Medical Command on the left shoulder of their battle dress uniforms and not, for example, the 4th Infantry Division's ivy leaves. But their desks were not in the hospital; they were in the unit's offices.[141]

This realignment meant that NCOs and officers now dealt with an individual provider whom they grew to know and came to trust. As in Iraq, leaders responded well when embedded providers helped them deal with a struggling or problematic soldier.[142] They also appreciated when providers told leaders that a soldier's mental health issue was not a mitigating factor in a disciplinary matter or that the soldier could still deploy with a diagnosis.[143] Whatever advice a psychologist or psychiatrist might give in a particular case, providers' honesty and their role in helping units stay effective slowly changed many leaders' perspectives. Even those who remained skeptical of mental health as a whole, Ivany found, came to see the value of *their* soldiers getting help from *their* provider. "We are never going to change all of those long deeply held convictions," he reflects, explaining that "they could still dislike mental health as a whole but like Dr. Johnson right down the street, who sees all of their soldiers."[144]

The Embedded Behavioral Health program was thus fundamentally about building trust between unit leaders and mental health providers, soldiers from two parts of the Army who did not always see eye to eye. In particular, under the traditional system, the EPICON reported, there was a sense among leaders that "providers did not understand the impacts of a BH [behavioral health] diagnosis and the potential for manipulation by soldiers in order to avoid negative consequences" or that a "diagnosis makes Soldiers non-deployable."[145] Embedded Behavioral Health, an Army behavioral health official explained at a 2012 briefing, helped mitigate privacy issues because a trusted clinician could tell a commander that a soldier had a serious issue that made them undeployable without disclosing what that issue was and "the commander believes it because [they] have been working together."[146] The Embedded Behavioral Health program thus contrasted sharply with the chain-teaching approach that had preceded it. Whereas the latter imagined that stigma could be reduced and mental health care made more acceptable through an hour-long PowerPoint lecture, the former understood that progress was incremental and individual. As Ivany put it, "There's no magic PowerPoint in this. . . . You have to build a consistent and enduring working relationship with these guys."[147]

Embedded Behavioral Health addressed nearly every issue that the EPICON had identified. In a 2016 article, Charles Hoge, Chris Ivany, and others pointed out that, when the Army's Public Health Command studied the program's effectiveness, it "identified significant correlations between implementation of the embedded teams and reductions in off-post mental health referrals, inpatient hospitalizations, and soldier risk behaviors, as well as an increased proportion of soldiers considered psychiatrically fit for deployment."[148] Based on this evidence, Rebecca Porter sought "to get Army leadership to buy off on rolling it out across the Army."[149] And while Peter Chiarelli was reluctant to embrace the program until Jayakanth Srinivasan, an outside consultant from the Massachusetts Institute of Technology, had presented his own study showing that it worked—an experience that frustrated Porter—he did eventually approve an Army-wide expansion of Embedded Behavioral Health.[150] By 2016, Hoge and his colleagues reported, more than fifty "brigade-sized unit[s]" were being served by embedded providers.[151]

* * *

Soldiers and leaders were not the only people who had to be convinced to embrace mental health care. While many doctors in the AMEDD believed that behavioral health should be more prominent in military health care, that feeling was by no means universal. Here again, simply ordering people to do more screening and treatment wouldn't work. "People have this idea that in the military everybody is in line and you salute," Charles Engel remembers. "Especially on the medical side, it's a bit like herding cats. . . . There was a great deal of skepticism, and so it took some time for us to gain credibility."[152]

Engel was referring to his work on RESPECT-Mil, the Army's initiative to integrate mental health screening into primary care. The program, which began in 2005 with a trial at Fort Bragg, grew out of his work running the program for Gulf War Syndrome at WRAIR in the late 1990s.[153] A self-described "blue state" native, Engel had grown up in Seattle and started college when politicians and generals were lamenting the post-Vietnam "hollow Army," and, he readily admits, he "joined because I needed the money to get through medical school."[154] He never expected to stay for thirty-one years.[155]

Engel researched psychosomatic medicine—the relationship between mental and physical health—and he became interested in how physical illnesses or injuries produced mental health issues and, conversely, how mental health conditions could create physical symptoms. For example, people with anxiety often experience physical symptoms like shortness of breath or chest pains.[156] People with these kinds of psychosomatic symptoms, however, are more likely to end up on their primary care physician's examining table than on a psychiatrist's couch. "You have to have already accepted the perspective that you have a mental health problem and that going to see the specialist is a valid way of addressing it" to make such an appointment, Engel explains. In contrast, it is relatively easy to make an appointment with your primary care physician if some physical symptom is making you uncomfortable.[157] Primary care physicians, however, often are not well prepared to screen for and treat mental health issues. Instead, they are trained to employ what Engel terms a "medical model" in which they assume that physical symptoms result from a physiological ailment that can be treated and cured.[158] When this approach proves futile, Engel explains, doctors generally dismiss the patients' symptoms, or write them a prescription, or send them to a specialist, an approach that rarely resolves the issue but frustrates everyone.[159]

Engel's solution to this problem was straightforward. During the 1990s, he spent several years running a clinic for veterans who had been diagnosed with the mysterious Gulf War Syndrome. His background in psychosomatic medicine led him to think, first, that the debate over the causes of Gulf War Syndrome wasn't "as mysterious of a problem as some might suspect" because many physical complaints are often linked to larger behavioral health concerns and, second, that identifying the causes of Gulf War Syndrome was secondary to treating patients regardless of the cause of their symptoms.[160] Counseling patients to view their symptoms as treatable and offering them a structured physical therapy program even when the specific causes of their complaints remained enigmatic led to positive results, and it immensely gratified Engel.[161] More important, he came to believe that mental health needed to be more prominent in primary care settings. The Army, he reasoned, needed to "go where the problems are."[162]

Making that approach a reality, however, was fraught with challenges, and it illustrates that even within the AMEDD mental health care was not without stigma. Ironically, the wars began right around the time Engel reached twenty years of service and was contemplating retirement. As tempting as a second career at a hospital or a think tank might have been, the wars presented a unique opening. "All of a sudden," he explains, "I had this opportunity to do what I had been writing about, which was the whole primary care piece trying to improve the mental health services."[163]

Engel first publicly described the program to Congress in the same hearing that featured Stefanie Pelkey's heartrending testimony. Moments before he spoke, Pelkey testified to the seriousness of the very problem that RESPECT-Mil was meant to address. "He saw his primary care provider," she explained, but despite "seven or eight visits for chest pains, high blood pressure, erectile dysfunction and even noted depression," no one mentioned PTSD.[164] Engel picked up on this theme, explaining "the need to bring safe, accessible, and confidential care to service members rather than waiting for them to seek care" and asserting that "primary care affords an excellent opportunity" to do so.[165] Nearly every soldier saw a primary care physician, he explained, but very few saw a behavioral health specialist.[166]

Engel's testimony was based on the results of a pilot study that he had overseen among 82nd Airborne Division soldiers at Fort Bragg. There, Engel and his team used a civilian model created for depression care in which a psychiatrist "provides informal advice to the primary care clinician" who treated patients and "formally supervises the care facilitator," the person who monitors patients to ensure that they followed recommended treatment.[167] To ensure that the program met the needs of returning Iraq and Afghanistan veterans, "PTSD [was] added as another clinical condition," and the program was modified so diagnoses relied more heavily upon patients' responses to intake surveys rather than on clinician observation.[168]

The procedure at the heart of RESPECT-Mil is familiar to anyone who has visited his or her primary care physician in the last decade. While patients waited for their appointment, they completed a health questionnaire that included questions about depression and PTSD. They included yes/no questions like "Over the last 2 weeks, have you been bothered by feeling down, depressed, or hopeless" and "Have you had any experience that was so frightening, horrible, or upsetting that, in the past month, you tried hard not to think about it or went out of your way to avoid situations that reminded you of it?" as well as questions that asked soldiers to rate the severity of their symptoms.[169] The provider then evaluated these responses in order to diagnose, treat, or refer the patient.[170] Often, the plan included getting soldiers involved in hobbies, a healthy lifestyle, reconnecting with friends, "or just find[ing] a quiet comfortable place and say[ing] comforting things to yourself" as part of a "self-management" plan.[171] In keeping with Engel's emphasis on recuperation rather than diagnosis, doctors were told that gathering "detailed information about the traumatic experience[s] . . . is not recommended" and that they should "focus instead on current symptoms and circumstances."[172]

At the end of the sixteen-month study, Engel and his team found that "approximately 10% (404) of screen[ed patients] were positive for depression, PTSD, or both."[173] Of those who entered into the RESPECT-Mil program and thus were monitored by the care facilitator, more than 80 percent "had a clinically significant drop" in their score on the PTSD questionnaire.[174] This was exactly what Engel had hoped. "With basically only the added resource of a care facilitator," he wrote, "we were able to screen and identify soldiers with depression and or PTSD who would likely have gone undetected and untreated," and "the majority of soldiers enrolled demonstrated clinically significant improvement." This, he understood, indicated that "RESPECT-Mil seems both feasible and worthwhile."[175]

Surgeon General Kevin Kiley agreed. When Engel briefed him on the pilot, he ordered RESPECT-Mil's implementation on the fifteen Army posts that had the highest deployment rates.[176] Engel, as a scientist, suggested that they structure the expansion as a clinical trial, implanting RESPECT-Mil at half of the sites so they could measure its effectiveness beyond the pilot.[177] Kiley, more concerned with the immediate needs within the force, insisted on a complete, immediate implementation. "I'm the guy thinking about it from a scholarly perspective, trying to think about how we could do this in a way we could learn something," Engel remembers, "and he's like 'Let's just roll this thing out, man.'"[178] As a result, by 2007 the program was being implemented throughout the Army.

Two points are important about this exchange. The first is that however much Kiley was maligned for his management of soldier care at Walter Reed, he clearly sought to expeditiously address soldiers' mental health care and was, as Engel puts it, "really quite good and quite attentive to the mental health needs."[179] As well, this moment illustrates how senior leaders were often keen to implement promising protocols, even if they hadn't been clinically validated to the standard that WRAIR researchers would have liked. If Engel believed that "it was better to make tiny, small, incremental change without mistakes than it was to take off and try to change the world in one fell swoop" because errors could scuttle the entire program, he recalls that Kiley "grew impatient with us doing this, and he attributed some of this to my research background: The idea that we were going at this far too slow, far too analytically."[180]

Such differences in opinion were common during the wars. Recognizing that soldiers had immediate needs, senior leaders like George Casey and Peter Chiarelli were often unwilling to wait for rigorous but often lengthy clinical trials to conclude before implementing a treatment. Casey, for example, recalls that, when he was presented with the argument that more research was necessary to validate the effectiveness of the Comprehensive Soldier Fitness program, his attitude was "I accept your opinion, but we don't have time and I'm implementing."[181] "I needed something that would make a positive impact at the time I needed it," he recalls.[182] For AMEDD researchers, this tension was a fact of life. The Army's leaders, Amy Adler understood, "have to think about moving things quickly. They only have a few years in their position."[183] The researcher's job thus became giving the best advice they had at the time, while the senior leader's job became "absorb[ing] the risk" if an initiative failed.[184]

Engel threw himself into the work "with almost religious fervor to convince people that this was the right thing to do."[185] He and his team had already developed a training manual that outlined the significant number of mental health issues that returning veterans faced, pointed to the "gap between need for treatment and receiving it," and asserted that RESPECT-Mil "provides one step towards closing this gap by providing background needed for primary care clinicians to provide high quality mental health care that has a solid evidence base for effectiveness."[186] It also encouraged clinicians "not to miss the chance to try the model at your first opportunity."[187]

Despite this entreaty, many clinicians were cool to the idea, and some were outright hostile. Given that psychiatry had historically enjoyed relatively little prestige inside the AMEDD, this is perhaps not surprising. Moreover, most primary care physicians had little training in mental health, and many did not view it as part of their job. Gale Pollock recalls, for example, that when she attended suicide prevention training at Fort Benning in the early years of the Afghanistan Wars, many doctors and nurses had the view that "that's mental health stuff. We don't go there."[188]

Engel encountered this firsthand as he encouraged AMEDD staff to embrace RESPECT-Mil. "In the first year or two of this program, when I would go places, I would get nothing but resistance," he recalls. Some hospital commanders so resented his initiative that they didn't even offer a handshake.[189] At Fort Hood, the chief of mental health and the chief of primary care actively resisted him.[190] There were multiple reasons for this resistance: Territorialism among specialties, the sense that mental health was somehow further down the hierarchy of medical expertise, and a sense within the Army primary care community that they had already been asked to take on a significant burden after the creation of Warrior Transition Units and that RESPECT-Mil was yet another task that was being foisted on them.[191]

Making the project work was thus a long-term proposition. Just as stigma could not be eradicated through a one-hour PowerPoint presentation and a directive that troops wouldn't belittle peers who sought help, Kiley's operationalizing RESPECT-Mil across the Army wasn't going to make individual physicians embrace the program or, more important, take mental health seriously. Rather, the program's success came through efforts similar to those that Chris Ivany had embraced: building working relationships, getting individual buy-in, and allowing the process to grow organically. It was cultural change on a retail, not a wholesale level. Engel spent seven years traveling around the world selling the program at Army facilities, and it took a few years before he began to feel like most of the people he was talking to were willing to go along with the idea. Largely, this was the result of his having cultivated relationships with doctors who embraced the program at one location and then became evangelists for it elsewhere when they got promoted and transferred. "Because I'd been running the program for so long," Engel recalls, "they had switched jobs, two, three, four times and they would run into me in each of these places, and so they got to know me, I got to know them, there was trust established. Over time they realized that . . . the ideas that I had were basically right and could be done and should be done."[192] At Fort Hood, after a new hospital commander with whom Engel had worked at another post learned of the resistance of the chief of primary care and the chief of behavioral health to RESPECT-Mil, both were reassigned to deploy. And while Engel is careful to caution that it's not clear that the resistance caused the deployment, "the feeling was that these guys had been there too long and it was time to move them along."[193] In the end, Fort Hood became a model for RESPECT-Mil's implementation.

Increasing evidence that vindicated the program added to its success. By late 2011, nearly every primary care patient was being screened, leading a RAND study to conclude that "RESPECT-Mil is identifying a considerable number of service members who are reporting depression and PTSD symptoms."[194] By July 2012, the program was in place at every Army clinic, and the military publication *Stars and Stripes* reported that "each month 100,000 soldiers are screened."[195] "It was very edifying," Engel recalls, "I loved it, probably more than anything else I ever did in the military."[196] Getting to this point, however, exacted a tremendous toll. "I pretty much worked myself into the ground," he explains.[197] Traveling around the country and staying in hotels, he gained sixty pounds and spent little time with his family. "I was very absentee during that time," he admits, and "I worked my way out of my marriage."[198] When he retired after thirty-one years of service, his marriage, which had lasted nearly as long, also ended.

* * *

Successful efforts to reduce stigma and make care more accessible, like Embedded Behavioral Health and RESPECT-Mil, illustrate that changing the Army's culture happened from the bottom up—when an embedded provider helped a company commander understand why some specialist was showing up late to formation, or when a primary care physician screened a patient for depression and referred them to counseling. This is not to say, however, that Army's leaders' actions played no role in changing the climate surrounding mental health. During the same period that Ivany, Engel, and others were toiling outside the public spotlight, the Army's most senior leaders—Peter Schoomaker, George Casey, and Peter Chiarelli—became more vocal about the strain that repeated lengthy deployments were placing on the force, implemented the chain teaching program, and fast-tracked these new programs' implementation. In their congressional testimony, they increasingly called upon a culture that had been largely detached from the military and the wars to recognize that two protracted conflicts could not be fought on the cheap.

In April 2007, the Army extended deployments to fifteen months in order to meet the surge's demands, a decision that Casey's predecessor, Peter Schoomaker, had made with the intention of helping soldiers and families set expectations for a long deployment rather than confront an unexpected extension.[199] This decision contributed to a broader anxiety that the wars were reducing the military's readiness and that, as one op-ed by two Democratic congressmen put it, "The U.S. armed services are literally at the breaking point."[200]

In particular, there was growing concern about the impact that long and repeated deployments with little rest—in military parlance, dwell time—was having on the force. The research done by MHAT-IV, as well as other research, suggested that longer and more frequent deployments were damaging soldiers' mental health.[201] The fifth Mental Health Assessment Team, which Paul Bliese led in late 2007 and whose report was published in February 2008, reached similarly troubling conclusions. Although the mission was almost scratched because of internal concerns over the methodology of previous MHAT data collection, Bliese and Army psychologist C. J. Diebold did go, and they collected more than 2,500 surveys.[202] As in MHAT-IV, the results raised concerns. Soldiers' morale and mental health bottomed out after about seven months, while suicidal ideation peaked a month later; alcohol abuse, in contrast, hit its high point right around the one-year mark.[203] "The model predicts a three-fold increase in the number of male, E l–E 4 Soldiers that will be positive for mental health problems at the 15th month of the deployment," Bliese wrote, later adding that "the adjusted percent of Soldiers reporting mental health problems at month 15 is significantly higher than the percent reporting problems in the early months."[204]

Mental Health Assessment Team V was also the first to include data from a significant number of soldiers on their third and fourth deployments, with equally grim conclusions. "An NCO on his or her second or third/fourth deployment reports significantly more mental health problems than does an NCO on his or her first deployment," the report revealed, as well as increases in "stress or emotional problems" that affected work performance, divorce rates, and alcohol abuse.[205] That repeated lengthy deployments were causing significant mental health challenges was, unsurprisingly, a primary takeaway in the press. "More than five years of recycling soldiers through Iraq and Afghanistan's battlefields is creating record levels of mental health problems," Gregg Zoroya wrote in *USA Today*, before quoting Casey's admission that "People aren't designed to be exposed to the horrors of combat repeatedly, and it wears on them."[206]

Addressing frequent deployments and shortened dwell times was thus a central goal. The Army set a goal of lengthening dwell time so units would be spending three years at home for every year deployed but was having little success in doing so amid the protracted conflicts. The Army had reduced the ratio to one-to-one early in the Iraq war and then to 15 months deployed and 9 months at home when the 2007 surge began. "During my whole tenure," Casey recalls, "it was a fight to get the Army back to 1 year out, 2 years back. 3 [years at home for every year deployed] has got to be the norm."[207] One avenue toward accomplishing this was fighting with a bigger force. From Casey's perspective, such an increase would go some way toward rectifying the nation's failure to realistically plan the demands that the invasion and occupation of Iraq would entail. "The force wasn't big enough to do what the country asked it to do," he reflects, later adding that "we certainly weren't prepared to sustain a 150,000 person troop rotation for a decade or more."[208]

In 2007, Army leaders successfully lobbied the Bush administration to increase the authorized strength of the Army from 482,000 soldiers to 547,000 soldiers.[209] This increase was explicitly tied to deployments and dwell time. In 2007, Peter Schoomaker told the Senate Armed Services Committee that "growing the Army will allow soldiers to remain at least 12 months between deployments. . . . This is the best way to reduce the strain on the force."[210] Testifying before the Senate Armed Services Committee in February 2008, Deputy Chief of Staff Lt. Gen. Michael D. Rochelle explained that the Army was "on target to meet this goal by 2010" and that "Army growth will help us return to shorter deployments [and] increased time at home between deployments."[211]

Even as Army leaders touted a larger force as a means of addressing the strain that the wars had caused, they knew that it would not solve the problem immediately. "It will take time to recruit and retain the soldiers to meet these increased authorizations," Schoomaker had testified in 2007, and a year later Defense Secretary Robert Gates was promising "with our 'Grow the Army plan,' we should achieve 48 deployable Active Army [Brigade Combat Teams] by fiscal year 2011" but that the Army in 2008 was predicting that it would be able to "limit deployments to 1 year in theater with at least that same amount of time at home"—still a boots on the ground/dwell ratio of 1:1.[212]

Increasing dwell time as the Army increased its size was thus dependent on reducing the number of soldiers who were required to deploy. That was a point that Casey's number 2, Peter Chiarelli, made before Congress as he cautioned that the lengthy wars and lack of dwell time were damaging the All-Volunteer Force. In 2009, Chiarelli explained the imperative to the Senate Armed Services Committee by critiquing the "increased deployments, shorter dwell-time, and insufficient recovery times for our soldiers, their families, and our equipment."[213] Chiarelli then admonished a culture that had not sacrificed much in the two wars that it had decided to fight simultaneously but that had asked much of the military. "We simply cannot achieve desired [Boots on the Ground]/dwell ratios until demand is reduced to a sustainable level," he wrote, adding that, "unfortunately, the Army cannot influence demand, and the current level does not appear likely to improve significantly for the foreseeable future."[214] Chiarelli portrayed the Army's hands as tied; forced to fight two prolonged wars with a small force, it had no choice but to "[ask] a great deal from our soldiers and their families," on whom the strain had become nearly unbearable. "It is a resilient force. It is an amazing force. But I have to tell you, it's a tired and stretched force," he began, adding that, "to turn around and go back to either Iraq or Afghanistan just under 12 months or just over 12 months, and have it be your third or fourth long deployment, is difficult. It's difficult on soldiers and families." He then cautioned that "the key . . . is seeing demand come down as projected. But if it doesn't, we'll have some issues."[215]

Internally, army leaders were making the same point into 2010. In April of that year, Surgeon General Eric Schoomaker wrote an email to Chiarelli calling dwell time "the most pernicious element of the entire army challenge."[216] He argued that

> it is time to draw a line in the sand which outlines emerging insights into the requisite time required to restore the baseline state of [behavioral

> health/mental health], to restore Family and community connections and to recover from psychological & physical stressors of prolonged deployments and combat.[217]

In explaining why this was imperative, he likewise offered a subtle critique of politicians and a wider culture that didn't appreciate the human consequences of the long wars and, perhaps, didn't care to. The issue, he wrote, needed to be addressed "before the bean-counters conclude that 1:1 or less is 'tolerable' and that an Army in protracted war can be kept small or demand kept high, at a 1:1 ratio." "THAT," he concluded, "is one of the real public health threats."[218]

The anxieties acknowledged by Army leaders should not be taken lightly. These men had entered the Army in the 1970s, after the Vietnam War had nearly crippled the institution, and they had spent their lives working to rebuild it. At least privately, they must have shared Sen. Carl Levin (D-MI)'s worry, expressed in 2007, that the country risked returning to "the hollow Army of the 1970s."[219] Their complaints certainly reflected this anxiety, and it was about as explicit as generals could be in critiquing the wars and the country that sent them to fight them. They also reveal that, for all of the Army's efforts, the ability to provide soldiers with optimum conditions for restoring their psychological well-being lies in large part with the public and the political representatives who send troops off to war. Nor were these efforts wholly successful. Especially for troops with specialized skills, even a larger force did not mitigate the wars' demands on them. Chiarelli recalls that "my aviators, in 2012, they were home for 365 days and on the 366th day, they were back."[220]

* * *

Programs like Embedded Behavioral Health and RESPECT-Mil undoubtedly made a significant impact in many soldiers' lives, but they were not silver bullets. In the years that followed their implementation, the Army continued to struggle with stigma and even with whether soldiers were receiving the appropriate discharges as they left the Army. The sixth MHAT, which appeared in 2009, reported that more than half of soldiers surveyed who were in maneuver brigades worried about stigma.[221] Those numbers had improved only slightly by 2011, when MHAT-VII, which focused on troops in Afghanistan, was released.[222]

More troubling was an EPICON report on crimes committed at Joint Base Lewis-McChord in Washington that reached conclusions that could have been lifted from the Fort Carson report.[223] Lower-ranking soldiers reported "name calling, bullying, teasing, harassing, [and] humiliation in front of others," that they "were afraid to disclose their personal issues or [behavioral health] concerns to their [chain of command] due to fear of repercussions, consequences, and being viewed or treated poorly," and that the demands of the war made it difficult to actually get treatment.[224]

Joint Base Lewis-McChord also turned out to be the place where the question of whether soldiers were being appropriately diagnosed before discharge reemerged. In 2012, the *Seattle Times* reported that soldiers "complained that doctors . . . unfairly stripped them of the PTSD diagnoses, which would help qualify them for a medical retirement, and instead tagged them as malingerers."[225] In this case, Maj. Gen. Philip Volpe, the head of the Army's Western Medical Command, and Col. Dallas Homas, the commander of Madigan Army Hospital, had embraced forensic psychology as the best way of ensuring that troops got the proper diagnosis. However, as Rebecca Porter explains, the methodology of forensic psychiatry carries with it a predisposition that soldiers might be lying or malingering, and so many soldiers found that their diagnoses were changed just as they prepared to leave the Army.[226] Unsurprisingly, the popular perception was not that Volpe and Homas were well intentioned but chose a poor screening method; it was that the Army was seeking ways to deny soldiers their benefits.[227] Arguments to the contrary were not helped by revelations that one of Madigan's forensic psychologists had made a presentation about how much a lifetime of disability payments for PTSD would cost taxpayers.[228]

Because she was the behavioral health consultant, the entire debacle landed on Rebecca Porter's desk, and she was justifiably furious. "I was angry at some of what was put out by some of my colleagues, as far as saying things like these soldiers are malingering or its our job to keep them from trying to scam on the government," she recalls. "To me, that is arrogant and it misses what our job is."[229] Over the next few months, Porter flew back and forth between Washington, DC, and Tacoma, Washington, investigating what happened and arranging for soldiers whose diagnoses had been changed to be reevaluated at Walter Reed's forensic psychiatry training facility. In the end, almost all of the soldiers who had had their diagnoses reversed were returned to their original status.[230]

On top of this, the Secretary of Defense ordered the Army to evaluate every soldier who had been discharged with a behavioral health issue during the wars—nearly one hundred thousand records—to ensure that they had been properly screened for PTSD and mTBI, an order that was subsequently expanded to the other services, "so I was loved by my counter parts in the Navy and in the Air Force too," Porter wryly recalls.[231] These efforts, however, surprisingly earned approbation from one of the most vigorous critics of the military's and VA's handling of mental health. "Without question, these are historic steps in our efforts to right a decade of inconsistencies in how the invisible wounds of war have been evaluated," Patty Murray (D-WA) remarked on the Senate floor, adding that, "because of this outcry from veterans and servicemembers alike, the Pentagon now has an extraordinary opportunity to go back and correct the mistakes of the past."[232]

Between 2007 and 2011, the Army, faced with a growing population of soldiers who needed care, made important strides in making mental health care both more accessible and more acceptable. Although some issues, like personality disorder discharges, could be addressed relatively effectively though policy changes, the most effective change happened at the level of individual relationships among soldiers, clinicians, and researchers. The events at Joint Base Lewis-McChord, however, reveal that, for all of its success, the Army's efforts to reduce stigma and improve access were not universally successful. When the Army did make progress, it did so incrementally—sometimes haltingly—and at the grassroots level. Only when individual researchers and doctors, or soldiers and providers, built meaningful relationships did attitudes about mental health begin to change. And while those projects resulted in significant success and eventually spread across the entire Army, they did not fully succeed in eradicating stigma or ensuring that every soldier got the care they deserved, nor could they compete with the prolonged and profound demands on the force. The story of the Army's efforts to improve mental health care between 2007 and 2012, then, is both a story of success and failure, of a complex organization learning how institutional change takes place, and dealing with the ramifications of being both beholden to and part of a culture that enables war and stigmatizes mental health.

5

"Military Families Are Quietly Coming Apart at the Seams"

Managing Family Mental Health and Critiquing the Iraq War

On the morning that Americans awoke to learn that "the air campaign in Iraq began in earnest," they also encountered one of the first indications of the psychological toll that the war would extract on the home front.[1] In an article entitled "In Military Towns, War a Constant Presence," *USA Today* profiled several service-members' wives. Some were resolute, declaring themselves "used to this" and "doing well" and their children "nervous" but "really proud."[2] These were part of a broader celebration of military families that had begun even before U.S. troops deployed to Afghanistan. "Our military families have accepted many hardships, and our Nation is grateful for their willing service," George W. Bush told the nation two weeks after the September 11 attacks, a comment that implied that, if the few Americans whose partners and parents faced combat could resolutely accept the wars, so, too, could the rest of the country.[3]

Not every wife in this article was so stoic. Andrea Flournoy, whose Marine husband was stationed at Camp Pendleton, California, recalled that "when my husband first left I was very sad. I couldn't eat. I lost 15 pounds. I felt very isolated. I would cry for no reason. Then I got really angry. Not at him, not at the president, not at anyone really. I was just very, very angry. Even now, I still sometimes cry for no reason."[4] Between 2003 and 2009, stories like Flournoy's became increasingly common. Whether in newspapers or medical journals, congressional testimony or think tank reports, or in military wives' Internet forums, Americans increasingly read about wives in the grip of anxiety disorders and wracked with depression as their daily stresses multiplied and they worried over their husbands' safety. Children shared these worries. Forced to take on adult roles as they, too, worried about the absent parent and watched the one who remained struggle, they fell behind at schoolwork by day and slept fitfully at night.

Outside of the military, accounts of these struggles facilitated critiques of the policies central to the war and of a culture in which wartime sacrifice was unevenly distributed. Attention to family mental health highlighted the wars' increasing reliance on National Guard and Reserve troops and on repeated, lengthy tours as well as on the reality that only a small number of families were bearing the war's burdens.

For military leaders, struggling families raised immediate, practical concerns. Having realized after Vietnam that maintaining the All-Volunteer Force required maintaining healthy families, the military had devoted considerable attention to family wellness and entered the post-9/11 period confident that their programs would meet the challenge. The demands of the Afghanistan and Iraq Wars, however, quickly overwhelmed those resources. By 2007, some leaders were questioning whether the All-Volunteer Force was sustainable amid two expeditionary wars. Before Congress, Army leaders cited the impact repeated deployments with very little dwell time were having on families as evidence that the stress on the Army was too great and called on the nation to either provide more resources or ask less of the force.

As leaders voiced their concerns, they also worked to mitigate the strain that the families faced. To do so, they placed more demands on its Family Readiness Groups (FRGs), the unit-based, volunteer-led organizations that provided information and promoted wellness among Army families. The Army also introduced two new programs, the Army Family Covenant and Spouse BATTLEMIND, that sought to provide resources that would make families' lives easier and help spouses and children become more resilient in the face of deployments. These programs were well intentioned; despite the criticisms and frustrations of many wives, it is simply not true that Army leadership didn't try to help families to weather the war. Nonetheless, these programs proved less successful than either the Army or the families had hoped, partly because they aimed make families more resilient without alleviating the underlying source of stress. The Army's turn toward resiliency is, as Jennifer Mittelstadt has argued, part of a larger neoliberal shift in which institutionally provided care has given way to a doctrine of self-sufficiency.[5] The reliance on unfunded FRGs staffed by volunteer wives untrained to provide mental health assistance left many wives frustrated and unable to access the care that they needed. Meanwhile, Spouse BATTLEMIND's emphasis on resilience and self-sufficiency, although important, provided few resources for families already suffering—and by 2007, with some families on their fourth deployment, there were many of them. Instead, these programs often took the wars as a given and established that Army families needed to willingly accept and facilitate the mission rather than question its wisdom and prosecution. For all of their good intentions, that is, these programs often provided insufficient resources while denying families' potential to critique the wars. Moreover, their effectiveness was questionable; in the years following their implementation, both families' mental health needs and their dissatisfaction with the Army remained high.

* * *

In the first months of the Iraq War, a military wife looking for advice on her husband's imminent deployment would have found cold comfort in the most popular self-help books. Karen M. Pavlicin's 2003 *Surviving Deployment: A Guide for Military Families*, for example, counseled spouses to "have a small, private pity party and then stop the depression while it's in the healthy stage. . . . Think positive. Now."[6] Meredith Leyva's *Married to the Military* (2003) was similarly chastening, arguing that "struggling wives spend too much time wallowing in their own depression" and "spend a lot of time moping on the couch. If you're one of these wives, my advice is to get up and get busy."[7] Another book reminded wives that "when you have a week or two behind you, it is much easier to deal with the separation" before asking, "What do you have to lose—a pity party, a frowning face, mood swings, or wild thoughts that control your every minute?" "Life is too short to spend your precious time in this stage," Shellie Vandevoorde concluded in *Separated by Duty, United in Love* (2003): "Let's move on."[8]

These sentiments might seem dismissive and callous, but they reflected the prevailing understanding about pre-2001 military life: Deployments were inevitable, but they were rare and relatively brief. After the short deployments of the later Cold War, Operation Desert Storm, and the peacekeeping missions of the 1990s, the notion that American troops would deploy multiple times, often for more than a year, during wars lasting more than ten years was simply unimaginable. Equally inconceivable was the psychological impact that those deployments would have on those waiting at home.

As Americans realized that the Iraq War would be neither as quick nor as clean as Operation Desert Storm had been, however, the notion that wives might simply "get over" deployment-related anxiety and depression increasingly appeared woefully ignorant of the wars' realities. Even in the war's early months, newspapers across the country explained that wives might become anxious and children might act out.[9] By the summer of 2003, when Americans were realizing that early celebrations of a quick and successful war were premature, *USA Today* reported that "the wars in Afghanistan and Iraq, with longer and more dangerous deployments, have made military life even tougher for families left behind. Children's grades suffer. Schools counsel students on anger management. Base therapists are busier than ever consoling spouses. Wives 'at their wit's end' call counseling hotlines at all hours."[10] The article went on to describe wives who struggle to discipline their children, who "can't deal with the stress of being alone" and who are resolutely positive when their husbands call and then "put the phone down and just cry."[11]

These stories became more frequent as the wars went on. In 2004, the *New York Times* profiled Sherri Souza, the wife of a Tennessee National Guardsman deployed to Iraq, who "tries to stay positive in front of her two boys but admits the extensions have been rough on her, too. She says she takes anti-anxiety medication but still barely sleeps these days and smokes through a pack and a half a day"; when she becomes worried that her husband has been hurt, "'I just stay in bed, call in sick and put the phone under my pillow.'"[12] As the death toll increased, the fear of a spouse's death—and the very particular anxiety that surrounds being notified of that death—became more prevalent in mass media. In 2006, Stacy Arizmendi, whose husband deployed to Iraq twice with the 4th Infantry Division, told her local newspaper that, beyond struggling to pay her bills and help her children with schoolwork, she was rendered sleepless worrying about her husband.[13] She explained, "I go everyday, in the back of my head, 'Is he still alive? Am I going to get a call tomorrow or a knock on my door?'" and confessed to "tak[ing] prescription pills for anxiety and depression to help her rest."[14] In Red Wing, Minnesota, Cherie Fritz, the wife of a National Guardsman deployed to Afghanistan, confessed to emotions that nearly paralleled the five stages of grief: "I went from total disbelief, to fear, to sadness, to desperation, to anger and depression," and she explained that "you worry constantly that people are going to come to your door and tell you your husband isn't coming home."[15]

Of course, it was not only spouses who struggled. In 2005, researchers at Virginia Tech concluded that adolescents whose parents were deployed frequently exhibited "several signs consistent with depression including lost interest in regular activities, isolation, changes in sleeping and eating patterns, sadness and crying," that "adolescents report a wide range of negative emotional responses to parental deployment, including feelings of fear, loneliness, anger, worry, and confusion," and that children in Guard and Reserve families were particularly vulnerable to having their "limited coping resources [tasked] beyond their capacity."[16] Young people reported worrying about their deployed parents' safety, being anxious when the phone rang, and worrying about the remaining parent's emotional well-being.[17] One child, interviewed in 2007, reflected that "I quit being a kid four years ago" and described turning to self-mutilation when both her mother and stepfather were deployed.[18] Parents described children acting out at school, developing sleeping problems, and having other issues.[19] The psychological stress associated with a deployment was so great, California's *Contra Costa Times* suggested, that it may be the families who have the greater struggle: "Rarely is mention made of the family they left behind, or of the hours, weeks, and months of anxiety the separation has caused them. It is sometimes more stressful for the families than for their loved one who is deployed. At least their loved one overseas is busy with his or her job."[20]

Mass media accounts of struggling spouses and children were the tip of the iceberg. On the popular Internet forum *MilitarySOS*, wives and girlfriends revealed identical complaints, often in more explicit and heart-wrenching terms that illustrated that claims that families were stressed but weathering the war were hardly true. "We had an idea Christopher was going to Iraq but got the official word a couple weeks ago," one wife wrote in June 2006. "I have been having dreams about him dying—in the dreams it's just me living life after his death, knowing he's gone. The dreams have turned into thoughts during the day as well, i don't want to say obsessive but it creeps up often."[21] Five months later, after Army Sgt. Willsun Mock was killed by a roadside bomb in October, another poster wrote "omg . . . that soldier that died, died so quick. he didnt know it was coming. it could happen at any time. im so afraid for [my boyfriend] if he ever gets deployed to Iraq. . . . and im crying as if i already lost him . . . this feeling i have, just the thought of something happening to him makes me sick to my stomach."[22] Yet another woman, whose husband "is in Iraq, then when he gets back he is headed to Korea for a year," asked how to interpret "a dream of being held hostage" and explained that "the Lexapro seemed to be helping. I went to the Dr. because I thought I was going to have a nervous breakdown, and he said he would call what I have transitional/adjustment depression."[23] Two years later, one woman confessed to suicidal ideation, writing that "sometimes I don't even want to live. I don't want to live without him that is. I think of all the horrible things going on in the world, all around us. And I just don't want to be a part of it."[24]

That despair was increasingly mixed with anger at both the military that had sent their partners into dangerous places and a public that did not share their sacrifices. In early 2008, one woman admitted, "I'm struggling in a big, big way," and

> Now? Now I'm angry. Because the Army controls him, they have control over my life too and that is killing me. . . . I'm angry at the military, but I end up venting it to him and we end up upset with each other. . . . When I think about the fact that our 6 month old son won't know him when he gets back, I feel rage. When I think about the fact that my 7 year old who loves this man just like a daddy is scared of the dark and has to spray his cologne on his pillows in order to sleep, I feel rage. When I think about the fact that our lives are put on hold and our relationship is put under this enormous stress, I feel rage. . . . When I think about the fact that they are lying to these men and telling them that they will only be gone for 12 months when we all know it will be 15–18, how do I not fly off the handle?[25]

Other wives and girlfriends were angry with the broader public, which they accused of having no understanding of their plight. Partly, it was because the experience of having a loved one deployed was relatively unique. One complained that "a lot of pepole [*sic*] I do know aren't military and don't understand and don't want to invite the sigle [*sic*] married girl over. . . . I just feel really depressed and it's wearing on me."[26] Another, who admitted that, "for the last few days I've cried myself to sleep again. Not just tears I mean chest hurting, barely able to breathe [*sic*] crying," explained that, while she was confident in her ability to persevere, "I just feel so alone lately. I'm so far from family and friends. I mean I've made a few friends here but I haven't really clicked with any of them and all of their husbands are around which makes it hard."[27] If these posts echoed what was being reported in American newspapers, they did so with a rawness that suggested that the psychological struggles of military families might be even worse than most Americans realized.

As families' deteriorating mental health became more apparent, it became a reason to critique the Iraq and Afghanistan Wars. In particular, articles focused on the impact that lengthy, repeated tours and reduced dwell times were having on families. In writing that the Arizmendis' story "began as a love story, but romance, marriage, and family quickly turned to separation, worry, and war," author Sarah Williams undermined portrayals of happy, stoic, and patriotic military families and posited the war as a disruptive force, noting that Stacy Arizmendi's husband Daniel "deployed to Iraq for his second tour of duty in December" and that "it's not clear when he'll return."[28] Their story, moreover, was presented as representative: "The Arizmendi's story is becoming more common as U.S. military forces dig deeper into the war in Iraq and more families are separated by tours of duty."[29] On the war's third anniversary, when nearly 60 percent of Americans judged the war a mistake, Williams linked military families' mental anguish to the continued commitment to the war.[30]

Around the country, other articles likewise portrayed repeated deployments and unscheduled extensions as the primary source of families' mental health struggles. In 2005, the *St. Petersburg Times* wrote that "military families are facing a period of continuous deployments" but that "the goodbyes don't get easier with practice. When a parent ships off to fight, children become the unintended timekeepers of a war that many are too young to understand. Some struggle to focus at school. Parents cry with them at night."[31] On Christmas Day, the *Providence Journal* profiled a wife who, because she "is missing her husband through his second deployment, [has] come to know her vulnerabilities and her triggers" and who warned that someone could, "for no reason, just say the wrong thing to me and the tears come."[32]

Newspapers portrayed these stressors as more profoundly affecting National Guard and Reserve families, who had not anticipated carrying such a heavy burden in these wars. As the wife of one of New Hampshire Guard member put it, "'We were totally blindsided when they said he was going."[33] In some ways, these families' stresses *are* greater; they typically live far from military resources, have less experience with military culture, and often endure increased financial stress during deployments because National Guardsmen often earn less on deployment than in their civilian jobs.[34] Moreover, the Iraq and Afghanistan Wars have consistently, and often unexpectedly, placed National Guard troops in combat roles. By 2006, the demand for troops had led the National Guard to provide troops much less dwell time than the five years that it had set as a goal.[35] By 2007, nearly half of a million Guard and Reserve troops—compared to about 1.2 million Active Component troops—had been sent to Iraq, and more than one hundred thousand Guard and Reserve troops—about 12 percent of the force—had been mobilized more than once for a post-9/11 mission.[36] These statistics led to significant worry about the Guard's ability to fulfill its domestic mission, but it also raised questions about their families' mental health.[37] The wife of a Missouri National Guardsman offered a bleak account of how repeated deployments and extensions damaged families' mental health: "'I came home and got an e-mail that said his unit had been extended again, and that's all I remember for a while. . . . I just hit that depression point where I couldn't go on. I had an ulcer. I wasn't sleeping. I wasn't eating. I lost 40 pounds over the last 14 months. I had chronic shaking and nightmares.'"[38]

* * *

If in the media the strain on military families helped facilitate anti-war critique, on Army posts around the United States that pressure was having more devastating effects. Although the Army had been investing in family well-being since the 1980s, when Army Chief of Staff Gen. John Wickham devoted considerable resources to family programs, those initiatives had served a peacetime Army.[39] Transitioning to serve families during wartime proved a challenge, however. "I think the Army at the beginning of the war was not quite there yet," explains Delores Johnson, who directed Army Family Programs in the 2000s. "We had probably lost some ground with the Wickham partnership with families."[40] At Fort Bragg, North Carolina, in 2006, the wife of one lieutenant colonel was discovering this firsthand. Kristy Kaufmann was, as she would later tell many audiences, an unlikely Army wife.[41] She had grown up in Upstate New York and gone to the University of California, Berkeley, where she had been a nationally ranked gymnast. No one in her family had ever served. But by happenstance—in Las Vegas, where, as she put it, "the only place Berkeley and West Point could possibly come together"—she met an Army officer.[42] Kaufmann and her husband, Reese Turner, married a few months before the September 11 attacks. She ended up at Fort Sill, Oklahoma, which Kaufmann called "a bit of a culture shock for me and for Oklahoma."[43] She didn't intend to be an Army wife for long; she figured that, after a couple of years, her husband would retire and she'd return to her career.

The wars in Iraq and Afghanistan changed her calculus. By 2006, she was at Fort Bragg, where her husband was attached to the 3rd Battalion of the 27th Field Artillery Regiment (3-27). A few days after Thanksgiving, 2006, Kaufmann turned onto her street just off post to find emergency vehicles outside one of her neighbor's homes. The next morning, she learned that sometime over the holiday weekend, Faye Johnson Vick, a thirty-nine-year-old mother of two whose husband, a lieutenant colonel, was in Iraq, had strapped her children into their car seats and then started the car while it was still in the garage, killing all three of them.[44] Investigators later learned that Vick had been suffering from post-partum depression.[45]

Vick's story was tragic, but Kaufmann worried that it wasn't isolated. As the wife of a lieutenant colonel, Kaufmann was in charge of the Family Readiness Group for her husband's unit. These groups are volunteer organizations, usually led by a leader's wife, and their goal is to provide spouses with the support, the information, and the resources they need to thrive during a deployment.[46] In her own FRG, Kaufmann was seeing the impact of repeated deployments and the high operational tempo. "We saw the impact all at once," she recalls, "and it was a shit show."[47]

Some wives were preparing for their husbands to deploy—again—while others were in the midst of deployments, and still others were dealing with husbands who had just returned. Several of the forty or so wives in the group, most of whom were in their early twenties, had contemplated suicide.[48] As they had in Vick's case, mental health issues remained hidden until the situation got dire. The wife of one soldier in Kaufmann's husband's unit seemed fine until halfway through her husband's deployment, when she pulled her kids out of school, reversed the lock on their bedroom door, and locked them inside.[49] "She had just lost it," Kaufmann remembers, but "we didn't know until the shit hit the fan."[50]

Getting help for those women turned out to be a challenge. The Family Readiness Group was stretched thin and underresourced. It "was me and two other wives trying to do the best we could," she recalled, and when she went to Army Community Services and asked for suicide prevention training, she was told that that wasn't in the purview of the Family Readiness Group.[51] Fort Bragg's leaders in fact discouraged FRG leaders from talking openly about Vick's suicide, calling it a private family matter.[52] For Kaufmann, this emblematized the Army's approach to military families. The Army "expected [FRGs] to tale care of mental health issues with no money and no skill set," she explained, calling the groups "the biggest unfunded mandate in the Army."[53]

Every unit is required to have a Family Readiness Group, and the Army simply "voluntold a bunch of wives to get involved."[54] There was no funding to professionalize the groups or significantly train the spouses regarding mental health concerns. In fact, the Family Readiness Group handbook from 2006 explicitly says that "the FRG does not serve as a professional counseling agency," but it did offer a modicum of advice for dealing with a struggling spouse or family.[55] The handbook acknowledged that FRG leaders might encounter a crisis, and instructed them to "assist when able and refer when a situation or issue is not your area of expertise. For example, do not handle individuals who are suicidal or may harm themselves or others, refer them to Chaplain, Mental Health or immediately call the Military Police."[56] This is not terrible advice; people faced with a suicidal individual should immediately assist them in locating professional help. But the reality of the Iraq and Afghanistan Wars meant that Family Readiness Group leaders were regularly encountering troubled families, and they were finding that this advice was insufficient to meet the needs of the FRGs.

Kaufmann was not shy, nor was she nor intimidated by traditions that often relegated Army wives to the role of demurring helpmates.[57] When she learned that Gen. George Casey, who had recently become Chief of Staff, would be visiting Fort Bragg, she wrote a ten-page memo outlining the problems inherent in the Army's reliance in FRGs to address family mental health issues. "The FRGs are being tapped in a way they were never meant to be tapped," she began, adding that "soldiers and their families are understandably starting to buckle under the incredible amount of stress and pressure."[58] Her critiques covered every aspect of the FRG system: Neither the wives nor the commanders had received appropriate training to organize and run them, the policies that governed them were unrealistic, and the events they held were not useful.[59]

Of the requirement that FRGs raise their own funds to sponsor events, she wrote, "No wife wants to spend her afternoon baking cookies so they can turn around and sell them to their husband in the motor pool. Wives are tired, apathetic and frankly, pissed off at the Army for sending their husbands into harm's way yet again. They don't want to be bothered and you can't blame them." Of declining participation in FRGs, she explained, "There's not much you can do to convince soldiers and their families to get involved if they have become overwhelmed with feeling[s] of bitterness and anger toward anything to do with the Army."[60] It was a memo that captured many of the frustrations that wives had voiced in newspapers, and it was not one that pulled its punches.

Although Kaufmann was frustrated that FRGs had been made the primary avenue for identifying and addressing mental health issues, she realized that they hadn't been designed to fill that role and thus had neither the structure nor the resources to accomplish it. Family Readiness Groups, she explained, were intended to be "an information conduit," but they needed to become more socially oriented because building trust among the members was a prerequisite to women feeling like they could reach out for help. "If wives are able to meet other wives within the unit and establish 'battle buddy' type relationships, they will be less likely [to] experience feelings of isolation during their soldier's deployment," she wrote, adding that, "if a wife feels like she has someone to rely on, someone who cares, she will be less likely to go off the deep end[,] especially during her husband's deployment."[61]

More serious was that FRGs were unprepared to address mental health issues when they did arise. "There is no guidance on how to deal with some of the specific and serious situations that may occur,"

Kaufmann wrote, including "handling a caller who says she is severely depressed or suicidal."[62] In fact, the FRG handbook instructs someone receiving such a call to either refer the person to mental health services or call the military police. However, Kaufmann explained, the problem was that, even when wives were comfortable reaching out to the FRG, volunteers encountered a host of issues around treatment and confidentiality for which they were entirely unequipped. "If we are going to put our volunteers in positions for which they should really have a degree in counseling," she suggested, "the very least we can do is provide them with specific and detailed instructions on what to do."[63]

Kaufmann was told that her memo would be delivered to Casey or, actually, to his wife.[64] However, neither Casey nor his wife got the memo. When Casey spoke to officers and their wives, Kaufmann therefore worked with others to ensure that he would be asked about military families' mental health. When the time came for questions after Casey talked, Kaufmann turned to see if the other wives—most of whose husbands outranked hers—would raise the question; however, they demurred, leaving Kaufmann to do it. The problems that she had outlined, she told him, "were not a 3-27 problem, not a Fort Bragg problem, it's an Army problem."[65] Her group "had just been recognized as the best FRG on post," she pointed out, asking, "If we're having these problems, what's happening everywhere else?"[66] "We need a social worker," she told the Chief of Staff. "We need mental health. I can't keep doing this."[67]

When Kaufmann finished speaking, the other wives gave her a standing ovation.[68] However, as Rebecca A. Adelman points out, such behavior is generally not welcomed in the Army, and "overtly angry military wives risk censure, ostracism, [and] the careers of their husbands . . . if they criticize the mission or the institution publicly."[69] Talking back to the Army's Chief of Staff, Kaufmann quickly learned, was not expected behavior for an officer's wife. An hour after they got home, her husband's battalion commander called and "absolutely killed" him, she remembers, the beginning of a three-week period in which he was "raked over the coals," repeatedly told that he should know that in the military, an officer's wife speaks for the officer.[70]

This moment, however, proved pivotal for Casey. If, as Robert Gates contends in his memoir, George W. Bush promoted Casey to Army Chief of Staff because "he did not want Casey, after all his service to the country, to leave with a cloud over his head because of the situation in Iraq," his promotion was auspicious for Army families.[71] There was probably no senior leader in the Army better poised to take their issues seriously. "He was absolutely the right leader," explains Delores Johnson, who was director of Family Programs for the Army when Casey was Chief of Staff. "General Casey has an intellectual curiosity that . . . allows him to be open to new ideas and be the right person to marshal change," but, more important, "he personally has managed adversity. . . . He had been at this place with the families. He understood what they might be feeling."[72] Indeed, unlike many Army officers, he had been on both sides of a deployment. "I understood what it was like to have a loved one deployed and when the car came" to tell you that that person has been killed, he remembers, referring to his father's service and death in Vietnam.[73] In particular, he recalled the isolation that his family had felt after his father's funeral. "When he didn't come back," he recalls, "It was our problem. I went into the Army . . . but my mother and sisters were in Arlington, Virginia, totally cut off from the Army."[74] And if this was true for a general's family, he thought, what must it be like for a private's wife?

Although he was "hearing it all over the place," Kaufmann's comment was, he recalls, "a crystalizing moment."[75] There was, he realized, "this perceived pressure for commanders' and sergeant majors' wives to step up, but we, the Army, didn't do much for them."[76] He assured Kaufmann that he was well aware of those issues, that the Army had taken them to heart.[77] But it was also, Casey remembers, the moment when he decided to commit more resources to military families, which led to his securing funds to add a Family Readiness Support Assistant to each battalion. An Army with a total budget of $254 billion that spent only $700 million dollars on family programs, he was realizing, had a problem. "We're saying families are important?" he asks. "We weren't putting our money where our mouth is."[78]

* * *

Pete Geren, the new Army secretary, was experiencing similar revelations.[79] In answering Sen. Carl Levin (D-MI)'s question during his confirmation hearing about the "major challenges that [he] would confront if confirmed," the former congressman reflected that "the Army is faced with many major challenges, including providing support to soldiers and families in time of war," listing it first among all of the other issues that Geren identified as important, such as modernizing equipment and "establishing strategic depth," noting that "the pace of operations has placed great stress on Army families."[80]

This issue remained a central topic two months later, when the Armed Services Committee held a hearing on *The State of the United States Army*. In his prepared testimony, Geren wrote that, "in order to sustain our All-Volunteer Force, we must do more to provide soldiers and families a quality of life equal to the quality of their service" and that "family support systems . . . designed for the pre-September 11 peacetime Army must be adapted to sustain an Army at war."[81] In particular, the secretary pointed out that, "with multiple deployments and dwell time filled with training for the next deployment, the stress on family increases."[82]

The notion that the war's impact on the family posed as existential threat to both the war effort and the force remained one of Geren's key themes. A few minutes later, he argued that

> we are in uncharted waters when it comes to family support. All-Volunteer Force, 7 years of war. Never have we done this before as a nation. The families are volunteers, just like the soldiers are volunteers. If we're going to maintain this treasure that is the All-Volunteer Force, we're going to have to do more to sustain these families. . . . If we continue in this era of persistent conflict, the stress on the families is going to be something that we have not addressed as a nation previously. . . . So this, to me, is one of the most important strategic issues that we face as a Nation. If we're going to retain the All-Volunteer Force, how do we retain the All-Volunteer Family?[83]

He went on to explain that "half of these soldiers are married. The success will depend upon those families hanging with us."[84] The notion that the wars themselves had worn families down to the point that retention was imperiled ran through his remarks. And while he praised the expansion of the Army by 74,200 troops as a "the clear recognition . . . [of] the need to reduce stress on soldiers and families related to the increasing and enduring operational demands," the clear message of his testimony was that the wars were not simply damaging soldiers and their families:[85] It was threatening the viability of the entire institution.

Geren and Casey were sincere in their concerns for families, and they worked together to make concrete improvements. "From the moment Casey came on board," Geren says, "we began adding more programs."[86] On April 19, 2007—around the same time as Casey's conversation with Kaufmann at Fort Bragg—Casey and Geren issued a directive that the Army conduct research to generate the Army Soldier-Family Action Plan.[87] This research included sending Army staff to installations to learn what family members wanted. According to Delores Johnson, who conducted some of these visits, the Army found that families were struggling in part because support structures that had been sufficient in peacetime or even as the Army began fighting in Afghanistan were insufficient given the demands of the Iraq War. Like Kaufman, she found that Army spouses and Family Readiness Groups were overtaxed.[88] What families wanted, though, was relatively simple. "I think families wanted to know that the Army was not going to leave them behind," she reflects. "It was this notion of partnership. Do you really care about what's happening to us? And the evidence of that has to be kind of a real leader engagement and understanding what we're going through."[89] When she returned to the Pentagon and briefed Casey, she told him, "The primary thing families needed to hear from leadership was that they were not alone" and that the Army needed to consider a covenant of the sort that Chief of Staff John Wickham had made with families in the 1980s.[90] She wasn't sure about actually calling it a covenant—"Kind of biblical, right?"—but in the end "it resonated with people" because it communicated that "you're indelibly connected to this institution and you add value to this institution."[91]

By September 20, that research had been completed, resulting in a 112-point Soldier-Family Action Plan that eventually became part of the Army Family Covenant and aimed at not only codifying the Army's commitment to families but also outlining what that commitment entailed in practical terms.[92] Assistant Secretary of the Army Donald J. James wrote that, "with this Charter, the Army is launching a comprehensive package of Family and [Morale, Welfare, and Recreation] initiatives to address existing shortfalls" and "[making] the Army Family Covenant a reality."[93] Previewing the concerns that Geren would voice a few months later before the Senate, Vice Chief of Staff Richard Cody cast this initiative as a means of sustaining the Army: "This plan will support all Soldiers and members of the Army Family, and, thereby help preserve the strength of the All Volunteer Expeditionary Army."[94] This was a point that Geren echoed when he introduced the covenant in October 2007 at the Association of the United States Army conference: "It is a strong Army. However, if we are complacent, if we ignore the lessons learned and warning signs of the six years of war, the seams that are exposed in the crucible of combat, in the field or on the home-front, our Army Soldiers and families will wear down."[95]

By January 2008, every Army facility had held a signing ceremony as a public display of the Army's commitment, and the army quickly pumped a hundred million dollars into programs.[96] Even before the covenant was announced, in fact, "the Director of Force Management [had] approved the concept plan to place 1,029 [Family Readiness Support Assistants] in deployable Active, Guard and Reserve battalions," and Casey had "authorized and approved $45 Million to fund FRSAs for all components."[97] By the end of 2008, the Army had nearly doubled the budget of the Family, Morale, and Welfare command and hired 1,079 new Family Readiness Support Assistants.[98]

These were meaningful steps forward, but FRSA positions were not aimed at providing better mental health care. To be sure, they were expected to "Inform the [rear detachment commander] of issues or problems reported by Family members that he/she needs to be aware of" and to "identify high-risk Families and work with the Commander to ensure those Families are assisted by the proper community agency," but there is no question that these were primarily administrative and clerical postings; as the FRSA handbook explains, the person hired into the position "performs a variety of clerical and administrative duties."[99] As well, many of the programs that the Army created, although aimed at improving quality of life, were not explicit mental health issue programs.[100] They were, instead, programs that included "a Holiday Ball," the "Wee-EIB and Spouse-EIB programs [that] allow children and spouses to experience being a Soldier for a day [by] competing in an Expert Infantry Badge–like event," and a Super Bowl party. If the second of these seemed more designed to cultivate admiration for the soldier and a desire to join the military than to assuage fears of what a deployed soldier might be enduring, the first was described as more of a distraction: "By allowing the teenagers to become involved in the planning and actualization of these events, they have kept them involved and assisted them in handling the deployment of their loved ones."[101]

There were some efforts, however, that more explicitly aimed at addressing mental health needs. By April 2009, for example, the Army had "increased the number of Military Family Life Consultants [MFLCs] to 180," and by June 2010 the Army was touting that the number of MLFCs had "exploded from 112 in FY05 to 620 in FY09."[102] These clinicians were stationed "in both on and off post locations and at schools," which made it easier for both children and the parent who had to pick them up to get counseling.[103] At U.S. Army Europe, Gen. Carter Ham was reporting that "children's mental healthcare improved with additional behavioral-health child psychiatrists in all medical treatment facilities."[104] The Army had also produced "200,000 video and training products to strengthen resilience in military children and families" and "expanded Battlemind psychological training to include spouses."[105]

The Army had also beefed up programs to sustain soldier marriages. In particular, the covenant expanded the Strong Bonds program, a chaplain-led initiative in which couples went on weekend-long retreats to reconnect and discuss their relationship and their shared goals. In 2008, the Army hosted "more than 1600" workshops for sixty-two thousand soldiers and family members.[106] The Army also "conducted 475 Premarital Interpersonal Choices and Knowledge (also known as 'How to Avoid Marrying a Jerk') programs for single Soldiers to enable them to make better life choices in their relationships."[107] In this context, "better life choices" meant not rushing into ill-advised marriages ahead of deployments or because doing so would provide them with more benefits or better housing.[108]

For Delores Johnson, programs like these were evidence of the Covenant's success. "It was really tangible to people. The medical community got something, the garrisons' [Army Community Service] centers and the Child Development Centers got employees and twenty-four spaces for childcare. . . . So as you went and talked to people, you were able to say in those communities, 'This is what the Covenant is bringing you that was important,' and I think that it didn't fall on deaf ears."[109] These resources did mark significant advances, but they also defined soldiers' spouses in ways that facilitated the perpetuation of the wars and impeded any opposition to the Army or the wars. Like the 1950s-era advertisements that Donna Alvah argues promoted Cold War militarism by "conflating the nuclear family, the home, the Air Force, and personal as well as national security," advertisements for the Army Family Covenant, which appeared as posters, movie theater advertisements, and tray liners at dining facilities identify military wives as acquiescent helpmates to the Army itself.[110] With the exception of one dual military couple, the women all appear as civilians, while the men appear in uniform, a portrayal that defines the husband's primary identify as soldier and other identities—husband, father—as secondary. The advertisements also militarize the wives, describing them as supportive caregivers "serving together" with their husbands. Ingrid Murray's child sit on her lap as her husband leans in from the right side of the frame, and Jamie Gordon stands behind her seated husband, Sgt. Lincoln Gordon, her arms around his neck with her wedding band prominent. The posters featuring the Gordon and Murray families also explicitly identify the wife's role as enabling the soldier's successful completion of the mission. Jamie Gordon, for example, tells viewers, "For us, the Army Family Covenant means that I have the support and services I need so I can support my Soldier and he can focus on the mission." The question of what that mission is—and whether it is worthwhile—is, of course, elided.

These ideas are also inherent in Spouse BATTLEMIND, which is part of the Army's broader effort to make spouses more resilient. Indeed, as a 2009 email about the covenant makes clear, resiliency is the entire goal: "The plan provides Soldiers and families a supportive environment to promote resiliency during the more frequent, less predictable separations of an Army at war."[111] The turn toward resilience, both in the military and in American culture more generally, has been the subject of increased scholarly critique that views the concept as the apotheosis of the welfare state's decline because it privileges individualism and self-reliance over the maintenance of the social safety net.[112] Also important, though, is that this training denies military wives the opportunity to critique the Army, the mission, or the war.

Spouse BATTLEMIND was an adaptation of the program that Amy Adler and Carl Castro had developed in 2005 for use with soldiers.[113] In fact, its development was motivated by soldiers' requests. Dennis McGurk, an Army research psychologist who as a major had worked with Castro in the program's development, recalls that

> we found that the sort of stressors that are part of a deployment were important to the entire family. A lot of times when we gave Battlemind, what soldiers reported was, "I wish you would give this [to] my spouse." So we did. Some of the time they wanted to do that so they could say, "Look, this is why I'm being like I'm being. You see what they tell you." But that's not how we developed it. We developed it to help the spouse, predominantly the spouse, prepare themselves for deployment [and] the same stressors of becoming single parents in some cases, the stressor of the service member being away.[114]

Castro, McGurk, and their team set about conducting focus groups with Army spouses to learn about wives perceptions of stressors and resources for dealing with them; unsurprisingly, wives were forthcoming with critiques.[115] The program that emerged focuses on helping spouses develop useful coping and communication skills.[116] But the training also endeavors to build unquestioning acquiescence to the demands of military life and, with it, a subordination of the possibility of critique. In doing so, these programs transfer the obligations of soldiers—who are legally prohibited from questioning orders—to the family, who occupy a liminal space, not quite civilians but also not quite soldiers, *of* the Army but not *in* the Army. The Army is quite clear about this reality. Soldiers are asked to "show your Spouse how much you appreciate his/her hard work and encourage him/her to pursue his/her own interests." And while this last instruction is intended to be egalitarian, it is crucial to remember that BATTLEMIND conceptualizes "independence" as "The capability to have a fulfilling and meaningful life as part of an Army Centric Family."[117] And lest there be confusion about what that means, leaders of Spouse BATTLEMIND workshops tell audiences that "what we mean by the Army Centric Family is that the Army is not just composed of soldiers; Spouses and children enhance and support the soldiers' careers" and that "sometimes the needs of the Army must be placed first."[118]

In seeking to build resilience with the primary goal of enabling soldiers to meet the mission, the Army Family Covenant, Spouse BATTLEMIND, and the Family Readiness Groups position spouses and children as unequivocal supporters of the Army's missions and prohibit them from identifying, or being identified, as victims of a disastrous foreign policy. These programs are thus another cultural location in which, as Rebecca A. Adelman puts it, "the political subjectivities of military wives are largely inaccessible . . . because of the institutional mechanisms that constrain their expression and dictate the terms of their appearance."[119] They thus also continue a long history in which military organizations have appropriated women's labor in support of militarism, demanding that wives accept and support deployments and the wars of which they are a part. As the feminist theorist Cynthia Enloe explains, "Patriarchal militaries need feminized military wives" who "have been persuaded that they are 'good citizens' if they keep silent about problems in their relationships with male soldiers for the sake of their fighting effectiveness."[120] More specifically, Jennifer Mittelstadt explains, in the All-Volunteer Force's early years, the Army's conception of the family restricted it to "serving both husband/soldiers and the army through their loyalty and care work."[121] These ideas continue to permeate the twenty-first-century force and implicitly deem questioning missions or policies inappropriate. And, as Adelman explains, the contemporary military wife "can be tearful, occasionally overwhelmed. . . . But she must always bear up. Anything less might be read as weakness, disloyalty, or worse."[122]

The Army's efforts to promote psychological well-being among family members during soldiers' deployments thus occupy a complicated cultural space. On the one hand, they illustrate that the Army did invest significant resources in attempting to help families cope with deployments. On the other hand, they also reveal that that assistance demanded acceptance of the deployments, the wars, and the Army itself. In that sense, these efforts did not do much to address the complaints that Army families had been making throughout the war.

There were several practical reasons why these programs didn't solve the problem. One was that the Army had no way of ensuring that families that needed treatment got it. While the Army could guarantee that every soldier sat through a PowerPoint briefing on PTSD or suicide prevention, it had no control over what the soldiers' families did. "The Army cannot order a spouse to do something," Dennis McGurk explains. "They're not in the Army."[123] As a result, Johnson explains, leaders became "very creative in how they build that network of inclusion so that those families feel comfortable now to come forward."[124] At Fort

Carson, Jimmie Keenan was finding that plenty of families were taking advantage of the programs that were available but that the programs were not reaching the ones who needed it most. As Chief of Staff of the program that had set up wounded warrior battalions, Keenan had seen a lot of families in crisis, with all manner of responses. "We had spouses that would come to us and say, 'I am so scared, I have no idea what's going on with my spouse, [when] he comes home from his appointment, he is angry, he won't talk to the kids, he won't play with the kids,'" she remembers. Other spouses, she found, "were taking the medication of the soldiers from them and taking it themselves. In some cases they were selling it."[125] Still others, overwhelmed by a severely wounded partner and the growing awareness that the lives they had known and the futures that they had planned had been severely disrupted by their partner's new reality, were "saying things like, 'I'd be better off if he died . . . [He's] just like my 3rd child.'"[126]

Shortly after she became hospital commander at Fort Carson, Keenan encountered a suicide in which that wish seems to have led a wife to talk her husband into taking his own life; when she arrived at the scene of the death, the spouse seemed most upset that her husband had used her new pistol to shoot himself, and she was so indifferent to the death that she interrupted her meeting with the chaplain and Keenan to declare that she and her friend were hungry and going to go out to eat.[127] "That was when, having seen other family issues when I was Chief of Staff, I said, 'Oh my gosh, we have got to start screening family members,'" she recalls.[128] The problem, though, was that the ones who were seeking help were also the ones who already had the most coping skills: "We had some family members who were engaged in everything, and if there was a seminar or a program with the chaplain, Strong Bonds to help build a marriage, they were at every appointment." In contrast, "the ones who really need it weren't there."[129]

Figuring out how to reach out to families was thus a constant project. In Europe, the 1st Armored Division reported that "we make every effort to contact families within 72 hours of arrival to begin FRG support."[130] In Vicenza, Italy, spouses of soldiers assigned to the 173rd Airborne Brigade Combat Team could get "childcare for FRG Meetings."[131] At Fort McPherson, Georgia, "Two FRG meetings are conducted weekly, at noon and in the evening hours, to foster participation. Food, child care, and guest speakers are provided to bring about a robust Family Readiness Group."[132] Keenan took a somewhat more creative approach, using what she called "the condo hook," a play on the supposedly free weekends that resort complexes offer while requiring guests to sit through a sales pitch for timeshares:

> We would identify who was at risk . . . and we said, "Hey, we realize this is hard on you, too, we'd like to give you some training, but the other piece of this is we are going to get you a facial, a mani, a pedi, free child care, we are going to give you a nice lunch, we are going to have it in a nice space here in Colorado Springs, would you do it?" . . . Once we started to do that, what we found was many of them just didn't even have coping skills. They'd come from families where just nobody talked to each other. If something went wrong, somebody left. . . . We were able to get them in, and then we were able to set up more intensive programming for each of them.[133]

The Army thus sometimes struggled—and had to be creative—to get families to even seek care. But even when they did, a second problem presented itself. By the time the Army began taking family mental health seriously, it was 2007, and it was 2008 before many of the resources had been deployed. That was often, in Kaufmann's words, "too little too late."[134] After six years of war, giving someone skills so that they could bounce back seemed insufficient to some Army spouses. For many wives, Kaufmann argues, the Army's approach came across as, "'We know it's been tough, and we're going to teach you to be stronger.' Is there ever an acknowledgment that everyone has a breaking point? We had to put our heads down and get through it."[135] This approach seemed to Kaufmann to ignore the issues that families were already facing. Proponents of resilience would disagree, of course. McGurk, for example, argues that resilience training produces "the ability to bounce back after you've been through a difficult time."[136] Kaufmann demurs, remembering one wife who told her, "It's like teaching me how not to get raped after it happens."[137] Here again, the imperative to meet the wars' immediate needs, rather than to address the root causes of the stress, prevented the Army from fully addressing military families' needs, and it left some wives feeling alienated.

Another issue was that, although family programs were popular, they never benefited from the research that other programs got that would empirically demonstrate that they helped. The BATTLEMIND program for spouses, for example, was implemented but never subjected to a clinical study, as the original BATTLEMIND program had been.[138] More troubling was that stories in the popular press suggested that, despite the military's efforts, the Iraq and Afghanistan Wars—and the policies through which the Bush administration prosecuted them—continued to produce a widening circle of psychological damage. Again, these difficulties were explicitly tied to the ways that the wars had been fought. In 2009, the *Christian Science Monitor* reported that "experts and commanders say 15-month tours are too long because they compound mental-health problems and other issues at home."[139] The *Philadelphia Inquirer* explained one expert's comment that "plenty of military families fall apart for a lot of stress" with the comment that "tours in Iraq have been extended to as long as fifteen months, three months longer than the one-year battlefield stint for troops in the Vietnam War. Less time passes between coming home and serving again."[140]

The reliance on Guard and Reserve troops, and the sense that the sacrifices were being borne by a small fraction of Americans, also remained at the forefront of stories about families' mental health. The *Buffalo News*, which profiled a National Guardsman's wife who explained, "I cry a lot, and I lose my temper easily [and] there's a lack of sleep and a lot more coffee drinking," explained that strain by noting that "citizen soldiers . . . have become a mainstay in the all-volunteer military. The toll on families and relationships back home has been steep, with casualties including crumbling marriages, financial hardship and frayed child-parent bonds."[141] The *St. Petersburg Times* piece quoted a chaplain who complained that, as much as the military could add additional resources, the problem was cultural: "The nation as a whole, he said, has not had to sacrifice in this war. So the hardships fall squarely on those who volunteered to serve and their loved ones."[142]

Among the many articles that highlighted families' still-deteriorating mental health was an op-ed that Kristy Kaufmann published in the *Washington Post*. Kaufmann had arrived in Washington in the summer of 2008, when her husband was transferred to a staff position at the Pentagon. Nearly a year later, she was wondering why she was still so upset about the issues she had seen at Fort Bragg and increasingly "felt morally obligated to speak for those who could not be heard."[143] Eventually, at the urging of journalist Tom Ricks, she decided to work on an op-ed about the needs of military families.[144]

This was not an easy decision. Kaufmann's husband was on the verge of being promoted to colonel, and she worried what her outspokenness might mean for his career.[145] Her husband was supportive and assured her that she had to get her message out, but a piece critical of the Army's handling of family well-being in the paper of record for the nation's capital is guaranteed to ruffle some feathers, and Kaufmann had already learned that stepping out of the circumscribed role of waiting wife was not without risks.

The op-ed, entitled "Army Families under Fire," appeared on May 11, 2009. Overall, it reprised the complaints that she had made privately in her 2007 memo. "Too many military families are quietly coming apart at the seams," it began, before asserting that "the response from Army and Defense Department leaders has been haphazard, sluggish and widely ineffective."[146] However much the Army claimed to care about families, and however much they had invested in wellness initiatives, Kaufmann asserted, the results remained inadequate—there weren't enough resources, officers weren't trained to appreciate the wars' tolls on families, and volunteers were burnt out.[147] The op-ed was written with the same directness she had shown at Fort Bragg. "New gym equipment and child-care facilities are great," she wrote, "but expanding and implementing mental health services so a soldier's child doesn't have to wait six months to see a psychiatrist is more important."[148]

The *Post* published Kaufmann's email at the bottom of the piece, and over the next few days she received hundreds of emails from other military wives. While some told her that she had stepped out of line, most agreed that the Army was not providing families with enough resources.[149] Another FRG leader told Kaufmann that "I've seen what happens to them (us) during deployments. Things that I never imagined that I would see—mental illness, abuse, infidelity (to include pregnancies) and emotional breakdowns. I felt like I needed a degree in psychology to deal with all the issues I faced. The resources were not there for me to help these families"; she felt guilty for leaving behind "a trail of broken and damaged families" when her husband was transferred to a new post.[150] Another described her experience running an FRG as being "in WAY over my head with the issues that I was having to deal with and there was really no one to turn to" and admitted that "it was pure hell just trying to get through the day raising our own three children, feeling so scared and worried about my husband . . . but then to have to take care of other people's problems, which I truly DID want to do, but was truly ill-equipped to do it, was too much."[151] Another poetically concluded that "I neither expect nor want formal recognition of the sacrifices I make for my family, husband, and country. But, I do not want to have to pretend that I'm not suffering, that it is not a sacrifice, nor that I am doing it willingly. . . . If they're going to make me dance, they should at least provide the proper accouterments and stop pretending that shoving me into a dark room with a static-filled radio is worth the price of admission."[152]

Other wives complained that the Army was not taking family mental health seriously, and the disgust and anger in these messages is striking. "The family covenant is lip service at its best," one woman wrote. Another described her "life [as] one horror story after another since December 2004" in an "army [that] doesn't care." Yet another admitted, "I am angry because I feel the Army has colluded with military wives into accepting this horrific lifestyle as a badge of honor" and described a culture in which there was to be "no complaining back at the home front, because God knows how much worse our husbands have it down range." And a fourth summed up her impressions in a statement that dismissed any sense that leaders like Casey and Geren truly cared about soldier and family welfare: "The thought that troops are expendable, so are the families, is pervasive in the entire hierarchy."[153] These emails offer a sharp counterpoint to the itemized lists of programs described in Army memoranda and which the Army touted publicly, and they reflected the degree to which, as Kenneth T. MacLeish explains, "many soliders and spouses bristle at the earnestness, superficiality, and heavy-handedness of the Army's 'recognition' and 'commitment' as well as the promises of support that might sometimes stand in relief against real material assistance."[154] These were not wives who were grateful for the Army Family Covenant or who had benefited from resiliency training. Rather, they were women whose families continued to struggle despite the provision of these resources and who resented the demands the Army placed on them.

A few days after the *Post* op-ed ran, Pete Geren emailed Kaufmann. "I want to assure you of my personal commitment to provide yours and all Army families the support you need, deserve and have earned through your service during these 7 plus years of war," he began, before declaring himself "most troubled by your perception of the response of leaders to concerns expressed by you and others" and requesting Kaufmann's "continuing input."[155] In the email exchange that followed, Geren asked her for examples of disengaged leaders and expressed his disappointment that she had encountered some. He also asked for her assessment of the "Army Family Action Plan," which Kaufmann described in her response as "the perfect example of a well-intentioned Army program that looks a lot better on paper than it is actually working" and as "poorly marketed and misunderstood" and suffering from a "lack of transparency."[156]

This exchange led to a meeting on June 2, 2009, between Kaufmann and Geren in which she asked for audits to ensure that medical treatment facilities were keeping appropriate records and that in turn those figures were driving the hiring of enough providers.[157] The op-ed had also gotten the attention of the Obama administration, however, and prior to meeting with the Army Secretary she was invited to the White House for a May 18 meeting with a staffer named Trooper Sanders.[158] The presentation that she gave was a characteristically direct assessment of the problem and its causes. "Behind every soldier in uniform today is a family that is struggling with the stresses of war," she explained, condemning the lengthy war and the repeated deployments that it had required as "result[ing] in an overuse of a volunteer army and the volunteerism of families" and the military's lack of foresight regarding family well-being: "No one thought about mitigating the family risks and designing programs in Year 2 of the war."[159] The Army's efforts, she explained, were "well-meaning," but "the fact is, they are not working" and were, rather, the equivalent of "slapping Band-aids on a gushing wound."[160] She spoke of poorly designed programs and spouses reluctant to seek care and shared harrowing quotes from military spouses.[161] Like Geren, she worried about how the strain on families affected the sustainability of the All-Volunteer Force.[162] And she called for an executive order and a presidential task force to address these issues and demand measurable improvements.[163] It took three years and the advocacy of many others besides Kaufmann, but the Obama administration did eventually issue such an executive order.[164]

The responses to Kaufmann's op-ed reflected the broader debate over military families' mental health during the Iraq and Afghanistan Wars. Leaders like Pete Geren were deeply concerned about family well-being and were engaged, reflective, and open to criticism and new ideas as they sought to address the issues plaguing families. Under Geren's and Casey's leadership, programs like the Army Family Covenant represented a considerable commitment to meeting families' needs. As Delores Johnson asserts, "For this generation of Americans and army personnel and their families, this was our first close-up look at what [war] was like, and I think we got [so many things] right."[165] But Johnson is also clear-eyed that those successes weren't universal. Reflecting on Kaufmann's work, she says "Kristy probably was accurate in [her] critique that we were in the midst of making things happen and they weren't uniformly visible to everyone," especially because the Army didn't always succeed at connecting with younger spouses. "I take her critique as real."[166]

In the end, family mental health represents one area that was overwhelmed by the wars' demands and duration. As the wars wound down, research bore out that those efforts were still struggling to mitigate the war's impact.[167] A 2013 Institute of Medicine report, for example, concluded that "given the exceptional demands that deployments to [Operation Enduring Freedom] and [Operation Iraqi Freedom] have placed on military families . . . there continues to be a need for military leaders to gain a better understanding of the needs of families and to use that understanding to implement more-effective coordinated programs and services."[168] This conclusion would hardly have surprised many military families, who continued to struggle and to feel as though the military's approach was insufficient.

Ten years after she met Reese Turner, Kristina Kaufmann co-founder and Executive Director of Code of Support Foundation, a veteran family support nonprofit. Her own marriage, however, had not survived the wars. When news of David Petraeus's affair with Paula Broadwell broke in 2012, Kaufmann was a guest on National Public Radio's *Talk of the Nation*, where the topic was "Infidelity on the Homefront." "I get asked all the time if it's because of war and deployments," Kaufmann said of her own divorce. "And I just don't have the perspective to answer that. The only marriage I've ever known has been a wartime one."[169]

6

"The Limited Science of the Brain"

Traumatic Brain Injury and Scientific Uncertainty during Wartime

By the time the Iraq War began, Jonathan Jaffin was used to seeing casualties. An Army colonel, he commanded the hospital at Walter Reed Army Medical Center, then the best-known and best-equipped hospital in the Army medical system.[1] Every day, injured soldiers arrived in ambulances or on white school buses that proclaimed "The Wheels of Army Medicine" after they had been hurt in Iraq or Afghanistan and stabilized at Landstuhl Regional Medical Center in Kaiserslautern, Germany.[2] That there were so many was a testament to the soldiers' equipment and the medics' and surgeons' training; because of the body armor that they wore, the clotting factor that they received, and a hundred other advances, U.S. soldiers wounded in these wars were more likely to survive than their counterparts in any earlier conflict.[3]

What did surprise Jaffin, though, were how many patients were being diagnosed with traumatic brain injury. By the end of 2003, almost two-thirds of the soldiers whom researchers at Walter Reed had determined might have been at risk of a brain injury had been diagnosed as having one.[4] Three things puzzled Jaffin about this increase. First, the diagnosis was being given long after the injury. The soldiers had been out of the fight for several days, which was usually enough time for concussion symptoms to clear. Second, many of these soldiers had not hit their heads; their TBIs were attributed to having experienced an explosion's blast wave. Finally, soldiers' symptoms weren't those associated with concussion; instead, it seemed to Jaffin, that it was "clearly a PTS [post-traumatic stress] type issue."[5] He worried that the rush to diagnose soldiers with brain injuries rather than stress reactions was not scientifically valid. "There wasn't any good information, and data was hard to come by," he says, recalling that "the data on how far people were from the blasts, how the blasts were effected, how big the blasts were—all of that was really lacking, and yet [doctors] were claiming these serious injuries without really having done their homework on the scientific side."[6] In particular, he was concerned that, "because of the stigma attached to the PTS, they are much more likely to want to call it TBI."[7]

Around the same time, an Army contractor named Leanne Young was reading research about which she was similarly skeptical. In late 2002 or early 2003, she learned that Deborah Warden, a neurologist at the Defense and Veterans Brain Injury Center (DVBIC), a research institute that had grown out of Department of Defense (DOD) and Department of Veterans Affairs (VA) research after Vietnam, had encountered soldiers who had symptoms of a traumatic brain injury but who had never had a physical injury; they had only experienced the force of the explosion. When Young read this research, it seemed "completely improbable"; all of the research focused on how explosions injured gas-filled organs like the lungs and intestines, and thus it seemed "completely improbable from the perspective of all the researchers, because . . . [y]our brain is not a gas containing organ."[8] Because of Warden's suspicions, though, researchers with whom she was working on another blast study decided to examine the brains of pigs who had been exposed to blasts in a study meant to test the efficacy of body armor. What they found surprised Young. "Even though the blast levels that we were using on these pigs were completely survivable if you were wearing protective gear," she recalls, there was "neuronal damage in [their] brains."[9] This evidence raised more questions than it answered. A pig's brain wasn't as advanced as a human's, after all, and the fact that there were physical changes to the brain didn't indicate whether there would also be cognitive and behavioral changes or how severe they would be. Still, Young thought, "This Dr. Warden may be on to something."[10]

In Iraq, an Army surgeon and obstetrician-gynecologist named Christian Macedonia was rethinking his assumptions about the effects of blast from a different perspective. Like all Army doctors, he had been trained in the principles of combat stress control. So when "guys who were really close to an explosion and said they couldn't concentrate and had difficulty with recall" but were seemingly physically unscathed came into his clinic, "we were taught to minimize. Don't let them perseverate or it might turn into PTSD."[11] His perspective changed when he found himself dangerously close to a high-explosive mortar round and survived the blast wave.[12] Having a blast wave run over you, he says, "is like a ghost has run through your body, and you get this metallic taste in your mouth."[13] The experience changed the way he thought about how a blast might injure the brain. "I remember that evening," he says, when he looked in the mirror and saw "the stunned, bloodshot expression on my face, that face that I'd seen in other people brought to me."[14] Macedonia had had near-death experiences before. He had served as the medical officer on a NASA expedition to Mount Everest, where he had fallen while ice climbing. He had crashed a motorcycle. He had been shot at. He had, in short, been afraid for his life before.[15] But when he looked in the mirror, the person looking back wasn't scared. That person's face was blank. As an NPR story in which he related this story concluded, "Macedonia was pretty sure . . . the blast had injured his brain."[16]

Alongside post-traumatic stress disorder, mild traumatic brain injury became the "signature wound" of the Iraq and Afghanistan Wars. Popular narratives have featured soldiers returning with brain injuries that leave them unable to conduct their daily lives and of a military that was failing to screen, diagnose, and treat them, and some scholarship has questioned whether traumatic brain injury is even real. The sociologist Jerry Lembcke is among the most vocal in this regard, who refers to the "constructionist properties of the TBI/PTSD formulations," argues for "the meaning of TBI being at least as much derived rom the sociocultural context of the war as from anything diagnostic," and asserts that "the press, and not medical science, [was] emerging as the difference makers in the legitimating of traumatic brain injury."[17] Lembcke is correct that TBI became significant within larger debates over the war, but his emphasis on the cultural narratives that surrounded TBI does not engage the larger debates inside the Army and the Army Medical Department (AMEDD) over the causes and consequences of traumatic brain injury.[18] To the extent that, as he puts it, "TBI was still a virtually blank page in the medical literature" in the midst of the wars, it was a page that clinicians, researchers, and Army leaders filled with competing narratives that led to vigorous and often acrimonious debates.[19]

As the wars became more violent and more troops were exposed to improvised explosive devices (IEDs), both the paucity of scientific knowledge of the brain and the shortcomings of applying civilian protocols for concussion treatment in a battlefield context became increasingly apparent. Over the next decade, a host of questions emerged: Was the concussive wave of an explosion in and of itself capable of damaging the brain? If so, how big did the explosion have to be, and how close did a person have to be to it in order to be harmed? How could mTBI be distinguished from PTSD when the events that produced a concussion or blast injury were also very often traumatic? Was differentiating the two conditions even important? Could soldiers be effectively screened before and after deployments to detect brain injuries?

Answering these questions proved to be among the most complicated and controversial challenges that the Army confronted during the Iraq and Afghanistan Wars, and they generated some of the most heated debates of these conflicts. Inside the medical community, there were competing theories of the causes and consequences of traumatic brain injury. As researchers debated them, they not only disagreed but were at times also shocked by their peers' arguments.

These internal debates also put AMEDD into conflict with Army leadership and the Army's congressional overseers. Leaders like Vice Chief of Staff Peter Chiarelli were eager to implement policies that would protect deployed troops and speed the recovery of those who had been injured, and they were often frustrated that the scientists' debates and disagreements did not yield clear prescriptions that they could immediately apply in the combat theater. As well, members of Congress who viewed TBI as further evidence that the war was a disaster sought to implement protocols to better screen and treat troops, including initiatives like pre- and post-deployment screenings and bariatric pressure treatment, implementations that frustrated AMEDD officials who were skeptical of protocols that had not been rigorously validated.

Although every leader, clinician, and researcher was committed to providing soldiers with the best care possible, the internal and external debates over diagnosing and treating TBI were marked by competing priorities and investments. Army Medical Department officials insisted on hewing to a model of scientific inquiry that embraced a problem's complexity, maintained a healthy skepticism of any hypothesis, and insisted that any conclusion be rooted in scientifically valid evidence. Some Army leaders and much of the general public, in contrast, viewed that approach as too deliberate in the midst of wars that were injuring soldiers every day. Efforts to address traumatic brain injury were thus marked by robust and sometimes acrimonious debates about what kinds of knowledge, inquiry, and evidence mattered most, and these tensions among the various stakeholders' perspectives reveal how efforts to care for soldiers afflicted with mTBI, then, turned on questions that were at once medical and cultural.

* * *

"Traumatic brain injury" is an umbrella term for a variety of injuries. A bullet that penetrates the skull and injures the brain causes a traumatic brain injury. The collision of the head with some other object—a car's dashboard, the ground, a shot put—likewise causes one.[20] They are startlingly common. In the United States, civilians suffer nearly two million of them each year.[21] One article on concussion protocol for youth athletes reported that "in sports; 300,000 concussions occur each year,"and the CDC discovered that "TBI is a contributing factor to a third (30.5%) of all injury-related deaths in the United States."[22]

Despite their frequency, however, concussion diagnosis is an imperfect science. While a penetrating head wound or a closed head wound that results in a coma provide ample evidence that the brain has been injured, the damage caused by a less vigorous blow to the head is harder to identify.[23] In 2003, the *Journal of the American Medical Association* lamented that "clinicians and researchers are humbled by how little is truly known about mild TBI."[24]

Even as late as 2007—four years into the Iraq War—scholars complained that TBI wasn't taken seriously in American culture, that effective diagnosis remained inconsistent, and that knowledge of effective treatment was still lacking.[25] Perhaps most strikingly, it was not until mid-2006 that the American College of Sports Medicine settled on a definition of what constituted a traumatic brain injury.[26] The guidelines suggested that in most cases athletes still only needed a brief exam and to be asymptomatic before they could return to the field, and recovery was still expected relatively quickly.[27] Moreover, patients arriving at an emergency room after a fall or an accident were not—and in many cases still aren't—guaranteed screening or treatment, regardless of their symptoms.[28]

"Civilians were nowhere," Geoff Ling, an Army neurologist who did extensive work on blast-induced TBI, says of concussion diagnosis, which posed a problem because, "when we go to war, we do the best that civilian medicine has to offer."[29] Indeed, in the wars' early years, military clinicians relied upon two criteria: whether there had been an event that could conceivably have caused a TBI, and whether in the event's immediate aftermath there were symptoms consistent with concussion.[30] The concussion's severity was measured by several factors, including patients' ability to move their eyes, communicate verbally, and follow commands—the scores for which constituted the Glasgow Coma Scale—as well as whether they had lost consciousness, for how long, and whether they had amnesia.[31] Army leaders and clinicians also relied upon the American Academy of Neurology's return-to-play guidelines.[32] First published in 1997, they stipulated that an athlete who had not lost consciousness "may return to contest if mental status abnormalities or post-concussive symptoms clear within 15 minutes"; more significant screening or rest was required only if symptoms lasted longer than that or if the person lost consciousness.[33]

These methods were designed to identify concussions caused by a traumatic impact. From the conflicts' outset, however, the potential for a explosive blast to damage the brain was central to the construction of the war's trauma and the military's inattention.[34] Newspapers reported that "a blast injury is like no other wound, a war unto itself," that in an explosion "the sound waves alone can cause concussion," and that "concussive blasts that rock the brain around inside the skull . . . leav[e] soldiers with "the sensation of having had their brains sucked out."[35] Just before the Iraq War's second anniversary, *USA Today* referred to "serious brain trauma" as "the emerging signature wound of the Iraq War" on par with Agent Orange.[36] These articles also portrayed these injuries as insidious and nearly undetectable. The article quoted a VA clinician who explained that "they think they've just had their bell rung. . . . But when they get home, they find out they have a lot of problems with daily functioning."[37] The next day, another article explained that "months after being hurt, many soldiers may look fully recovered, but their brain functions remain labored."[38]

Stories like these contributed to the same anti-war narrative that accounts of veterans with PTSD had constructed. Rather than being part of what Lembcke calls a "home-from-war-and-hurt narrative [that is] tasked with distracting Americans from the war itself," newspaper coverage of TBI presented suffering soldiers as bright, dedicated, successful members of society who had been reduced to dependency and infantilized because of their service.[39] Readers encountered, for example, a veteran for whom "putting on shoes is an enormous task," one who "is easily distracted, and he struggles to remember appointments and schedules"; another who "couldn't match items like socks" and who "couldn't figure out how to put on his sweatpants"; and yet another who remembered every detail of the incident in which he was injured but couldn't answer "a question requiring an answer with a date or time, such as when he enlisted in the Army" and who was "learning again how to talk, to read, to drive and, ultimately, to care for his wife and 17-month-old daughter" and "work[ing] part time at the hospital delivering mail, a job that gives him a sense of independence and a $200 paycheck every two weeks."[40]

Newspapers also pointed to the military's paucity of knowledge about mTBI. In June 2006, *USA Today* ran an article arguing that the military lacked the capacity to easily diagnose concussions and was unsure about the impact of repeated concussions and that, as a result, many concussed soldiers stayed in the fight.[41] Other articles were more critical, pointing out that Army doctors initially failed to diagnose soldiers' TBI without noting that the same error often occurred in civilian hospitals.[42]

Unsurprisingly, mTBI, and the government's seeming inability to address it, quickly became part of anti-war arguments. "For our soldiers, it's not only deaths but lost limbs, disfigured faces, shattered spines, traumatic brain injuries and suicides," the *Buffalo News* argued, condemning the Bush administration for "racing to war in Iraq while unprepared for its aftermath."[43] Legislators were equally vigorous. Calling the war "utter folly," Rep. Mike Honda (D-CA) argued in February 2007 that "each Member of this House has tales of constituents whose lives will never be the same because of the Iraq war" and pointed to his own example, a constituent who "upon returning to work . . . found that he had difficulty concentrating as a result of his head injury" suffered in Iraq.[44] Ending the war would be "the right decision to make," Rep. Sheila Jackson Lee (D-TX) argued two weeks later, "when you . . . see the mounting numbers of 22,000, 23,000, 25,000 severely injured troops, many of them with brain injury."[45] In April, Rep. Carolyn Cheeks Kilpatrick (D-MI) argued for a spending bill that included a timetable to end the war and pointed to "untold numbers of women and men who have been affected by traumatic brain injuries that we are just discovering, and will suffer for decades from post traumatic stress disorder."[46] Other editorials and congressional speeches condemned the Bush administration for cutting funding for, or vetoing bills that funded, TBI research.[47]

* * *

It was in this climate of growing criticism that the Army began more aggressively studying mTBI. To be sure, there were already a range of programs across the Army. At Fort Carson, for example, a reservist named Heidi Terrio had instituted a program to screen returning troops, and the Defense and Veterans Brain Injury Center (DVBIC), which had been operational since the end of the Gulf War, was by 2006 evaluating the case files of every returning patient at Walter Reed.[48] A year earlier, Ling recalls, the Army had "looked at the civilian section and said they had no answers here. None. Zero. So we decided to create our own."[49] This included better tabulating the number of TBIs. In 2005, DVBIC reported a total of 12,834 mild TBIs across the entire Department of Defense; in 2007, it reported 19,721.[50] Ling argues that this was not because the number of actual injuries increased but because there was better surveillance. "Once we started counting," he recalls, "we realized that there was a lot bigger problem here."[51] By the spring of 2005, DUBIC had also developed the Military Acute Concussion Exam, or MACE, an eleven-section test that fit on a pocket-sized card and that medics and doctors could use in theater to evaluate soldiers.[52] Little of this, however, was systemic. In 2006, the Army's instructions to commanders regarding concussion were relatively vague: "Commanders and leaders at all levels should be alert to concussion and have soldiers evaluated . . . as the tactical situation permits," a July memo explained, providing a list of symptoms and noting that anyone with them "should see a doctor immediately."[53]

The demand that the Army develop a more consistent and coherent approach landed on Donald Bradshaw's desk. Just after Thanksgiving 2006, as news of TBI became increasingly prominent in the news, the brigadier general, who headed the Southeast Medical Command, received a call from Surgeon General Kevin Kiley, who selected him to lead a task force to study the current state of knowledge on mTBI and recommend improvements for diagnosis and treatment. The call came as something of a surprise. Bradshaw was not a psychologist or a neurologist; he specialized in family medicine. His experience with concussion was largely limited to an ice skating accident that had knocked him unconscious for twenty minutes during his senior year of high school.[54] For the one-star general who was already overseeing the Army's medical operations in eight states, the Caribbean, and Central and South America and who had two school-age children, it was an additional task.[55]

Bradshaw constructed a team that included members of the Army Medical Command and representatives from each of the other services, the VA, and DVBIC. They quickly found that, while people in the Army were increasingly aware of mTBI, they had very few clear answers on some of the issues related to it. How much of a danger was blast injury? What happened when soldiers were repeatedly concussed? How long would a soldier be affected by a concussion, and what did that mean for his ability to participate in a mission? This last question illustrated that these medical questions had significant tactical implications that, in turn, affected the safety of other soldiers. "If you get your bell rung, can you turn around and go out two hours later and do it again, or not?" Bradshaw remembers the group asking, knowing that the issue was that "you have that balance of, you've got a ten-person squad, they go out, something happens, are you going to sit three of them down and send the other seven out?"[56]

To answer these questions, the Traumatic Brain Injury Task Force looked at brain injury care across the entire spectrum of operations. They reviewed in-theater policies for screening and treating soldiers, went to Landstuhl and military hospitals around the country to learn how mTBI was dealt with as soldiers moved from the combat zone and into the military hospital system, and to VA and civilian facilities to learn what happened after the transition from active duty. They conducted focus groups with soldiers affected by TBI and their families. They worked to adapt the American College of Sports Medicine's TBI definition for military medical use.

Bradshaw's report, delivered on May 15, 2007, was a clear-eyed but troubling assessment of exactly how little the Army knew about TBI and how fragmented its approaches were. "Many best practices were identified but were inconsistent with the Regional Medical Command (RMC) and between RMCs," the report pointed out, explaining that no policies required that providers be trained or that soldiers be screened after deployment.[57] Although there was a strong suspicion that "Traumatic Brain Injury, from mild to severe, is currently the most common injury in OEF [Operation Enduring Freedom]/OIF [Operation Iraqi Freedom]," the screening of potentially injured troops was inconsistent, there were no universally accepted guidelines to follow, and many soldiers with mild TBIs were likely simply returning to duty.[58] There were likewise no questions about TBI on the Post-Deployment Health Assessment (PDHA) or Post-Deployment Health Reassessment (PDHRA), and in hospitals across the Army, the screening of injured soldiers was inconsistent and often insufficient.[59] In fact, the Army hadn't even settled on a definition of what constituted a TBI.[60]

The Army began implementing some of the task force's recommendations even before the report was published.[61] Even so, military and VA leaders were increasingly readily acknowledging the uncertainties that surrounded TBI and the difficulties that they faced before Congress. Army Chief of Staff Peter Schoomaker, for example, responded to a question from Sen. Susan Collins (R-ME) by admitting that he relied on his brother, who would soon become Surgeon General, to help him understand TBI and acknowledging that, "if we do not understand how to measure it and we do not know what it looks like and these things have delayed effects, then we have got to get smarter pretty quick here because of the nature of how we are operating."[62] On Valentine's Day 2007, Michael J. Kussman, the VA's undersecretary for health, told Congress that "I think what we are learning is that it is more complicated than anybody thought. . . . We all know what happens if somebody has a gunshot wound to the head, or significant TBI. The challenge is undiagnosed or minor TBI."[63] In March, Navy Surgeon General Donald Arthur—who had himself suffered a TBI in a motorcycle accident in 2005—offered a similar message: "Traumatic Brain Injury is an especially difficult entity to define right now," he explained. "We are just coming to the realization that blast injuries, the concussive injuries, especially multiple concussive injuries, have an affect on the brain . . . and we have to now define what that is."[64] The message from military and VA leadership was simple: Despite their concern, civilian and military science hadn't yet provided any clear answers.

* * *

Developing solutions would prove controversial. On January 31, 2008, the *New England Journal of Medicine* published an article in which some of Walter Reed's most prominent researchers—Charles Hoge, Carl Castro, Charles Engel, and Dennis McGurk—reached a conclusion about TBI that stoked a debate that was as much cultural as political. Examining the records of more than 2,500 troops who had returned from Iraq, the researchers found that about 15 percent had suffered a head injury that met the criteria for mTBI.[65] Unsurprisingly, the surveys also found that many of them "reported significantly higher rates of physical and mental health problems"—they saw their doctors more frequently, they called in sick more often, and they complained of symptoms as varied as fatigue, chest pain, and diarrhea.[66]

These results confirmed anecdotal reports of mTBI's impact on the force. What was startling, though, was what happened when Hoge and his colleagues controlled for PTSD. Just over 40 percent of the soldiers who had had the most significant mTBI's—the ones who had lost consciousness—and about a quarter of those who had endured less severe concussions also screened positive for PTSD.[67] When those soldiers were removed from the sample, the number of health problems significantly decreased. "The analyses suggest," Hoge and his colleagues wrote, "that the high rates of physical health problems reported by soldiers with mild traumatic brain injury 3 to 4 months after deployment are mediated largely by PTSD or depression."[68] The point was not, the authors emphasized, to discredit mTBI as a diagnosis, but it did challenge its ascendency as the war's defining injury.[69] It wasn't, Hoge recalls, "that the concussions weren't important, but that they were only one relatively small piece of the puzzle of why a lot of service members come back from war experiencing generalized and fairly severe and chronic health concerns"; rather, he explains, "it really has a lot more to do than just TBI. . . . PTSD and depression . . . had a much stronger impact on general health."[70]

Differentiating between PTSD and TBI was both a clinical and a cultural issue. The question of how mTBI and PTSD were related has a lengthy history, and debate over it became more pronounced as the wars went on.[71] In the first place, there was the sense that the symptoms overlapped. Newspapers regularly mentioned the two together as the two "signature wounds," in framing that implicitly conflates them by pointing out, for example, that "about 20 percent of the troops serving in Iraq and Afghanistan can expect to come home with behavioral problems from either traumatic brain injury or post-traumatic stress disorder."[72] Others mentioned that they often have the same symptoms, a point that Col. Elspeth Cameron Ritchie, then the Surgeon General's chief psychiatry consultant, had explicitly told Congress in 2006: "Many of these symptoms are similar to post traumatic stress disorder, especially the symptoms of difficulty concentrating, and irritability."[73] The medical literature evinced similar debate, with some research arguing that brain damage and amnesia might insulate veterans from PTSD and other research speculating the opposite.[74] Other research pointed out that it was unclear whether TBI generated PTSD or whether the same event produced both.[75] And certainly, it was common sense that an event violent enough to cause a TBI might also generate PTSD.[76]

As a result, there was broad concern in Congress, in the media, and in the Army that the differences between the two weren't being sufficiently parsed and that soldiers were being misdiagnosed. Which diagnosis a soldier got mattered for both clinical and cultural reasons. On the clinical side, there was the question of ensuring proper treatment. Geoff Ling, who calls Hoge's research "meticulous," explains that "misdiagnosing somebody with TBI when they actually have PTS or vice versa is a clinical tragedy. . . . They require two different kinds of treatments."[77] At the same time, TBI and PTSD signified differently in the culture. The most public concerns over misdiagnosis worried that what was really mTBI was being labeled PTSD. Bradshaw's task force, for example, had noted that "symptoms may be misattributed to other diagnoses such as ASD [Acute Stress Disorder] or PTSD," while Sen. Susan Collins (R-ME) told Peter Schoomaker in January 2007 that her constituents had concerns that they were being diagnosed with PTSD instead.[78]

In parts of the AMEDD, however, the suspicion was exactly the opposite—that soldiers were being diagnosed with TBI when they were in fact suffering from PTSD. Traumatic brain injury "was more culturally acceptable, I think," Jaffin recalls, "and so the numbers got way, way, way inflated."[79] Eric Schoomaker, who became Surgeon General in 2008, agrees. Like Jaffin, he had visited soldiers shortly after they had arrived at Army hospitals and found them alert and able to describe the event that had injured them; then, when he returned on a subsequent visit a few months later, he found that many had been diagnosed with mTBI. "Concussion and physical injury had a greater credibility to it, for some reason," he explains. "It was physical, and people could grasp that as something that they were okay with for some reason. . . . Having a psychological response to combat somehow didn't have the merit or cachet."[80] These comments illustrate a point that Lembcke presciently makes: "With 'traumatic brain injury' mixed in," he asserts, "the PTSD package now becomes nominally biological, relieving it of its otherwise 'only psychological' stigma."[81] In a culture that stigmatizes mental health issues, attributing symptoms to a physical wound may have been more palatable for both soldiers and the wider public.

Jaffin's and Schoomaker's experiences aligned with Hoge's conclusion that many of the complaints from soldiers who had been diagnosed with mTBI were more associated with PTSD.[82] Because the vast majority of mild concussions heal within a few days, many of those diagnosed with TBI, Hoge thought, were better described as soldiers who had *had* a TBI. "The typical person who gets labeled as being a TBI patient is a soldier who has come back from deployment who has had a couple of concussions during their deployment," Hoge explains. "But he's considered to have TBI *now*, because he is having headaches or concentration or memory problems, or fatigue, or sleep disturbance."[83]

The problem with that diagnosis, in Hoge's view, is that it was made too long after the event to be valid. "The definition of mTBI only includes symptoms that occurred at the time of the injury, not the range of symptoms that may persist in some patients," he explains.[84] Soldiers who complained of such symptoms months after returning from Iraq or Afghanistan thus could be living with the persistent post-concussion syndrome, but they might instead be suffering from post-traumatic stress or depression.[85] Screening soldiers months after an event to find out whether they'd been concussed and what symptoms they were experiencing potentially constructed a false correlation.

Hoge's *JAMA* article seemingly clarified some of the uncertainty that shrouded the diagnosis of mTBI, but it proved controversial. On the one hand, it validated the views of people like Jaffin and Schoomaker, who worried that the stigma surrounding PTSD was leading to the overdiagnosis of TBI and that misdiagnoses might be causing soldiers to not receive the proper treatment. It also led the AMEDD to increasingly adopt the approach that Chuck Engel had long championed, in which patients would receive treatment for their symptoms even if those symptoms' precise etiology was unclear.[86] This included co-locating combat stress treatment and concussion care in the same clinic, a recognition that the two conditions often co-occurred.[87]

Hoge's article, and the AMEDD's embrace of it, also led to changes in the questions about mTBI on the PDHA and PDHRA. Asking veterans whether they had experienced a concussive event seems, on its face, like common sense, and the notion that the Army had not been doing this had in 2007 suggested to some that the military wasn't doing enough to identify and care for soldiers with mTBI.[88] In June 2007, the *Washington Post* ran the story of Joshua Calloway, an Army private who "had been exposed to multiple bomb blasts in Iraq, but after seven months at Walter Reed he had not been tested for traumatic brain injury" and was ultimately retired from the army with a disability rating for PTSD.[89] In July, Sen. Patty Murray (D-WA), one of the war's most vocal critics and the senator most concerned with mental health, inveighed that when she had visited wounded soldiers at Walter Reed who "talked about not knowing they had traumatic brain injury even a year and a half after they had been wounded and came home . . . they knew something was wrong, but no one had taken the time to ask them what they had seen on the ground in Iraq or what they had been involved with that might have caused a brain injury."[90]

Ironically, Murray's speech came six weeks after the Army had begun working to incorporate questions about TBI into the PDHA and PDHRA. On June 15, Gale Pollock issued a memo telling commanders that "accurate documentation of blast exposure is mandatory to optimize medical management and minimize soldier risk," explaining that "recording blast exposures in this fashion allows encounters to be part of the soldier's longitudinal medical record" and announcing that questions about blast and concussion would be added to the PDHA and PDHRA.[91] Those questions were fully incorporated into the two questionnaires in January, and by the summer of 2008 the Army had refined instructions for in-theater screening, requiring that "soldiers involved in a blast, fall, vehicle crash, or direct impact who have an alteration or loss of consciousness" undergo the Military Acute Concussion Evaluation.[92]

Although screening questions had been in limited use for several years—at Fort Carson, for example, Terrio had been using them since

2005—their utility had only been validated in a single DVBIC study that the authors subsequently realized was methodologically flawed.[93] As a corrective, they asked Hoge to include the questions on his own study.[94] "They may have regretted their decision to give us those questions," he recalled, because "we ultimately found that there was a very complex interaction between those questions and outcomes that we were interested in." Post-deployment screening questions, Hoge found, "identify [that] the concussion may have happened in theater, but they don't do a good job of helping you attribute whether symptoms are caused by concussion or not"; rather, "what really predicted post-concussive symptoms was the depression and PTSD, and not the injury itself."[95]

This result, as one might imagine, hardly pleased the DVBIC researchers. Over the next three years, they and Walter Reed researchers engaged in an ongoing debate over whether the questions needed revision. Terrio's subsequent research helped the Army's case by calling into question the validity of the latter two questions, which asked whether an individual had experienced a potentially concussive event and had had or still had any symptoms. Only the first two questions, which asked whether a potentially concussive event had happened and what symptoms had been present immediately following, effectively determined whether the person had had a TBI.[96]

Although her intent was to make sure that soldiers who weren't exhibiting symptoms still had their concussions noted and received follow-up care, her conclusion echoed Hoge's—the current screening questions needed to be revised, and it would be better if the questions about the incidence of concussion were matched with a question that asked what kinds of symptoms followed.[97] Separating questions about the incident from the subsequent symptoms was Hoge's goal as well. The 2011 revision moved the questions about symptoms to a separate section of the assessment, thereby breaking any association that they might have had with the event.[98]

The debate over revising the Post-Deployment Health Assessment and Post-Deployment Health Reassessment illustrates not only how much knowledge of TBI and its relation to PTSD was evolving in the moment but also how the AMEDD's evidence-based approach conflicted with popular assumptions about diagnosing mTBI. Hoge's initial research, and Terrio's subsequent work, showed that simply asking soldiers what had happened to them and how they felt wasn't clinically valid. In this way, the AMEDD's embrace of a scientific approach challenged popular wisdom about screening and denied an easy resolution to critiques that the primary problem with mTBI was that the Army was failing to effectively screen its troops.

Hoge's research also generated debate when it came to research on blast-related TBI. Young, who had been working with Ling to study the effect of blast on the brain, recalls that when she read Hoge's article she "was appalled" and "very bothered." In her view, because Hoge was so well respected, particularly inside the Surgeon General's office, the AMEDD embraced his perspective to the point that it "felt like the shutting down of important research" on blast.[99] This sentiment reflects a long-standing debate among researchers. When Young had first heard of Warden's research, she recalls, "There were at this time, and continued to be for a while, some individuals who were just adamant that there is no such thing as a blast traumatic brain injury. It doesn't exist. It's all PTSD. And . . . that camp continued to be out there for quite a long time."[100] Macedonia agrees, recalling that some in the AMEDD "continued to push the idea that it was 100% PTSD."[101] He categorically disagreed, pointing out that some troops had witnessed many potentially traumatic sights—for example, seeing a squad-mate decapitated by a fifty-caliber round—with no symptoms but had significant symptoms after exposure to a single blast. Rhetorically, he asks, "What makes the blast so scary and the fifty-caliber round not?"[102]

Importantly, researchers like Hoge and Schoomaker never dismissed the significance of TBI, nor did investigators like Ling and Macedonia argue that all post-deployment symptoms were caused by it.[103] Rather, the debate revolved around which symptoms were attributable to which conditions and thus what treatments a soldier needed. How vigorous and at times acrimonious this debate was became more evident shortly after Peter Chiarelli became Vice Chief of Staff. Journalist Yochi Dreazen has called Chiarelli "an unusually compassionate and emotional officer."[104] In 2004, when Chiarelli was commanding the 1st Cavalry Division in Baghdad, his biographers Greg Jaffe and David Cloud write, he so deeply felt the Iraq War's toll on the young men and women fighting it that "he began carrying index cards in his breast pocket with the names, hometowns, and parents' names of every soldier in his division killed in action, a stack that grew and grew during his year in Iraq, eventually numbering 168. He went to all of their memorial services, and when he had a spare few minutes, he'd study the cards."[105] Inside the Army, his regard for his soldiers is legendary, and taking care of their problems dominated his career. At the ceremony marking his promotion to Vice Chief of Staff in 2008, Secretary of Defense Robert Gates told the audience that "he has been described by troops under him in many ways: as a father figure, a health advocate, a career advisor, and even a marriage counselor" and that "as long as there is a single soldier in harm's way, as long as there is a single Army family in need, Pete will never rest."[106] Almost any officer or civilian who worked under him offers a similar assessment. Christopher Philbrick, who as a colonel served on Chiarelli's staff and managed the Health Promotion, Risk Reduction, Suicide Prevention Task Force from 2009 to 2011, asserts that the surest way to enrage Chiarelli is to question his commitment to his solders' well-being.[107] That sensitivity, matched with a keen intelligence, an aggressive personality, and a nearly monomaniacal commitment, shaped Chiarelli's aggressive approach to solving problems. The *Washington Post* called him "an iconoclast."[108] Bruce Shahbaz, who served on the task force with Philbrick, calls him "a bull in a china shop."[109]

Learning about the prevalence of PTSD and mTBI in the force shocked Chiarelli. "On the 4th day of becoming Vice in 2008, somebody came into my office and showed me we had more individuals with post traumatic stress and traumatic brain injury by a factor of over three times . . . [than people] who had lost arms, or legs, or multiple limbs," he remembers, "and I looked at the chart and said, 'my God, what's TBI?'"[110] His briefer explained to him that a TBI was synonymous with concussion, but he admits, "I had no idea what post-traumatic stress was."[111] He recalls that he had been "brought up right" as a junior officer, and so when he got to Iraq he followed all of the combat stress control guidelines that he had been given. He explains, "I gave them priority for aviation, . . . and I made sure they got to locations where they were needed after losses."[112] What he didn't realize, and had never been told, is that some of the people being treated by those teams would end up with post-traumatic stress.[113] Looking at the slides, though, was a revelation. "I didn't know a hell of a lot," he admits. "But when you're the new Vice Chief of Staff of the Army, you're paid . . . to look out for those things that are going to be problems. . . . And it was clear that this was going to be a problem."[114]

Although he was a tanker who "had spent decades aching to lead troops in combat" and would have preferred returning to Iraq to being named Vice Chief, Chiarelli had also taught social science at West Point and at one point thought of staying on the faculty permanently, and he encouraged debate and dissent among junior officers.[115] That openness, however, did not translate into endless patience with the Army Medical Command, and particularly not with the Surgeon General, Eric Schoomaker.

The AMEDD is a unique subculture within the Army. If the big Army of infantry, armor, and artillery units tends to value decisiveness, the AMEDD is by nature more deliberative.[116] It's filled with MDs and PhDs and was headed during Chiarelli's tenure by Eric Schoomaker, who had one of each degree, specializing in internal medicine and earning a doctorate in genetics with a dissertation on sickle cell disease at the University of Michigan. If Chiarelli was a bull in a china shop, Schoomaker was more of an academic. To some Army leaders, this was precisely the demeanor that the moment required. He was "particularly well-suited for the challenges that we faced," former Secretary of the Army Pete Geren recalls, because "some in the medical profession are fairly rigid in their approach to medical protocols. Eric is open to new ideas . . . he's a thoughtful, creative, open-minded, highly principled guy."[117]

Perhaps chief among those principles was an unwavering commitment to the scientific method. During the 1991 Gulf War, he had watched with dismay as leaders made decisions that he felt weren't scientifically defensible because of the pressure that the conflict created, a failure that he felt scientists were ethically compelled to avoid. "It's always been a hallmark of what we do, at least in my career, that . . . we seek the best evidence-based approach and that we not be satisfied with anecdote," he explains, because "too often something that looks like its going to work, when its subjected to a really rigorous investigation [it] doesn't work out. . . . It's unethical to apply things to people without knowing firmly that they are the best approach."[118] When he became Surgeon General, Schoomaker and his staff "adopted a very hard line that we were not going to do things simply because they looked good on the surface when in fact . . . there might be a better explanation or a more evidence-based approach."[119]

For Schoomaker and others in AMEDD, teasing out exactly what constituted an mTBI, how it affected soldiers, how the co-occurrence of mTBI and PTSD could be usefully parsed, and which treatments were most promising called for precisely this kind of evidence-based rationality. This approach, however, frequently put them at odds with Chiarelli, who wasn't sure Schoomaker was taking traumatic brain injury seriously enough. "I had some senior, senior people in Medical Command look at me and say, 'You know, concussion's not a problem,'" he remembers. "I had a very, very senior person in Medical Command—the most senior person—look at me and say, 'You know, I was concussed five or six times, I lost consciousness playing rugby, and look at me, I'm a surgeon.' So even within Medical Command there was a kind of a pooh-poohing of these injuries."[120]

Schoomaker's comment was perhaps less a dismissal of concussion's seriousness than a claim in line with Hoge's argument: that concussion alone might not be the major issue that soldiers faced and that PTSD might be a significant co-occurring factor. Like Hoge, Schoomaker found illogical arguments that concussion alone could account for all of the symptoms that soldiers were having months after their injuries. And his comment that he had been concussed playing rugby but not suffered any long-term cognitive effects reflected his understanding that there were millions of concussions in the United States every year, but civilians didn't complain of post-concussion symptoms of the sort, or at the same rate, that soldiers did. "There's something important contextually about how you get concussed," he would later explain in an interview with NPR's Daniel Zwerdling:

> When you're concussed on the rugby pitch or you're a Ben Roethlisberger and you're a quarterback of the Steelers, and you wake up from your concussion, or you recover from your daze, and you've got a trainer by your side, and the crowds are cheering, this is a different context from when you wake up after an IED blast or in combat where people have been killed, and friends may be badly injured, and they may still be shooting at you.[121]

This reasoning illustrates two tenets of the AMEDD's thinking on concussion. The first is that a blunt force concussion in combat is, physiologically, no different from one that occurs in the civilian world but is, psychologically speaking, freighted with an entirely different set of implications. The second is that acknowledging and treating those psychological implications seems critical to returning soldiers to health.

The insistence of psychiatrists like Hoge that post-traumatic stress, not mTBI, might account for many of the symptoms that were being dubbed "post-concussive" put him and his colleagues in direct conflict with Chiarelli. Having discovered the disputes that defined the field, the Vice Chief collaborated with Marine Vice Commandant James Amos to invite a neurologist from the University of California, Los Angeles named David Hovda to meet with Army specialists on mTBI and PTSD in fall of 2008. Hovda approached traumatic brain injury from an entirely different direction. He argued that even mild concussions had potentially catastrophic consequences for an individual's cognition, ranging from how quickly they were able to process information to the death of cells throughout the brain, and he had reached these conclusions through brain-imaging studies.[122]

The meeting didn't go well. "This food fight broke out," Chiarelli remembers. "It was crazy. It was mostly with my doctors disagreeing" about the nature of TBI and its relationship with PTSD; Hoge, in particular, thought that the potential impact of PTSD was being discounted because of the physical injury.[123] "Amos and I were disgusted," Chiarelli recalls. "We were action kind of people, and all we saw was this room full of experts and this guy that we really, really thought was fantastic, and everyone was arguing with one another, and nobody was saying what we were going to do to try to fix the problem."[124] Infuriated, the generals stormed out of the room.[125] Hovda followed them in, telling them that disagreement was productive, but Chiarelli fired back, "Yeah, but that disagreement is causing young men and women to pay the price."[126]

Both the *Washington Post* and in *Politico* have related this story, casting it as the moment when a maverick general stood up to a hidebound bureaucracy protective of its own interests.[127] The import of this moment, however, lies not in its revealing an unresponsive Medical Command but rather in its highlighting the competing investments and approaches of different stakeholders in the debate and of the conflicts that ensued as a result.

That the state of brain science meant that there was very little medical certainty about what caused post-deployment symptoms and how they could best be treated made it difficult for anyone to definitively give Chiarelli something tangible that he could do to solve the problem. This increasingly frustrated the general, and the Pentagon meeting was one skirmish in an increasingly acrimonious relationship between the Vice Chief of Staff's office and the Surgeon General's office. Chiarelli was frustrated by what he increasingly perceived as the Medical Command's refusal to take the issue seriously and by the diverging opinions throughout the medical community. "I saw this tremendous disagreement, everyone from my surgeon general, who was saying this wasn't a problem, to people who were saying it was a huge problem."[128] Schoomaker, who believes that "my greatest flaw and greatest strength was my skepticism" and insisted on being "the voice of dissent, self-described rationality and caution in the room," was equally exasperated that "our action and decision-oriented leadership were not tolerant of this caution."[129] Macedonia's assessment is more blunt. "Chiarelli felt like it was a cabal of military doctors trying to keep him in the dark, and he railed against Eric Schoomaker," he recalls. "They were like oil and water."[130] That Schoomaker's approach "was often interpreted as insubordination by Army Staff officers" led to several occasions in which the tension between the two men became so great that the Surgeon General offered to resign.[131]

Among Chiarelli's frustrations was that the pace at which rigorous scholarly research operated could not meet the immediate demands of the wars. Rebecca Porter recalls one meeting during which medical officers' assertion that researching the potential benefits of placing soldiers in a hyperbaric oxygen chamber to treat their TBI would probably take a decade led to Chiarelli becoming "kind of angry that here's a problem that's impacting soldiers, and you're telling me that you can't have an answer for more than ten years?"[132] Leanne Young also saw this conflict play out. "Research isn't fast, and the needs of the field are today," she reflects, explaining that "we did have people like General Chiarelli that were kind of pounding the table tops and saying 'You've got to give me something I can do,' and meanwhile we're going, "well, we're not sure.'"[133]

Convinced that he "wasn't getting a straight story from [his] docs," Chiarelli turned to outside experts like Hovda.[134] With Amos, he established in early 2009 a "Blue-Ribbon Panel" to develop a protocol for screening soldiers on the ground, particularly for repeated concussions, a decision that further inflamed tensions between his office and the Office of the Surgeon General.[135] To some, Chiarelli's decision to turn to outside experts must have seemed the latest example in a long history of the larger Army's dismissal of Army medicine expertise. Several researchers and clinicians express frustration at what they view as a pervasive sense across the Army that Army doctors must be substandard, otherwise they would have found more lucrative careers outside the Army. Rebecca Porter explained that "I've had people say to my face that the only people who stay in Army medicine are people who can't hack it on the outside," while Amy Adler, one of Walter Reed's research psychologists, admits that "we are very conscious that people just assume that you wouldn't be in the job you are in if you were good."[136]

Unsurprisingly, this is a view that those inside AMEDD roundly reject. Donald Bradshaw, for example, is adamant that people who serve in the medical corps "do it because they love this country. Most of my doctors could have gotten paid 200 to 500 percent more than they did, and slept every night with their family."[137] Adler likewise points out that "Charles Hoge, Carl Castro, Paul Bliese . . . are not people who would struggle to make it on the outside," while Porter argues that "we have some of the greatest minds in psychiatry and military mental health. . . . We certainly have enough expertise in house to inform the problem and the solution."[138] Given these competing views, that it seemed to Eric Schoomaker that the Vice Chief was "very resentful, very dismissive, and, quite frankly, very disparaging of us" and "felt that his own medical people, including me, I guess, were either not bright or not proactive enough" must have stung both professionally and personally.[139]

* * *

If Young was concerned that Hoge's article was directing attention away from blast-induced TBI, that was in part because there was mounting evidence that a blast wave could, in fact, damage the brain. Prior to the Iraq and Afghanistan Wars, the notion had never been considered. To understand why that is, it is first important to understand the basics of blast physics. An explosion releases gas at a significantly higher pressure than the surrounding air; that gas, in turn, forces the surrounding air molecules to compress, producing a wave of high pressure—a "blast wave"—that moves outward from the explosion.[140] The larger the explosion, the greater the pressure differential between the blast wave and the ambient air.

In addition to anything that it carries—pieces of the bomb, for example, or other shrapnel—a blast wave that comes in contact with human bodies has destructive power precisely because of that pressure differential. Organs like the lungs, the bladder, and the intestines are filled with air that is at the same pressure of the pre-blast surroundings; when the blast wave collides with that air, it displaces it, causing tissue damage as the organs struggle to adjust to the sudden introduction of the new air at a substantially higher pressure.[141] In close enough proximity to a powerful enough explosion, that tissue damage can be fatal.

However, a scientific law called Hopkinson's Rule explains that a blast wave is like any other wave: its strength diminishes over distance.[142] The question, then, was, how close did a soldier have to be to a blast to sustain organ damage? The Army conducted considerable research on this issue in the 1980s that included shelling armored vehicles in which researchers had placed body armor-clad pigs and sheep and found that, because "the intensity of the blast (the peak overpressure) decreases rapidly with distance from the detonation, personnel must be very close to an explosion" for the blast wave itself to have wounding power. At greater distances, a soldier was more likely to be wounded by shrapnel, as "conventional munitions travel in air far beyond the distance that the blast will cause injury."[143] Surprisingly, given what would come during the Iraq and Afghanistan Wars, though, the Army's textbook on blast injuries devotes little attention to the impact of blast waves on the brain, noting only that injuries to the lungs can create fatal brain embolisms.[144]

Because the brain is not an air-filled organ, there was initially little suspicion that it could be damaged by a blast wave.[145] "That is something which there was not a lot of research on," Geoff Ling explains. "Why would there be? Most people outside of the war who were getting hit in the head were getting hit in the head by soccer balls or punched."[146] As the wars went on and more soldiers complained of injuries from explosions, however, the possibility that a blast wave alone could be injurious became a critical concern. Accepting that physical force could cause a concussion compelled Ling and others to consider the potential effect of blast more seriously: "You say to yourself, 'But a shock wave is a physical force, just not one you and I can see.' So we just went ahead and said 'Okay, that's the debate.'"[147]

By the mid-2000s, the Army was conducting significant research into the implications of blast on the brain. In 2005, Ling created PREVENT, an acronym for Prevent Violent Explosive Neurotrauma, under the auspices of DARPA, the Defense Advanced Research Projects Agency.[148] This research was pivotal. In 2007, Ling, along with Leanne Young and others, conducted a study in which they had dressed Yorkshire pigs, which Young and Ling had chosen because their cranial structure is roughly similar to a human's, in the same body armor that U.S. troops were issued, anesthetized them, and attached instruments to monitor their vital signs.[149]

After being transported to the blast site by ambulance so their vital signs could be monitored, the pigs were placed inside a large tube, in a building or in a model Humvee, and oriented so that they would be in the path of the blast wave.[150] Young, as an engineer, was in charge of making sure, first, that the blast was the same strength as those that Americans typically encountered in Iraq and Afghanistan and, second, that only the blast wave affected the pigs by, among other precautions, placing the explosive in a cardboard box so there would be no shrapnel.[151] As Ling points out, having confidence that the blast alone had affected the brain meant ensuring that other factors, like the heat of the blast or the electromagnetic pulse of the explosion, were accounted for.[152]

After the pigs were exposed to the blast, they were driven back to the lab, where they were sacrificed and their brains were examined. What the researchers found surprised them. In a concussion that is sustained from blunt-force trauma like a fall or an automobile accident where the head hits the dashboard, the brain is bruised both at the point of impact and on the opposite side, where it ricocheted into the other side of the skull—a classic coup-contrecoup concussion.[153] The pig's brains, of course, didn't display this sort of injury, but they were damaged. When the brain was dissected, the researchers found inflammation as well as "dead axons and dendrites," the parts of neurons that transmit data.[154]

They also found increased astrocytosis, an indication of inflammation of the cells that connect neurons, which is another sign of neurological injury.[155] In layman's terms, the brain had been damaged on a cellular level. Shock waves separate neurons, Ling explains, which leads to "an inflammatory response" and then "programmed cell death."[156] Although the researchers were measured in their presentation of the data, the study seemingly confirmed that blast TBI was real.

What wasn't clear, though, was whether the damage was significant to the point that it would cause cognitive or behavioral changes in a human.[157] A pig's frontal cortex is, of course, not as complex, so it was impossible to gauge whether higher-order executive functions, memory, or emotional state had been affected.[158] And, of course, intentionally exposing human subjects to a blast wave wouldn't pass muster with an institutional review board.[159]

Finding human subjects who had been exposed to blasts, however, turned out to be simple. In 2007 and 2008, Young and a group of colleagues got permission from the Navy to conduct research on Marines at the Dynamic Entry School at Quantico, Virginia. "Dynamic entry," or breaching, is a military euphemism for blowing a door open with an explosive. Young and her colleagues would later write that "breachers are a unique population who, as part of their regular training, are exposed to series of controlled blasts under supervised conditions that minimize the risk of injury from debris, fragments, or whole-body translation."[160] The Marines had taken precautions to protect those taking the course from the potentially concussive effects of blast, but the researchers were interested in determining whether the repeated blast exposure caused neurological changes.[161]

In their experiment, Young and the others, led by a Walter Reed researcher named Walter Carr, had Marines who were enrolled in the course take a series of cognitive function tests before and after the course. Here again, the results were troubling. While there was not "clear evidence of neurological impairment," there was evidence "suggestive of blast-induced impairment in selected domains of cognition among individuals subject to sustained repetitive blast exposures," in particular "a reduction in accuracy of . . . responses on tasks that place demand on memory ability."[162] Perhaps more important, the research quantified how much exposure would lead to deleterious results. Using gauges attached to the breachers' helmets, Ling explains, the researchers determined that a blast wave of sixteen pounds per square inch was the threshold for neurologic injury. "If you stayed below a blast exposure of 16 PSI," Ling explains, "you had no evidence of neurologic injury, even if you were exposed repeatedly over a 2-week period."[163] In short, these results offered the first scientific validation of anecdotal media reports.

For the neurologists, the results of this study and the earlier swine study were all they needed to settle whether blast injury existed, whether it had long-term cognitive effects, and whether it might cause the symptoms that overlapped with PTSD. "It dispelled the discussion," Geoff Ling remembers. "I don't think anybody now has an issue [with the notion] that exposure to a blast shock wave causes TBI."[164] It took until 2016 for the results of the Quantico Breacher Study to be published in *Military Medicine*, but leaders were being briefed on it long before then, and leaders responded. By January 2008, the Army was field testing blast gauges in Afghanistan; by 2011, they were in widespread use.[165]

Researchers did increasingly accept that a blast wave could damage the brain, but questions remained. The most significant related to the reality that soldiers in the combat zone rarely experienced a blast wave separate from the other effects of an explosion. As Ling explains, "An explosive blast is pretty dramatic. There's a big fireball, there's a bunch of junk flying around, and you probably get hit by stuff and thrown off your feet . . . you might even hit your head," so determining exactly which factors injured a soldier was a significant challenge.[166]

Given this reality, some researchers argue that because a shock wave dissipates over distance, any soldier close enough to be seriously injured by the wave itself would also have significant physical injuries. For example, John Holcomb, the Army's top trauma surgeon, claims that, "if you are within 5–10 feet [of an] open space explosion, you may have this primary blast injury, but you are going to be shredded to pieces, literally, by the fragments"; thus, he explains, "guys look like they are suffering from the holes, not some magical blast over pressure wave."[167] Charles Hoge agrees, explaining that, "if a Soldier is within the radius of a primary blast wave, he or she may certainly suffer neurological damage" but "will most likely also suffer catastrophic physical secondary and tertiary injuries from shrapnel, blunt trauma, [and] burns."[168] These soldiers, he goes on to explain, would likely be medically evacuated and thus were unlikely to first complain of symptoms months or years later.[169] Similarly, while Hoge values the breacher studies for demonstrating the effect of repeated exposure to blast on the brain, what remains unclear is whether one or even a few blasts have a deleterious effect.[170] Macedonia, however, disagrees, arguing that "we had people who were in bunkers" or armored vehicles with open windows and thus shielded from shrapnel but who nonetheless "had contemporaneous medical evaluations consistent with blast overpressure injuries."[171]

These were thus significant questions that generated serious debate and resisted simple solutions. But what Ling and Young's research into blast did do was change the way the Army screened soldiers on the battlefield.[172] The growing acceptance of blast TBI coincided with growing concern over multiple concussions and, particularly, how the brain would be injured if a second concussion occurred before the first had healed. This was an issue that the civilian world, largely because of attention to the long-term effects of concussions among National Football League players, was beginning to appreciate as well, with several studies in the preceding years illustrating the deleterious cognitive effects of multiple concussions.[173] It was also, of course, one that Americans in and outside of the Army were beginning to realize affected soldiers. In 2008, for example, the *New York Times* profiled Indiana reservist Kevin Owsley, who had endured at least two potentially TBI-inducing events: "After each attack, he did what so many soldiers do in Iraq. He shrugged off his ailments. . . . Given that he never lost consciousness, he figured the discomfort would work itself out and kept it to himself."[174] Owsley had returned from Iraq suffering a range of symptoms, and the article pointed out that his experience was far from unique. "These are the first wars in which soldiers, protected by strong armor and rapid medical care, routinely survive explosions and return to combat," the article noted, pointing out that "these mild concussions, which do not necessarily lead to a loss of consciousness, are easy to dismiss, simple to misdiagnose, and difficult to detect. The injured soldiers can walk and talk."[175]

The potential challenges posed by undiagnosed and multiple concussions increasingly preoccupied the Army. In a clarifying moment for Chiarelli, Hovda had provided him with a slide of three brains—one of a person with normal brain function, one of a UCLA football player who suffered a hard hit in the first half but been cleared to continue playing, and the third of a patient comatose after a car accident. While the first brain glowed with neurological activity, the other two were nearly silent.[176] According to Chiarelli, the slide of the hit football player's brain was "the most important slide I ever received," and it convinced him that the failure to screen soldiers with mild-TBI had potentially catastrophic consequences. "I had a Surgeon General who didn't really believe in the effects of concussion," he explains, "yet I was seeing kids get a concussion on Tuesday, keep themselves in the fight, get a concussion on Thursday, show up at Landstuhl, and [I'd] have to medically discharge them out of the army because they had huge cognitive issues."[177]

To ensure adequate screening and prevent a second concussion, a team of outside experts and a group of Army doctors worked to set criteria that would define which service members would be automatically evaluated for a concussion and rested before returning to duty.[178] In 2009, Admiral Mike Mullen, the Chairman of the Joint Chiefs of Staff, chose Macedonia to lead the Department of Defense's so-called Gray Team of neurologists. Initially, it was not a job Macedonia wanted. "I was really skeptical of the commitment that flag officers and General Officers had toward getting to the bottom of things," he explains and thought, "If this dude is just trying to form a task force as a Band-Aid, I don't want to be part of his Band-Aid. So he better be the real deal."[179] An Army obstetrician-gynecologist was not the logical choice for the job, but he got it largely because he was willing to talk back to the chairman and because he, like Mullen, excelled at systems thinking.[180] He had also trained at the National Institutes of Health and worked at DARPA, where he developed skills in bioinformatics and become skillful at reviewing research studies.[181] Knowing that the task force would meet opposition from each of the services, it wasn't necessarily a role Macedonia relished, but it was one he felt obligated to take. "I didn't [take the job] because I enjoyed blood sport," he reflects, "but out of my love [for the Army] and some guilt that I sent some people back who shouldn't have been sent back."[182]

Macedonia's concept was to "Red Team" the military's approach to traumatic brain injury by "walk[ing] the dog all the way from the point of injury back to the U.S. to look at where we were doing it right and where we're sucking wind."[183] Because the concept of Red Team—which in Army parlance implied looking for flaws and weaknesses —"will get everyone's hackles up," Macedonia reasoned, it was better to call it "Gray Team" because of the color of brain tissue and, self-deprecatingly, the relative age of those selected to serve on it.[184]

In Afghanistan, the team met with an Army doctor named Jennifer Bell and two Navy medical officers, Tracy Skipton and Doug Nasky. Although they were stationed in different regions of the country, they had independently arrived at an effective way of ensuring that troops who had potentially been concussed recovered before returning to duty. The treatment was surprisingly simple and drew upon the principles of Combat Stress Control that the Army had been using throughout the wars. Soldiers who had potentially had a concussive event in the field would be sent to a clinic near the front so they could maintain contact with others in their units. Medical personnel would treat their symptoms—Tylenol for headaches, sleeping pills for insomnia—and provide activities like crosswords and Sudoku to keep their minds active without having them look at digital screens.[185] "Jennifer and Tracy and Doug got remarkable results," Ling remembers. "Most of their troopers went back to duty within a couple of days, and when I went out with the Gray Team and found these remarkable clinicians and what they did, we said we have got to make this become the basis of a new way of treating our brain-injured soldiers."[186] That was not all they did, however; even as they pursued standardizing TBI treatment, they also argued for increasing forward-deployed behavioral health resources.[187]

The Gray Team's research convinced Chiarelli and Amos to begin implementing this sort of concussion care throughout Iraq and Afghanistan.[188] To convince soldiers that they needed rest if they had potentially been concussed, Chiarelli conducted video conferences with deploying units from his Pentagon office in which he displayed the slide Hovda had given him of three brains (one with normal brain function, one of a football player who had a hard hit and continued to play during a game, and one of a comatose car accident patient). "That slide was worth millions of dollars in [terms of] making the stigma go away," Chiarelli remembers, because soldiers could see the physical consequences of untreated concussion.[189]

Still, the Gray Team faced opposition. "Each one of the service medical departments fought like crazy to keep the status quo," Macedonia recalls, and in his view "a vocal but powerful minority fought at every step of the way not to study the question" of blast TBI, whether it was putting blast monitors on soldiers' helmets or sending MRI machines to Afghanistan.[190] This reluctance frustrated him. "Since when in medicine," he complains, "have we been afraid to record data about a patient and try to understand a disease process?"[191]

In the end, formalizing this process for screening service members required that Undersecretary of Defense William J. Lynn issue Directive Type Memorandum 09-033 on June 21, 2010.[192] This order set a much stricter surveillance regime, in which "any service member in a vehicle associated with a blast event, collision, or roll-over" and "any service member within fifty meters of a blast" received mandatory screening and rest.[193] For Ling, this was a crucial step forward because "it obligated the leaders to ensure that, if somebody was at risk of having a traumatic brain injury, they [would] be screened by a medical provider."[194] Moreover, the protocol proved effective. "Over 90% of [soldiers who were screened and rested] returned back to duty without having to leave Afghanistan or Iraq," Ling explains. "And then the last few that they found that did not show recovery, almost all of them had mental health issues," most likely post-traumatic stress.[195] And because the Army was increasingly taking a multidisciplinary approach to medical care, TBI treatment and Combat Stress Control were located in the same facilities, which meant that soldiers could receive both types of care simultaneously.[196]

Like the research that showed the effects of blast, though, the fifty-meter rule was not universally embraced. Some in the room thought the radius should be larger, perhaps one hundred meters, a distance that Chiarelli and others felt would have generated too many false-positive results.[197] Even fifty meters seemed "pretty conservative" to Schoomaker, who felt that "even at fifty meters the risk is fairly low for the type of blast we were seeing."[198] Hoge argues that "the fifty-meter rule is completely arbitrary, but it's a reasonable rule," and calls it "commonsense concussion care that is not unlike what's done on a football field."[199] Macedonia mildly disagrees, calling the distance "quasi arbitrary" because it was "based on the average explosive being used by the enemy," which was a "forty-kilogram C4 equivalent charge large enough to overmatch an uparmored vehicle." As he explains, "We exploded forty kilogram C4 charges and saw at what radius you're at risk to get tympanic membrane damage, and that was about thirty-seven meters," adding that "tympanic membrane damage has been shown to correlate with blast concussion injuries."[200]

The truth was that, because the size of bombs and the paths of blast waves varied, there was no way to make a precise prescription. Still, as John Holcomb put it, "At least a decision was finally made."[201] For Schoomaker, this was perhaps what was most important about the decision because it resolved a long-standing frustration. "One of the things that drove me nuts," he recalls, "was on the sports field or in a motor vehicle accident, at the moment of injury we start managing people. We don't wait twelve months later to ask them, 'Were you concussed?'"[202] The fifty-meter radius "forced us into doing something at the point of injury so that we could then say later we've done everything in our power to save [the soldier] from the concussive injury and possibly the overlapping psychological injury."[203]

The relatively large orbit in which soldiers were to be screened served an additional purpose beyond making sure soldiers got screened and treated. It was also intended to help researchers get better information about the relationship between proximity to the blast and the likelihood of injury. "When we wrote the protocols for taking care of TBI," Jonathan Jaffin explains, "one of the things we wanted was [to] . . . really figure out, if we screened everybody within this radius, how many people, really, right after the fact, had symptoms of a TBI?"[204] Thus the June 2010 memorandum specified not only that each soldier within the fifty-meter radius be examined but also that their distance from the blast be noted, with the goal of eventually generating a large data set that could predict which soldiers were most likely to suffer a blast-induced TBI.[205]

This goal, however, was never effectively communicated to the Army as a whole. Instead, well-meaning commanders, or at least commanders who faithfully executed their orders, screened everyone within the radius and reported that they had done so. "The only numbers that they ever reported," Jaffin recalls, "were how many people they screened for TBI, and people started assuming that everybody in this big radius that we picked should be treated as having had a TBI."[206] This was perhaps because the memorandum also specified "a 24-hour rest period for all exposed personnel involved in a mandatory screening event," which effectively mandated that everyone who *might* have been affected be treated as if they definitely *had been* affected.[207] As a result, the Army never got the data it wanted, and determining precisely what kind of blast was likely to produce a TBI remained a challenge.

The institution of the fifty-meter radius thus represents a different way in which the practical needs of the moment came into conflict with the AMEDD's research-based approach. The decision was somewhat unscientific, and a *Politico* story has called it "a rare victory in [Chiarelli's] fight with the doctors," who "usually told him his ideas didn't make sense—politely, of course—or that they couldn't be implemented because of a lack of equipment or trained personnel."[208] However, Army doctors were comfortable with it for two reasons. First, it took a conservative approach to soldier safety, and implementing it had, effectively, no clinical downside because it had a clear benefit—giving more soldiers rest and screenings—but wouldn't expose soldiers to a treatment that hadn't been validated. It would adhere, that is, to the dictum of *primum non nocere* ("first, do no harm") that Schoomaker and others embraced when they advocated for evidence-based treatments.[209] Second, it provided an opportunity to conduct the kind of research that they viewed as essential to making meaningful progress on identifying and treating mTBI. The larger Army, however, failed in that second effort. That lapse might be excusable, given the tactical demands that commanders faced, but it certainly represents a lost opportunity to better understand blasts, and it may have contributed to an overdiagnosis of mTBI.

* * *

Perhaps the most dramatic example of the tension between popular demands for better TBI diagnosis and treatment and the AMEDD's insistence on evidence-based approaches came when the Army was legislatively mandated to implement a screening procedure that some in the AMEDD believed was not clinically valid. Despite Mike Kussman's 2007 testimony that there was not yet a single test for mTBI, members of Congress and some inside the military remained interested in the possibility. In 2006, for example, the Armed Forces Epidemiological Board had argued that "a baseline screening tool to enhance utility of post-injury neuropsychological testing . . . would be most effective if implemented upon entry into military service. At a minimum, implementing baseline testing should be considered pre-deployment and in military occupations at high risk for blast or impact injury."[210] The Army's TBI task force similarly called for "a baseline (pre-deployment), post-deployment, and post-injury/exposure neuropsychological evaluation."[211]

Likewise, in a set of questions that she submitted after a February 2007 hearing, Sen. Hillary Clinton (D-NY) argued that "knowledge of a soldier's or a marine's normal memory and cognitive skills prior to deployment could potentially improve post-deployment detection of mild TBI" and asked both the Army and the Marine Corps about pre-deployment testing, which Marine Commandant James Conway again explained didn't exist.[212] Nonetheless, a provision for the "documentation of the cognitive (including memory) functioning of each member of the Armed Forces" prior to deployment in order "to facilitate the assessment of the cognitive (including memory) functioning of each such member upon returning from such deployment" made it into a Clinton-authored bill, the 2007 Heroes at Home Act, and subsequently became law in the 2008 National Defense Authorization Act.[213] Faced with this requirement, the Department of Defense chose the Automated Neuropsychological Assessment Metrics (ANAM), a program that had been originally developed in the 1980s by Defense Advanced Research Projects Agency (DARPA).[214] By the 2000s, it had been used to evaluate the recovery of stroke patients and the incidence of concussion in West Point's boxing program and the National Football League.[215]

This requirement is certainly well intentioned, and at first glance, the idea of evaluating a person prior to and after a deployment seems like a simple and useful means of determining whether they'd incurred any cognitive changes. After all, doctors use such tests all the time to monitor other changes in physical health; it's why routine physicals include blood tests for liver function or cholesterol and blood pressure screenings. By monitoring changes over time, physicians are able to identify the onset or progression of disease.

Inside the AMEDD, however, the notion of screening all soldiers was met with some skepticism, for three reasons. First, as the military and VA made clear, is that there simply wasn't the technology to conduct a test that will accurately identify a TBI. In particular, the effectiveness of the ANAM had been called into question in 2004 and 2005, particularly with regard to its usefulness "as a concussion detection tool."[216] Second, as the authors of the Traumatic Brain Injury Task Force report pointed out, the information generated by any test wouldn't reveal much because, "in operational settings, performance on these tests when administered after an acute event . . . could be expected to have a significant number of false-positive results."[217] Moreover, the results were dependent on whether the soldier took the test seriously, which could not be guaranteed.[218] This point relates to the third issue, which is that creating a baseline cognitive function for an individual is difficult. Unlike cholesterol or blood pressure, for which there are, despite differences between people, accepted ranges of what is considered healthy, cognitive ability varies widely among individuals. As a result, gaining a baseline score from an individual cannot be achieved through a onetime test. Much as a patient's fear of the doctor might increase his blood pressure or a Boston cream donut might skew her blood test, an individual soldier's mental state during the test will shape that soldier's ANAM result. As Charles Hoge explained,

> A person can sit in front of a computer and do the test and then it will generate a score, but it doesn't mean anything, because that person may have gotten drunk the night before, they may have been sleep deprived, or because they were anxious about deployment, or they may have gotten into a fight with their wife, or they may have just been perfectly on point that particular day.[219]

Hoge and others thus believed that developing a usable baseline score requires averaging together the results of several tests. "ANAM baselines do not capture a true baseline," a 2011 Army history of the program complained. "Traditional ANAM research studies found ANAM to be effective by giving the subject multiple practice runs in order to establish their baseline."[220] In fact, the history explained, a 2009 study had recommended that subjects take the test twice "and [that medical officials should] use the second run as the baseline for future comparisons."[221] This presented the Army with a logistical problem in that having every soldier take a cognitive exam twice before deployment was probably not feasible, given all of the other tasks and screenings required to render a soldier deployable and the short turnaround time on which units were operating. In Hoge's view, however, there was also a problem from the standpoint of good science, which is that the more a soldier takes the test, the higher his score is likely to be.[222] The ANAM was adopted, then, over the Army's skepticism of its value. The Traumatic Brain Injury Task Force's report, for example, noted that "the provisions forwarded in this bill may not be valuable, or even possible."[223] In the end, Hoge's assessment of the ANAM was even more blunt: "Basically, the test is completely worthless."[224]

Nonetheless, the Army was by 2008 bound by law to conduct the testing. In August, Schoomaker issued Operations Order 08-65, announcing that "USAMEDCOM supports establishment of ANAM pre-deployment baseline testing for comparison to post-injury testing as one of the tools available for the assessment, management, and return-to-duty decisions for Service members with a traumatic brain injury/concussion."[225] By the end of September, the Army had built the infrastructure to conduct baseline testing of deploying troops and to retest those who had been injured, with a plan for more complete testing beginning in October.[226] But in issuing the order for the regional medical centers to begin standing up the training, identifying units that should be screened and so on, the AMEDD also made clear that, although it would dutifully carry out the initiative, its skepticism remained intact: "The ANAM has not been validated for use as a screening tool for pre-deployment or post-deployment populations. The results serve as stand-alone data to be made available for comparison with future neurological assessments."[227]

That skepticism may have informed Schoomaker's November 2008 order directing the Army to cease post-deployment testing.[228] Over the next few years, this decision would be criticized by researchers and in the media. To Leanne Young, for example, "It just seemed unconscionable that we would not be doing pre-emptive screenings on everybody and if there's a change in cognition, putting it in the record so at least it's noted. That's the very least the military should do."[229] She, like others, felt that the ANAM was better than nothing and that in rejecting it Schoomaker was letting the perfect be the enemy of the good.[230]

The Surgeon General, for his part, agreed with Hoge that using the program to screen returning troops wasn't scientifically valid. In written responses to a set of questions that National Public Radio's Daniel Zwerdling submitted in preparation for a 2011 story, Schoomaker explained that "virtually all experts in concussive injury recommend that diagnosis and management of mTBI/concussion occur as soon as possible after a potential concussion. Prompt recognition and treatment provides the optimal chances for full recovery and prevents a repeat concussion during the vulnerable period of healing. Waiting until a Soldier redeploys—in many cases months after an injury—does not represent optimal care for that Soldier."[231] Here it is clear that Hoge's view that symptoms would likely dissipate within a few days, and certainly by redeployment, still held sway.

This 2011 answer is informed by research that during 2009 and 2010 confirmed some of the suspicions that surrounded the program. In January 2009, Schoomaker sent a team of AMEDD researchers to Iraq to study the ANAM's implementation. The results weren't inspiring. While the tests accurately pointed out "neurocognitive dysfunction within 72 hours of a concussive event," within a few days its utility waned, and the researchers pointed to "a lack of utility as a diagnostic screening test beyond the first ten days following a single, uncomplicated concussion."[232] Using the ANAM to test retuning troops, it seemed, would do little good.

Other initial results of the ANAM's implementation did little to assuage this anxiety. By 2009, the Army had collected information on more than 107,000 soldiers—about 10 percent of the force, including the Guard and Reserve troops. From that data, the University of Oklahoma researchers who had developed the program culled the information for about ten thousand, all of whom had deployed from Fort Campbell, Kentucky, and who took the ANAM during their first week back. Of them, about two thousand reported that they had experienced an mTBI. The report found that almost all of the soldiers who had had a TBI performed about the same as they had prior to the deployment and that the soldiers who did perform worse were those who still complained of symptoms.[233] The test seemed to confirm the dangers of a second concussion, but otherwise it didn't do much.[234]

This was not a ringing endorsement of the ANAM. In fact, and as the researchers pointed out, the test largely revealed what the science had already agreed on: that in the vast majority of cases the symptoms of an mTBI dissipate quickly and, therefore, that a test given weeks or months after the exam was unlikely to reveal much about how the concussion had affected the soldier. Nonetheless, the Army remained legally obligated to continue screening soldiers with the test and by 2010 had gathered baseline data on about six hundred thousand troops.[235]

This was not to say that the ANAM had no utility. The Army had during this same period determined the test that could be "used selectively as a post-injury management tool in conjunction with a focused clinical examination to guide return to duty recommendations."[236] The difference here is important. Rather than comparing pre- and post-deployment screens in hopes of recognizing a previously undiagnosed mTBI, which was ineffective given the likelihood that an individual's concussion would have healed by the time she or he returned home, the ANAM could indicate when a soldier's injury had healed to the point that she or he could return to duty. To do so, a psychiatrist or psychologist would compare the soldier's score to either his or her pre-deployment baseline—by fall of 2008, most deploying units had been screened—or an average score of Army soldiers.[237] By 2010, the Army had constructed three dozen screening centers in Iraq and Afghanistan, and every deployed psychologist had a laptop loaded with the ANAM software; the results of that second screening could be nearly immediately compared to the soldier's baseline score, which the provider could request from a stateside repository.[238]

The AMEDD's—and particularly Schoomaker's—understanding of how the ANAM could and could not be used effectively collided with a growing public sentiment that the Army was short-changing brain injured soldiers. Zwerdling's 2011 NPR story in particular savaged the Army's decisions: "Troops have little reason to feel reassured" from having taken the test, Zwerdling reported. "Evidence shows that military officials have made poor decisions about the testing program, preventing it from helping many troops who have brain injuries."[239] The article went on to refer to Schoomaker's decision to not rescreen low-scoring soldiers prior to deployment as "unethical" and assert that he had made the decision because, "if troops who score low on ANAM are referred for medical evaluations, it could postpone their deployments."[240] In short, the report argued that the AMEDD was more interested in expediency than in caring for soldiers.

The story didn't sit well with the Surgeon General, who wrote to NPR three days after the report. Calling himself "deeply dismayed," Schoomaker asserted that "the reporters failed to understand and accurately report the magnitude of efforts" to address mTBI.[241] The letter amounted to a defense of evidence-based research and the scientific method. "All of the Armed Forces utilize scientifically-credible guidelines for early recognition, rapid management, and post-injury recovery, and rehabilitation following concussions," he wrote, adding that "the specific guidelines have been validated by internal DoD subject matter experts, external reviewers, and peer-reviewed scientific literature."[242]

The ANAM, Schoomaker insisted, was thus only being used in ways that were scientifically defensible, that "all [neurocognitive assessment tools] have limitations largely due to the limited science of the brain," and that "based upon empirical evidence of the ANAM's limited ability to detect concussion . . . the Army does not use ANAM or any NCAT [Neuro-Cognitive Assessment Tool] as a stand-alone screen or diagnostic test."[243] And then, to drive the point home once more, he added, "This limitation has been validated in the scientific literature by leading subject matter experts in brain injury and neuro-cognition."[244]

Zwerdling's story and Schoomaker's response encapsulate the debate that surrounded the diagnosis and treatment of traumatic brain injury in the midst of the Iraq and Afghanistan Wars. When the wars began, brain science was not sufficiently advanced to answer the important questions that would emerge as U.S. troops increasingly encountered improvised explosive devices in Iraq and Afghanistan: Could a blast wave alone damage the brain? If so, how and how much? How strong did the explosion have to be, and how close did a soldier have to be to it? How long did it take for a concussed soldier to recover, and what happened if the soldier got a second concussion? What was the relationship between PTSD and TBI? Was there some way to effectively screen soldiers before they deployed so a TBI could be recognized when they return?

These questions resisted clear answers, and efforts to find them were marked by vigorous and at times acrimonious debate. Although they often disagreed with each other, Schoomaker and the majority of those in the Army Medical Department were deeply committed to the scientific method and evidence-based solutions, even as they sometimes disagreed with each other over the etiology of symptoms. Chiarelli, for his part, was sometimes frustrated by the scientific uncertainty as he aggressively sought to develop protocols that would help troops in the field. As the leader of Mullen's Gray Team, Macedonia was frustrated by what he saw as an insular and defensive AMEDD that "stymied Chiarelli over, and over, and over."[245] As a result, AMEDD clinicians and researchers often felt besieged by the Vice Chief of Staff; although they never questioned his commitment to the troops, they sometimes felt that he was dismissive of their expertise and too quick to seek solutions outside the Army. They also chafed against congressional demands that they adopt unproven protocols and public perceptions that they were not doing enough or, worse yet, were intentionally failing soldiers.

The story of the Army's efforts to address traumatic brain injury between 2006 and 2011 is thus the story of the tension between scientific knowledge and popular assumptions, between the pace of research and the immediate demands of war, and of the competing ways that different parts of the Army solve problems. It reveals that, as much as the different parts of the Army are committed to the same goals, they have different visions of how to achieve them, which sometimes produces conflict when they collide. All of these tensions would play out in another area of behavioral health after 2007, as Army leaders realized that the active-duty suicide rate had crept above the national average.

7

"Leaders Can Once Again Determine the Kind of Culture the Army Is Building"

Active-Duty Suicide and Anxiety over Army Culture

As a two-tour veteran of the Iraq War who had been diagnosed with PTSD, Doug Hale could probably have been forgiven if he had chosen to forgo a fireworks display. The notion that combat veterans are troubled by such noises have become something of a trope in U.S. culture, and it is a phenomenon that the Army's BATTLEMIND training program warns veterans to beware of.[1] Nonetheless, the twenty-six-year-old seemed at ease as he spent July 4, 2010, with his family and girlfriend, who later recalled "a huge smile on his face as they set off fireworks."[2] Hale's apparent happiness made all the more shocking his decision two days later to send his mother a text message telling her that he loved her, that he was sorry, and that "I hope u and the family and god can forgive me" before shooting himself in a restaurant men's room just outside Fort Hood's gates.[3]

By 2010, deaths like Doug Hale's were increasingly common. He was one of twenty-two soldiers who took their own lives at Fort Hood that year. During the last week of September alone, four soldiers—all veterans of Operation Iraqi Freedom or Operation Enduring Freedom—committed suicide on post.[4] These men and women were among the 156 active-duty suicides that Army experienced in 2010. These numbers were marginally better than those of 2009, when there were 160 suicides and 1,713 suicide attempts.[5] But, as they had been since 2007, soldiers were killing themselves more often than their civilian counterparts. Ten days after Hale's suicide, *USA Today* reported that "soldiers killed themselves at the rate of one per day in June making it the worst month on record."[6] For many Americans, active-duty suicide was further evidence of the trauma that the unpopular wars were visiting on the men and women who had volunteered to fight them. Newspapers regularly published profiles of soldiers like Hale, men and women haunted by their wartime experiences and who took their lives, often after receiving little help from the Army.

Inside the Army, though, the *Washington Post*'s conclusion that the rising suicide rate "suggest[ed] that, after nine years of combat, the Army is showing some serious signs of strain" was resonant.[7] That more soldiers were taking their own lives raised questions about whether the wars were damaging the institution itself. Peter Chiarelli, George Casey, and Eric Schoomaker were among a generation of leaders who had spent their career rebuilding the institution from its post-Vietnam nadir into one of the most respected institutions in the country, an organization that advertised itself as not only defending the country but improving the men and women who served.[8] Amid relentless deployments and the tactical demands of two prolonged wars, though, it now seemed that the Army that had long prided itself on caring for its people was to struggling to understand and meet soldiers' needs, and soldiers were killing themselves as a result.

In 2008, Army Secretary Pete Geren and Casey tasked Chiarelli with developing a plan to lower the Army's suicide rate.[9] The Vice Chief's Health Promotion, Risk Reduction, Suicide Prevention Task Force and his suicide council embarked on a sometimes contentious process of studying why soldiers took their lives and how suicides could be prevented. Suicide prevention efforts became a forum in which leaders, medical officers, and soldiers debated what kind of institution the Army was, how the prolonged wars had changed it, and what could be done to recuperate it. If these debates, like those that surrounded TBI, often pitted the Army Medical Department's deliberative approach against the action-oriented Vice Chief's office and Chiarelli himself against the larger army, efforts to prevent soldier suicide also created spaces for critiques—some subtle and some less so—of the wars themselves.

In 2007, when the Army's suicide rate passed the demographically corrected civilian rate, its suicide prevention program was nearly a quarter century old.[10] Dave Marlowe, the godfather of the Mental Health Assessment Teams (MHAT), had been working on suicide prevention since the 1980s, but a key moment came in 1984 at the direction of then Chief of Staff Gen. John A. Wickham.[11] Wickham believed that divine intervention had saved his life in Vietnam and that his survival was an injunction to make Christian moral principles central to Army life.[12] Wickham empowered the chaplain corps, and suicide prevention became one of its responsibilities. Because soldiers tended to take their lives less frequently than other Americans, though, the program was never at the forefront of Army leaders' concerns. On the eve of the September 11 attacks, Army writings about suicide prevention training emphasized the chaplains' role and often counseled little more than sending a troubled soldier for spiritual consultation.[13]

The delegation of suicide prevention to the chaplaincy raised few questions inside the Army, an organization heavily influenced by evangelical conservatism. A survey of more than three hundred troops by John J. South, a reservist chaplain completing his strategic studies project at the Army War College in Carlisle, Pennsylvania, found that 90 percent of soldiers wanted chaplains to take a greater role in counseling and that the same number felt that religious belief was a powerful factor in preventing suicide.[14] Suicide prevention, South concluded, offered "a great opportunity to mentor and disciple soldiers who obviously want and believe in God's direction and encouragement."[15] For most soldiers, chaplains were their only exposure to suicide prevention. Gale Pollock recalls, for example, that the Army's concern about suicide "generated a number of training programs, but it seemed to be more delegated to the chaplains. It wasn't something that I saw . . . the Army Medical Department investing very much time or energy in."[16]

As late as 2007, the new Army Suicide Prevention Resource Manual, which had been updated for the first time since the wars began, emphasized spirituality even as the Army acknowledged both the rising suicide rate and soldiers' increasing dissatisfaction with chaplains' suicide prevention work.[17] Soldiers learned, for example, that "connectivity to the Divine is fundamental to developing resiliency that allows one to deal with disappointments"; trainers were told to "emphasize the importance of spiritual health, connectivity with a faith community, and a relationship with God"; and the training included a video featuring former Pittsburgh Steelers quarterback Terry Bradshaw, in which he was "very open about his faith in God and his relationship with his church" and that illustrated that "spirituality is an invaluable ingredient in his battle with [depression]."[18] Chaplains conducting the training were expected to "openly advocate behavioral health as a resource," just as mental health professionals were told to "openly advocate spirituality and religiosity as resiliency factors."[19]

Beyond this, the training did not do much to suggest how soldiers could keep from becoming suicidal or how leaders could help them. The manual suggested little in the way of creating a positive mental health climate and instead taught soldiers and officers to recognize depression and encourage help-seeking behavior and to intervene when a soldier was on the verge of suicide.[20] Perhaps most surprising, almost no consideration was given to what impact the wars might have been having on the troops.

This manual appeared just as the suicide rate was rising above civilian levels. Historically, the Army had had fewer suicides that the rest of U.S. society, and that mattered deeply inside the Army.[21] This figure began to shift in 2004, with a steady upward trajectory from fewer than ten suicides per one hundred thousand individuals to over twenty by 2008.[22] That rise caught the attention of the *Washington Post*, which in January 2008 reported on "an alarming phenomenon in the army's ranks: Suicides among active-duty soldiers in 2007 reached the highest level since the Army began keeping such records in 1980."[23] The story featured Lt. Elizabeth Whiteside, an Army reservist who had made two suicide attempts, one in Iraq and a second at Walter Reed. Whiteside, who was facing a court-martial for charges stemming from her first attempt, was for *Post* writer Dana Priest emblematic of the consequences of an "Army [that] was unprepared for the high number of suicides and cases of post-traumatic stress disorder among its troops, as the wars in Iraq and Afghanistan have continued far longer than anticipated", and noted that the Army had found that "the current Army Suicide Prevention Program was not originally designed for a combat/deployment environment." Even worse, the institution was unsympathetic; Whiteside's own commander judged that she was using her psychological issues as "'an excuse' for her actions."[24]

Priest's article was one of the first to acknowledge that the Army's suicide rate had crept higher than the civilian average. Over the next two years, other articles as well as congressional attention would join it to construct a narrative that linked active-duty suicide to the wars, the lengthy deployments that they required, and the trauma that they produced. Indeed, the day that the Whiteside article was published, Patty

Murray (D-WA) took it to the Senate floor to make precisely this point: "Our great service-members who face deployment after deployment without the rest, recovery, and treatment they need are at the breaking point," she began, adding that "many of them have seen their best friends killed. They have seen other untold horrors. Yet we still are expecting them to head back to the battlefield to perform unaffected by what they have seen or gone through."[25] For the senator, active-duty suicide was further evidence of the Bush administration's failures. The president, she inveighed, "needs to lead by example and show he understands what these never-ending deployments are doing to our troops."[26]

Murray was hardly alone in connecting the wars' unconscionable length to the troops' declining mental health and increasing suicidality. In April, a *Providence Journal* editorial began by noting that "suicide rates in the Army have spiked alarmingly, suggesting that U.S. deployment strategies urgently need reappraisal" before adding that, "stretched thin by the conflicts in Iraq and Afghanistan, the Army has found itself unprepared to deal with the stress on individual soldiers."[27] Five days later, Rep. Debbie Wasserman Schultz (D-FL) critiqued the rising rate as "an astronomical jump"; asked, "When do we say that we care about these troops as people, not as fighters, not as defenders of America, but as people? And when do we recognize that there is a limit to their ability to hold down their lives and to be able to return to a quality of life that they had before they left?"; and criticized "the refusal of this administration to recognize that there is a cost and a toll that is being taken on these families, [and] on the individual troops."[28]

On the surface, these critiques made sense. By 2008, anti-war sentiment rested in part on ample evidence that the wars had deleterious effects on the troops' mental health, and pointing to high the incidence of PTSD, TBI, and suicide allowed the war's critics to critique that war and its policies while remaining insulated from charges that they were not supporting the troops. And yet, these assertions made the connection a bit too quickly and cleanly; they were, to use the scholar Rob Nixon's terminology, "in thrall to speed and spectacle."[29] Writing about the environmental destruction of war, he argues that much media coverage "lacks the attention span to follow war-inflicted catastrophes that take years or generations to exact their toll" and, thus, fails to fully appreciate "major shifts in the ways that contemporary wars kill."[30] He urges attention to what he calls "slow violence[,] . . . a violence that occurs gradually and out of sight, a violence of delayed destruction that is dispersed across time and space, an attritional violence."[31] This argument about environmental destruction also applies to active-duty suicide. Although the notion was that soldiers who had deployed much and had been traumatized were likely to take their lives, it turned out that soldier suicide constituted a form of slow violence. The rising suicide rate was, Army researchers concluded, connected to the wars, but it was not a direct result of deployments and trauma. It was, rather, the result of broader structural changes that the wars had wrought on the Army.

* * *

Indeed, that nearly two-thirds of soldiers who took their lives had deployed to Iraq or Afghanistan was something of a red herring. By 2007, it was a rare soldier who had been in the army for any length of time and had *not* deployed. What was more significant, and surprising, was that soldiers who had been in combat accounted for fewer than a quarter of the suicides, and soldiers who had deployed more than once accounted for only 10 percent.[32] Thus, as Charles Hoge pointed out in a 2008 Senate Armed Services Committee hearing, while deployments and dwell time did affect mental health overall, and that mental health issues in turn contributed to suicide, "we haven't been able to make a direct link" between deployment and suicide.[33]

Hoge was probably referring to a review of all of the Army Suicide Event Reports that Elspeth "Cam" Ritchie, then the Surgeon General's behavioral health consultant, had completed in January, released internally in March, and made public in May.[34] The *Army Suicide Event Report* was a compilation of the data gathered after any suicide attempt, completed suicide, or incident of severe suicidal ideation, which were reported on individual forms of the same name. Ritchie's report tallied 108 suicides across the Army and 166 attempts in Iraq and Afghanistan, and it revealed the shortcomings of the Army's suicide prevention program.[35] Three-quarters of the soldiers who took their own lives had not given any indication that they were contemplating suicide, and in fewer than half of the cases was there a clear reason why a soldier had chosen that path.[36] A similar percentage had been diagnosed with a mental health issue.[37] Perhaps more troubling, "trends suggest that Soldiers who complete suicide in OEF [Operation Enduring Freedom]-OIF [Operation Iraqi Freedom] may see Chaplains prior to the event more frequently than non-OIF-OEF soldiers who completed suicide" and nearly half of soldiers who took their own lives "had been seen in at least one of the programs/clinics within 30 days of the event," data that suggested that a suicide prevention plan that emphasized the role of the Chaplain corps and that focused on crisis management was insufficient.[38]

The report's data did little to change the conversation about the causes of suicide. In *USA Today*, Sen. Patty Murray again pointed to repeated deployments as the primary cause of suicide.[39] So did the *New York Times*, which, despite noting that "26 percent [of suicides] had never been sent to either conflict," quoted Ritchie's remark at a news conference that suicide was "mainly [caused by] the longtime and multiple deployments away from home, exposure to really terrifying and horrifying things, the easy availability of loaded weapons, and a force that is very, very busy right now."[40] An Alabama newspaper called the report "a statistical gut-punch to the outgoing Bush administration's war-making decisions," and railed, "That American soldiers are killing themselves at record rates says a great deal about this nation's decision to wage year after frustrating year of war, especially in Iraq. It's clear that the strain on America's fighting men and women is undeniable, and unbearable for far too many."[41] In Massachusetts, the conclusion was that "officials need to get out of denial and deal with this preventable act."[42]

The connection between combat trauma, deployment, and suicide remained a frequent emphasis throughout 2008 and 2009. Most common were portrayals of soldiers who had been traumatized by their deployments. The *St. Paul Pioneer Press*, for example, based an article on suicide entirely on a profile of Benjamin Miller. This was another story of a promising life undone by the trauma of combat. Miller "became a man" in the military and "had a sense of direction," but combat's violence was his undoing, as the article's first paragraph makes clear: "The charming and outgoing Army Sgt. Benjamin Miller seemed different when he came home to Minnesota this month on leave from Iraq. . . . Family members say he seemed frustrated and detached. He spoke of cleaning up the carnage of suicide bombings and the unmistakable smell of death" before "he just snapped" and shot himself.[43] The *Akron Beacon Journal* portrayed Derek Hendon in much the same terms, attributing the suicide of a man who remained his ex-wife's best friend to his having been "changed by the war."[44]

Portrayals offering vivid detail about the lives and deaths of soldiers who took their own lives because of combat-related PTSD continued to dominate coverage of soldier suicide in 2009, another year that began with the announcement that the Army suicide rate had again broken records.[45] *USA Today* began the year with the account of Josh Barber, an Army cook who "took a job . . . that was devoted to serving and feeding other soldiers" but was pressed into service as a machine gunner and witnessed a suicide bombing that killed almost twenty-five soldiers.[46] Like previous accounts, it is not only the violence that he witnessed that is rendered in chilling detail but his death as well; the article described Barber as arriving at Fort Lewis, Washington, heavily armed but that he "took only one life that day. He killed himself with a shot to the head."[47]

Other articles didn't present veteran suicide victims as potential dangers, but they did construct them in detailed, empathic terms. Ivan Lopez, a Pennsylvania National Guardsman who "was always a happy-go-lucky person" but "never came back the same" hanged himself in February.[48] In May, Roy Brooks Mason shot himself in his car in San Jose, California, after he "called emergency dispatchers . . . and told a dispatcher a dead body would be in a red Chevrolet Cobalt there, . . . [and he] asked that someone 'clean up the area' before children see anything amiss."[49] In August, the *New York Times* described the death of Jacob Blaylock, who had had a troubled childhood before witnessing a roadside bombing that killed two of his colleagues in Iraq: he became "heavily intoxicated" in December 2007 and "lifted a 9-millimeter handgun to his head during an argument with his girlfriend and pulled the trigger."[50] On Memorial Day weekend, the *Washington Post* ran a story about "Marine Gunnery Sgt. James F. Gallagher, a 19-year veteran, [who] returned in 2006 from a combat deployment to Iraq in which his unit lost a dozen members. Back at Camp Pendleton, Calif., Gallagher hanged himself in the garage of their home, where was found by [his wife] Mary and their daughter Erin, then 12."[51]

Articles like these offered detailed accounts of promising lives undone by a traumatic deployment to Iraq, and these affective, empathic stories establish these deaths as the most significant suicides of the war. The Army, however, was finding that deployment and suicide didn't correlate. Some articles made this point. A few articles emphasized, as one chaplain quoted in the *Salt Lake Tribune* explained, that "the trend we've seen . . . is that these suicides are relationship-based" in that, "when you take people who may already be on the edge of crisis—maybe they have family relationship issues, divorce, health problems—and then you add a deployment on top of that . . . you see people who may have been teetering before are now pushed into more of a crisis."[52] In all of 2008, only one article—in the relatively minor *El Paso Times*—details such a suicide, and then all but anonymously.[53]

As 2009 promised an even higher suicide rate, the *Christian Science Monitor* noted that "there is not yet a clear cause-and-effect relationship between suicides and deployments to a war zone. At Fort Campbell in Kentucky, for example, seven of the 18 confirmed suicides this year involved soldiers who had never deployed. Eight of the soldiers had been deployed once, two had been deployed twice, and one had been deployed four times." The *Monitor* quoted Chiarelli's comment, "As I look across all the factors, from the number of deployments individual brigade combat teams have gone through, to everything else, I cannot find a causal link that links anything."[54] A widely reprinted wire story quoted Chiarelli's comment that "I've scrubbed the numbers ever way I possibly can. . . . I cannot find a causal link."[55]

The juxtaposition of intimate portraits of some soldiers and the mere mention of others and the anecdotal stories of wartime trauma leading to suicide alongside Chiarelli's comments that the statistics didn't bear that correlation out illustrate how media coverage of active-duty suicide remained, to use Nixon's term, "in thrall to speed and spectacle": Stories of a decent person driven to suicide by the horrors of war made sense and were compelling in their dreadfulness. From a public health standpoint, however, such stories were problematic because they obscured attention to the broader patterns of violence that the wars were producing: their slow violence.

* * *

Hoge's comments before the Senate, Ritchie's report, and Chiarelli's public statements all demonstrate that the Army quickly understood that the problem was more complex than anecdotal stories suggested and thus precluded a simple solution. As Eric Schoomaker recalls, "The problem that we all were concerned about is, What does it say about the whole force? What is under the surface of the iceberg, the point of which is suicide, but underneath that is a lot of growing, suffering, and dysfunction that we can't even see?"[56]

Answering this question required a shift in how the Army thought about suicide. Although the Army's more nuanced understanding of suicide emerged rhizomatically, one key moment in this evolution came when Chiarelli requested a briefing on suicide prevention from Elspeth Ritchie and the brigadier general who was in charge of suicide prevention and realized with frustration that the program essentially amounted to a single PowerPoint presentation.[57] In Ritchie's estimation, that moment marked the beginning of Chiarelli's commitment to reducing the suicide rate.[58]

In March 2008, Chiarelli formed the Suicide Prevention Task Force. Within weeks, it was renamed the Health Promotion, Risk Reduction, Suicide Prevention Task Force, a gesture that Schoomaker and others applauded as it reflected the Army's growing awareness of the problem's complicated dimensions.[59] Its organization explicitly denied the pigeonholing of suicide as a chaplain's issue or even an Army Medical Department (AMEDD) issue and enabled the task force to look across the Army at the range of factors that might provide insight into soldier suicide. Chiarelli and others realized that medical information alone was insufficient to understand and prevent suicide, and the AMEDD lacked the authority to access information from across the entire force.[60] Thus, while the Surgeon General's office contributed people and information, the task force was explicitly designed to bring together diverse components of the Army, each of which had its own information stream, responsibilities, and authority.[61] This organization was partly in response to the recognition that as much as the Medical Department could develop a set of recommendations on suicide prevention—which, in 2008, Ritchie had—it couldn't order their implementation across the Army.[62] But Chiarelli could.[63]

The extent to which the task force brought suicide prevention outside of the Medical Department was evident in Chiarelli's choice of Brig. Gen. Colleen McGuire, a military police officer who, among other commands, had run the military prison at Fort Leavenworth as its leader.[64] In a sense, she was an unlikely choice. "I'm not a medical officer. I'm not a counselor. I'm not a chaplain," she explains, and unlike many commanders, she'd never experienced a suicide in one of her units.[65] But as a military police officer, she was well versed in how the Army investigated them.[66]

McGuire decided that the task force should include representatives from across the Army: people from human resources with access to personnel files; from the military police who could review criminal histories; from the medical command who could determine what kinds of physical, psychological, or substance abuse treatment soldiers had been receiving; from the chaplaincy; and from the Guard and Reserve.[67] What that amounted to, McGuire explains, was having "the permission of the Vice Chief of Staff to look in everyone's closet."[68] The primary representative from the Office of the Surgeon General was Bruce Shahbaz, a retired lieutenant colonel who had served as a medical service corps officer for two decades and earned a reputation for thriving in complicated situations and solving challenging problems.[69]

The task force quickly determined that media coverage and political rhetoric that emphasized a direct link between deployment, trauma, and suicide oversimplified a more complex relationship between the wars and suicide. In fact, no factor had sufficient explanatory power.[70] Deployments, as Shahbaz put it, "may have been no more contributory to the suicide death than the fact that they had brown hair. It was just something that happened over the course of their military career."[71] The same was true for combat exposure; as Charles Hoge explained, it was easy, but inaccurate, to assume that combat exposure was related to suicide because "the rising suicides were really only seen in the Army and Marine Corps, not in the Air Force and Navy. So the increase of suicides really only affected ground combat forces." This seemed revealing, but it turned out that, "within those ground combat forces, the rise of suicide was proportionally the same whether or not individuals had deployed."[72] When the task force looked at the records of those who had taken their lives, Eric Schoomaker recalls, it found that "we had plenty of people who had fought in worse circumstances that didn't commit suicide."[73] And even though soldiers who committed suicide had often had endured the end of a relationship, that, in and of itself, didn't indicate that a person might be at risk for suicide. "In America, half of the marriages fail," Shahbaz explains, "but we don't have a 50 percent suicide rate."[74]

At times, the task force's increasingly complicated understanding of the issue frustrated Chiarelli. Despite clearly appreciating the complicated nature of suicide, he was skeptical that deployments weren't a primary driver of suicide and eager to implement a solution, and he sometimes became aggravated by the task force's nuanced answers. Shahbaz recalls a briefing in which the Vice Chief was "approaching it from a tanker, armor officer perspective of, 'Stop trying to confuse the issue. Does it or doesn't it?,'" leaving the retired lieutenant colonel to tell a four-star general, "Boss, it isn't as easy an answer as that."[75]

It was not that Chiarelli was unwilling to embrace a complicated answer. He understood quite well that it was not war-related trauma so much as the generalized strain that war placed on the force had created the conditions under which a host of personal problems that contributed to suicide manifested. Before the Senate Armed Services Committee in March 2009, for example, he testified that suicides were the result of "a combination of factors" for which there was "no simple solution," but he also told Sen. Lindsay Graham (R-SC) that

> you can look at the numbers and try to make yourself feel like it's not totally dependent on that stress, by looking and saying that one-third of those individuals don't have any deployment history at all. . . . But I just don't think that's the case. I think it's a cumulative effect of deployments that can run from 12 months to 15 months. . . . If you were to ask me to identify one thing that has caused the spike, that is, in fact, it.[76]

It was a point Chiarelli would reiterate in July, when he appeared before the House Armed Services Committee.[77] In Chiarelli's estimation, the links between failed relationships and suicides were attributable to the short dwell time that soldiers and families endured. Discussing his visit to six Army posts in the late spring of 2009, he told the panel, "Spouses were telling me their husbands were not reintegrating with the family. They just realized that that was too hard to do in the short period of time they had, and they would back off from the family, which creates the relationship problem."[78] He likewise told me that "I've always believed there was a linkage there. I think it's absolutely ridiculous to have a benchmark in 2001 that just so happens to coincide with when we got into these wars. It doubles by 2008, and you're going to say that had nothing to do with it? The fact that you were sending people for twelve months and bringing them home for twelve months?"[79] Indeed, there was agreement throughout the Army that the wars themselves were somehow responsible for the increase. "I think it would be naïve to think it wasn't," Schoomaker reflects. "We were doing the majority of the fighting." But it was not simply fighting, Schoomaker explains: "Our families were subjected to the majority of the issues, we knew there were other factors like deployment length and frequency and especially dwell. . . . We knew that this was lubricated by alcohol and other drugs. We knew that three-quarters of the time when we went back and looked at it, it was related to the loss of a close relationship"; what they didn't know, however, was "what the cause and effect was."[80]

The task force's most important suicide prevention initiative, the 2010 *Health Promotion, Risk Reduction, Suicide Prevention Handbook*, or the Red Book, reflected the awareness that if it was next to impossible to identify a single cause, suicide itself was rarely the first sign of trouble. Soldiers had often had other issues—medical, disciplinary, legal, financial, or familial—that preceded a suicide attempt, and the handbook concluded that addressing those issues early on could prevent soldiers from spiraling into suicidal ideation.

The Red Book presented this trajectory as a maze with suicide at the center, surrounded by increasingly severe levels of what it termed "high-risk behavior": suicide attempts, substance abuse, crime or misconduct, medical conditions, and even activities like motorcycle racing or unprotected sex. This conclusion emerged from the task force's review of individual suicides, which showed that, when soldiers became addicted to drugs, faced a criminal charge, or suffered a physical injury that threatened their career, their risk for suicide increased.[81] If these factors pushed soldiers toward the center—that is, toward suicide—the opposite was also true: "Escape from the maze will generally require help-seeking behavior and/or leader intervention to arrest the spiral toward the center."[82] To illustrate this, it cited the example of a soldier "stuck in the maze," having hit nearly every possible issue that leads to suicide: a failed relationship, multiple deployments, a PTSD diagnosis, criminal behavior, drug abuse, "financial difficulties," "depression symptoms," and "suicidal ideations daily" who had been "found dead from an apparent self-inflicted gunshot wound to the head."[83] This soldier was clearly an extreme case, and he was likely chosen precisely because his tragic story illustrated the task force's primary conclusion: that there were very few people who suddenly become suicidal and that there is almost always a pattern of troubling behavior that precedes a suicide attempt.

What's notable throughout the Red Book, however, is how little attention it gives to deployment, combat experience, and PTSD as causes of suicidal ideation. The distancing of the war from the suicide epidemic is most explicit in the report's "vignettes," which emphatically emphasize domestic difficulties while nearly ignoring combat stress. For example, one vignette describes "a Staff Sergeant [who] had a hard childhood" and who suffered "a very violent improvised explosive device (IED) Attack" that led to "difficulty sleeping and . . . nightmares" and whose "parents stole his identity and incurred a large debt in his name."[84] While both of these events likely contributed to this soldiers' suicide, the report concludes only that "indicators visible to unit leadership may not identify stressors in personal or Family life," effectively eliminating the mTBI as a factor worthy of consideration.[85] Another vignette describes a soldier who "started drinking more than usual," who "had nightmares about OIF and . . . slept with a gun under his pillow," and who "was depressed about his friends leaving the unit and . . . was not happy about going to Drill Sergeant School" and similarly concludes only that "transitions can be particularly stressful."[86] Other vignettes follow a similar pattern, for example, describing a suicide by a soldier who "had previously deployed twice [and] had no documented behavioral health issues" or where the focus is on relationship and financial stress.[87] Throughout, the Red Book marginalizes what might seem to be obvious combat-related mental health issues while highlighting domestic and quotidian stresses. Only two vignettes could be construed as linking suicide to the effect of combat.[88]

This is not say, however, that the report dismisses the wars as significant factors in soldier suicide. Rather, the report identified a changing Army population and changing standards of leadership as having created a population of soldiers more disposed to suicidal behavior and, in doing so, critiqued the wars as detrimental to the Army as an institution.

This critique was rooted in assumptions about what kind of organization the Army had understood itself to be at the beginning of the twenty-first-century conflicts and what those conflicts had in turn done to that understanding. In the aftermath of the Vietnam War, the Army had put considerable effort into rebuilding itself to the point that it could reasonably promise, to use a phrase from the Army's 2001 Posture Statement, that it was prepared to fulfill "its non-negotiable contract with the American people to fight and win the nation's wars decisively."[89] More than this, though, as Beth Bailey explains, after the Cold War "the army . . . presented itself as a force for good in civilian society. It was a provider of education, of training, of discipline."[90] It was a place, in short, where families and communities could be proud to send their children, a vehicle of what Lauren Berlant terms the "upwardly mobile good-life fantasy" that dominates neoliberal U.S. culture.[91]

The twenty-first-century wars tested each of these assertions. By 2010, it was clear that neither the Iraq nor the Afghanistan War had ended with a "decisive" victory. In addition, the prestige that the Army enjoyed seemed at risk of being diminished as a result of declining enlistments and a set of scandals that included the torture of prisoners at Abu Ghraib, the killing of Afghan civilians by a self-fashioned Army "kill team," the cover-up of Pat Tillman's death in Afghanistan, and the horrid conditions at Walter Reed Army Medical Center.[92] The task force was certainly aware of the damage that such events had done; the 2012 sequel to the Red Book, *Army 2020: Generating Health and Discipline in the Force ahead of the Strategic Reset* (the Gold Book), asserted that the actions of Calvin Gibbs, Nadal Hassan, and John M. Russell had "significantly eroded the Army's reputation."[93] For individuals who had been in the Army through this rebuilding period and who loved the institution, the Iraq War thus must have seemed, to use Berlant's phrasing, "a moment of extended crisis, with one happening piling on another," of which the increasing suicide rate was another example.[94]

The Red Book's explanation for why the suicide rate was increasing reflected the anxieties about the wars' impact on the institution. First, it argued that their duration and nature had changed recruitment and the shape of the force. Soldiers who enlisted in the midst of war, the report argued, were "a population inherently attracted to risk" in that they "enlist fully knowing that they will serve in a combat zone," and "soldiers who deploy may even be more comfortable accepting high levels of risk and uncertainty in their lives."[95] This "inherent attraction," in the Army's view, also occurred outside of deployments, in the form of substance abuse, sex, fiscal irresponsibility, and so on, that could set a soldier on the path to suicide.

The report and its 2012 successor also argued that the Army was taking in less-qualified soldiers who were potentially more inclined to substance abuse and crime. Indeed, as the wars went on, the Army faced a recruiting shortfall that it made up for, in part, by enlisting recruits who would previously not have met its standards.[96] Exactly how those soldiers fare in the force has been the subject of some debate, and in 2014, the Army STARRS study (Army Study to Assess Risk and Resilience in Servicemembers) reported that these soldiers were not particularly disposed to suicide.[97] The Red Book, however, argued that "we are retaining sub-standard trainees in the generating force and moving them quickly into the operating force," while "decreases in separation actions combined with current crime and alcohol/drug rates indicate an increase in tolerance, if not acceptance, of high risk behavior."[98] Two years later, the Gold Book maintained that crime committed by soldiers "impacts Army and unit readiness in a variety of ways," including "the cost to the Army's reputation and the sacred trust we owe the nation."[99] The presence of these soldiers was thus an indication that the wars had damaged the Army.

Second, this shifting enlisted population had occurred alongside a shift in leadership. That soldiers were engaging in these behaviors was one thing; that leaders were allowing them to do so was quite another. Both reports focused on failures in leadership as they described why this was so, asserting that senior NCOs and officers had lost or never developed the capacity to lead and mentor soldiers. This claim generated some criticism both inside and outside the Army when the Red Book was published—the *New York Times* reported that "the report put a large part of the blame on commanders who either failed to recognize or disregarded high-risk behavior among their troops"—but in fact the report is less a condemnation of leaders then the circumstances under which they were forced to lead.[100] The report is perhaps best read as a critique of the wars themselves and the damage that they had done to the Army.

In the post-Vietnam Army that prided itself on producing good citizens, maintaining good order and discipline was not only paramount but a primary measure of a leader's effectiveness. As Lt. Gen. Raymond

Palumbo, who when the Red Book was published was commanding U.S. Army Alaska, recalls, "I was in the Army nine years before we went to Desert Shield and Desert Storm, and we stayed for less than a year. . . . So I was brought up on garrison training." What that meant in practice was that promotions were dependent on things like "how your equipment was maintained, [or] how many DUIs you had" in your command.[101] Rebecca Porter, who by 2010 had become the Surgeon General's behavioral health consultant, had begun her career as a military police officer in Germany before returning to graduate school; she agrees: "When I was a platoon leader in the eighties," she recalls, "it was expected that I would go . . . and make sure that [my soldiers'] living conditions were satisfactory."[102] Part of the Army ethos, she explains, is that "all those things that seem so inconsequential and stupid actually have great importance. Shining your shoes . . . is important because reflects discipline, it reflects that you're taking care of your equipment, that you are paying attention to detail."[103] Colleen McGuire, who had spent her entire career in the Army's law enforcement system, likewise remembers being held accountable for disciplining her subordinates and that "when I was doing this book, I was appalled at the number of NCOs who were still in the Army after being charged with Article 15's"—the Army's term for a command-issued punishment.[104] As a commander, she had disciplined her soldiers much more rigorously. "I did what was required to ensure that we had quality soldiers in the Army," she explains, and "it was apparent that many leaders did not enforce these standards."[105]

The Red Book repeatedly argues that those institutional capacities have been lost because of the unique demands that the Iraq and Afghanistan Wars placed on the Army. Chiarelli's opening letter in the Red Book, for example, explains that "we now must face the unintended consequences of leading an expeditionary Army that included involuntary enlistment extensions, accelerated promotions, extended deployment rotations, reduced dwell time and potentially diverted focus from leading and caring for Soldiers in the post, camp, and station environment."[106] The report's central chapter, "The Lost Art of Leadership in Garrison," took this point further, critiquing not the leaders themselves but the conditions that had defined their careers. "Today's operational tempo . . . has eroded the technical skills and experiential knowledge needed to lead and manage effectively in the garrison environment," the report explained, adding that "prolonged, recurring combat rotational requirements have resulted in young and mid-level leaders whose only command experience is meeting the demands of the deployment-to-combat-to-redeployment cycle."[107] In perhaps the report's most damning passage, Shahbaz and his colleagues wrote that "while our commanders and subordinate leaders are phenomenal warriors, they are unaccustomed to taking care of Soldiers in a garrison environment. Value of and appreciation for good order and discipline practices . . . have been lost." The problem, they added, "appears to be worsening as the requirements of prolonged conflict slowly erode the essential attributes that have defined the Army for generations."[108]

Porter, now in her role as the Chief of the Behavioral Health Division, had been asked to proofread the Red Book, and she thought it was "just a phenomenal piece of work."[109] In particular, the claims that leadership failures largely resulted from the demands of the wars resonated with her. "If you've never been in a garrison environment where those kinds of leader behaviors were expected," she explains, "it just feels like a bridge too far when you are exhausted and you don't realize the importance of it."[110] Palumbo agrees, maintaining that twenty-first-century junior officers had the attitude of "'I don't care about [garrison leadership]. I care about kicking the Taliban or Al Qaeda's ass' . . . and that's what they were getting rated on, frankly." This was not, however, a criticism of the new generation of officers; as Palumbo put it, they couldn't have understood the significance of those kinds of leadership skills because "we didn't train them to be . . . garrison commanders."[111] From McGuire's perspective, the Army had been trying to "conduct business as usual and fighting a two-front war. It really was too much for people."[112]

These passages and comments reveal a sense that because of the wars the Army had failed to maintain institutional capacities that were central to the post–Cold War Army's self-fashioning as a critical and effective societal institution. Crucially, however, these concerns amount less to a critique of the current generation of Army leadership than to a critique of the circumstances under which they have been asked to lead. The argument for why the suicide rate had gone up was, in fact, a critique of the wars themselves.

The task force thus imagined that stopping suicide meant a return to an Army populated by disciplined, committed soldiers and officers who were competent managers, and the Army's suicide prevention handbooks are jeremiads that call on Army leaders to recuperate the Army's pre-2001 culture first by removing problematic soldiers and second by returning to earlier standards of leadership. "When standards slip, we begin to see dangerous trends in behavior," the Red Book explains, again imagining once-high sets of expectations that have eroded during the war. "However, by enforcing policy in a fair, judicious, and equitable manner, leaders can once again determine what kind of culture the Army is building."[113] If this passage imagines leaders as having been essentially powerless over the direction of the military in recent years, it is also a call for them to recapture their earlier capacities by becoming more proactive disciplinarians and mentors.

The reports thus counsel removing problematic soldiers from the force and prescribe corporate models of efficiency, effectiveness, and professionalism. In 2010, Chiarelli was adamant that "we can't use these people up, have them develop a problem and then throw them away and not take care of them. There is no way. I can't be part of an organization like that. . . . Part of the reason they're having the problem is the situation we put them into."[114] This was in itself an argument about the Army's exceptional nature, and an admirable claim. But in his opening letter in the Red Book, the Vice Chief was equally adamant that "we must ensure that Soldiers who cannot adapt to the rigor and stress of this profession find sanctuary elsewhere for their own wellbeing and for that of the force."[115] Encouraging leaders to "understand when to mentor Soldiers and when to accept that they will not meet Army standards," he asserts that "these high risk individuals pose an unacceptable level of risk to themselves and the Army."[116] Later, the Red Book suggests that, while "on the one hand, there is opportunity to mentor and shape the career of a young Soldier[,] on the other, there is the need to remove a problem before it can take root and undermine the entire unit. Arguably the most important tool the leader has . . . is separation authority."[117]

The passages that emphasize removing those who are "a danger to the Army" or who might "undermine the entire unit" reveal that recuperating Army culture is a central project of the suicide prevention program. If at times that mission is implicit, there are moments when the authors seemingly call for ensuring the exceptionalism of the Army over meeting its public mandate. Arguing that "a larger number of individual Soldiers, who would otherwise have not been added to the ranks, have been retained due to the protracted conflict," the Red Book asserts that, "at some point, the decision must be made either to accept the erosion of Army standards (entry and retentions) for the sake of manpower or enforce regulations to maintain order and discipline."[118] It contends that, "by not fully complying with established policy, commanders are in fact communicating to their troops that Army standards of conduct are less important than the scope of the overall mission,"[119] a striking statement in an organization whose most explicit mission is "fighting and winning the nation's wars."[120]

The call for increased professionalism to recapture military exceptionalism became more strident in the Gold Book, which states that the Army "has an opportunity to select and retain professional soldiers to fill its ranks . . . [and] to deselect and separate those Soldiers who do not meet the professional standards of conduct required of an all-volunteer force."[121] More problematically, however, the Gold Book argues that retaining problematic soldiers amounts to something akin to a poor business decision because "the cost of behavioral health conditions is not restricted to financial expenditures. It also reflects loss of time and productivity" and because "caring for and properly disciplining these Soldiers consumes a significant portion of leaders' time. These Soldiers also end up costing the Army a great deal of money."[122] In these passages, the Gold Book seems less a call for a return to an Army that mentored young people to create good citizens and more reflective of an institution that has embraced neoliberal models of efficiency.[123]

* * *

In spite of the its neoliberal outlook on soldier behavior, the Gold Book also claims that the Army's future as an exceptional organization demands a repudiation of the cultural shifts that the long wars have necessitated. Indeed, the suicide prevention program's emphasis on the disciplinary, particularly in the Gold Book, coincided with increasing concern that misconduct had become a convenient way to discharge problematic soldiers without regard for their mental health. In 2012, the *Fayetteville Observer*—Fort Bragg's hometown newspaper—reported on the struggle of a depressed soldier to receive an honorable discharge after he'd been charged with misconduct.[124] In May 2013, the *Colorado*

Springs Gazette published the story of Jerrald Jensen, a sergeant who received an other-than-honorable discharge after a positive drug test, explaining that, "after a decade of war, the Army is discharging more soldiers for misconduct every year. The number kicked out Army-wide annually has increased 60 percent since 2006" and that "some of the discharged have invisible wounds of traumatic brain injury and post-traumatic stress disorder but are kicked out anyway."[125] The article went on to detail Warrior Transition Units at Fort Carson and other posts as more focused on discipline than on rehabilitation, and Fort Carson's commander came off as less than sympathetic: "You are still a soldier until you take the uniform off," Maj. Gen. Joseph Anderson said. "So you cut your hair, you don't smoke pot, you take care of yourself, you don't tell people to F off, you don't get DUIs, you don't go smoke spice."[126] Two years later, National Public Radio reported that "the U.S. Army has kicked out more than 22,000 soldiers since 2009 for 'misconduct,' after they returned from Iraq and Afghanistan and were diagnosed with mental health disorders and traumatic brain injuries."[127]

Correlation is not causation. A soldier with PTSD or mTBI does not necessarily engage in misconduct because of that condition. However, these increasing numbers suggest that, at least for some commanders and at some installations, the Red Book's lauding of "separation authority" and the Gold Book's call to "deselect and separate" problematic soldiers had trumped earlier advice like that found in a Walter Reed memorandum's post–Gulf War explanation that "in the post-combat situation the good soldier who suddenly starts getting into trouble has not become a bad soldier—he may be signaling a possible emotional problem."[128] The juxtaposition of these documents is illustrative of just how much the Gold Book in particular sought to recuperate the Army's reputation and culture while rejecting the aspects of pre-war culture that emphasized compassionate mentorship and instead embraced a neoliberal vision of corporate efficiency and that the emphasis on restoring "good order and discipline in the force" led in some cases to a shift from compassionate leadership to something more punitive.

These articles garnered congressional attention, and an amendment to the 2015 National Defense Authorization Act required the comptroller general to study "the impact of mental and physical trauma relating to Post Traumatic Stress Disorder (PTSD), Traumatic Brain Injury (TBI), behavioral health matters not related to Post Traumatic Stress Disorder, and other neurological combat traumas . . . on the discharge of members of the Armed Forces from the Armed Forces for misconduct," and it also mandated that soldiers who had been discharged for misconduct receive a more thorough review.[129] By 2016, congressional pressure had forced the Army to better document its processes when discharging troops for misconduct.[130] Still, a year later, the Government Accountability Office found that "a majority of troops discharged from the military for misconduct between 2011 and 2015 suffered from post-traumatic stress disorder, traumatic brain injury, or other mental illness."[131]

At the same time, however, the Army's call for a renewed focus on leadership was widely embraced. It was not without its skeptics, of course; George Casey, for example, feared that this emphasis on discipline might inspire micromanagement: "I didn't like the Army of the seventies. That's the Army I came into," he explains. "You were measured on the wrong things because those were what was measurable." Thus, while he embraced the need for leaders to do more, he was adamant about "not going back to a 'chickenshit' Army."[132] Further down the chain of command, however, the idea was enthusiastically embraced. When Chris Ivany was setting up the Embedded Behavioral Health units at Fort Carson, for example, "one of the main points that we tried to drive home was making sure that the junior level leaders knew their soldiers well, and they knew not just about who could shoot well . . . but [also] how many kids does that guy have? Is he married? On the first wife, second wife, third wife? Does he have financial problems? Is he overdrawn in his checking account? Is his mother sick?" The rationale was the same as the Red Book's: "You've got to understand who these people are, where they are from, what are the things that bother them, at the junior leader level so that you can sense when thing start to go wrong."[133]

Jimmie Keenan, who as the commander of Evans Community Army Hospital was Ivany's boss and whose commitment to suicide prevention was inspired in part by her father's 2001 death by suicide, had already taken this advice to heart. "I would talk to [our warriors in transition] and be honest with them and say, 'You know, some of you may think you are a burden, but I'm telling you, you're not. If you feel like you want to harm yourself, here's my number, please call me. Because I care. I don't want you to make that choice and then [have your family] be like I am, where I never got to talk to my dad again.'"[134] And the soldiers did call her, often surprised that a colonel cared enough about them to reach out. Even as Keenan prepared to retire from the Army at the end of 2015 as Deputy Surgeon General and head of the Army Nurse Corps, soldiers she had supervised at Fort Carson still called her when they were in crisis.

Both the Red and Gold Books reveal that suicide prevention was related to larger institutional anxieties that the Iraq and Afghanistan Wars produced. They reveal that the wars threw into crisis the Army's assumptions about what kind of organization it was, what experiences it offered those who served, and its place in the nation. If prior to the wars the Army had prided itself on, and been lauded for, transforming young Americans into fine citizens, the pace and demands of the war had created an organization more focused on immediate tactical needs. While this had produced undesirable outcomes from increased crime to suicide, by the Gold Book's publication it was also evident that the Army's imperiled reputation and place in the culture faced an equal or perhaps more significant danger. In response to these threats, the task force counseled a return to earlier visions of more responsive and empathic leaders who knew and mentored their soldiers while insisting that they meet high standards. These initiatives increasingly placed a heavy an emphasis on discipline, sometimes to the detriment of soldiers with behavioral health issues. All of these efforts, however, reveal that the Army's suicide prevention handbooks were not simply training manuals; they were cultural spaces in which the Army contemplated, and critiqued, the wars' institutional impact and sought to begin a process of recuperation.

* * *

The 2010 *Health Promotion, Risk Reduction, Suicide Prevention* report was well received inside the Army, but other suicide prevention efforts remained controversial. As the task force set about its work, it sometimes found itself in conflict with the Medical Command, which, as it did in the debates over TBI, felt that its expertise was being ignored and that it was unfairly singled out for critique. As well, some senior leaders were uncomfortable with—and sometimes openly resented—the changes that the task force proposed and the pace at which Chiarelli made them. These conflicts reveal another way in which the Army's efforts to address mental health issues during the Iraq and Afghanistan Wars were as much a matter of cultural change as medical innovation.

As the task force got to work in its Pentagon office, it quickly became clear that the Army was not organized to effectively share information about past suicides or at-risk soldiers. McGuire worried that, as in the civilian world, the Armed Forces Medical Examiner was probably underreporting suicides, for example by identifying some suicides as accidents.[135] As well, getting a complete picture of any suicide required input from at least seven different parts of the Army—the Surgeon General's Office, military police, human resources, Army operations, the Public Health Command, the Substance Abuse Program, and the chaplaincy—and that was for active component soldiers. If a soldier was in the National Guard or Army Reserves, additional layers needed to be plumbed. Quite quickly, Bruce Shahbaz and his colleagues on the task force realized that information sharing was not an Army strength and that often those responsible for caring for a soldier in one regard had no idea that a soldier was facing other issues. "We had so many programs . . . [and] everyone wanted to help," McGuire remembers, "but they would do what they thought the Army needed. . . . None of them were synchronized."[136] For example, a soldier could conceivably see a chaplain for marriage counseling, a therapist for post-traumatic stress, and be arrested by military police, and none of those entities would know about the soldier's interactions with the other two. "We were going through one case review, and the medical community had said that the individual had no substance abuse history," Shahbaz recalls. "And the law enforcement guy pulls up the individual's arrest records and says they got arrested six times for DUI in over eighteen months." The issue was that that soldier "had never been sent to the substance abuse clinic," so there was no medical record.[137]

This sort of disconnect might seem unremarkable in civilian life; it's easy to imagine talking to your pastor about one issue and seeing a therapist about another, but it posed two problems for soldiers. One was that, as in the civilian world, soldiers sometimes received conflicting advice that left them confused, sometimes with disastrous consequences.[138] The lack of information sharing also meant that commanders and others who could intervene often didn't have complete knowledge of the soldier's challenges. Here, it is important to remember, as Chris Ivany points out, the different nature of Army life. In the civilian world, he explains, "No one tells you what time to wake up, what time to go to bed, [whether] you can have alcohol," nor are civilians required to tell their bosses about their depression, alcoholism, or failing marriage. "But in the Army," he explains, "your whole world is controlled to a much, much greater degree."[139] This control had the potential to devolve into the sort of "chickenshit" that Casey had experienced in the 1970s, but it also meant that leaders "have much more opportunity to shape the soldiers' environment."[140]

Here the Red Book's emphasis on leadership overlaps with the question of information sharing. "If commands take care of soldiers," Ivany explains, "then the rates of mental illness in those soldiers is lower."[141] But taking care of soldiers required having a complete picture of their health and well-being. The Army, however, was not organized to deliver that information. It became evident that, in some cases, one part of the Army's regulations about privacy and reporting conflicted with one another, with some Army commands mandating reporting of certain issues and others mandating that information on those same issues remain confidential.[142] The Army's own structures, it turned out, were preventing it from conducting the kind of medical surveillance necessary to save soldiers' lives.

To address this, the task force aggressively changed policies. Because it reported directly to Chiarelli, it had the capacity to quickly make changes outside of the normal chain of command, which is how things were usually done inside the large, ponderous bureaucracy that is the Army. Typically, any decision at the Pentagon had to be approved by committees composed of each succeeding rank, beginning with a council of colonels and preceding through working groups at each flag rank before a four-star general rendered a decision.[143] Chiarelli cut out all of those middlemen, leaving only one layer of bureaucracy—the Army Suicide Prevention Council—between him and the task force.[144] "It was bureaucracy on steroids in terms of the speed at which things got done," Shahbaz recalls. "We cut through layers and layers of bureaucracy."[145]

At times, the information gleaned through the task force's reviews of suicide cases led to immediate changes. One of key findings, for example, was that soldiers often took their lives shortly after arriving in a new unit and that the leaders of that unit had often been unaware that that soldier had been receiving behavioral health care. At Chiarelli's direction, the Office of the Surgeon General developed a protocol through which behavioral health providers would take a stronger role in deciding whether soldiers could be moved to new units and in ensuring that their counterparts in the receiving units were aware of that soldier's issues. In a first for the Army, the company commander of the sending unit would brief his counterpart in that soldier's destination unit on that soldier's well-being.[146]

This sort of change seems commonsensical, but privacy regulations and other concerns meant that they were sometimes challenging to implement. As Rebecca Porter explained, Army psychologists had to live with "the pull between different masters, and who is your patient or who is your client," meaning that, on the one hand, they had the obligation to provide care to an individual soldier, and, on the other hand, they had the obligation to be transparent about whether that soldier could fulfill his role in the force.[147] In the stigma-prone Army, a soldier's fear that he might be belittled, not believed, or not promoted if he sought care made privacy perhaps a somewhat more significant issue. Thus, some initiatives aimed at enabling commanders to know more about their soldiers, like the Army One initiative—a commander's online dashboard that would allow a soldier's commanding officer to view his interactions with physical health, behavioral health, the chaplaincy, and other aspects of the Army—were complicated by the Army's need to respect the Health Insurance Portability and Accountability Act, which protects patients' privacy.[148] Commanders were thus sometimes frustrated by the fact that they couldn't get complete information about their soldiers. In the end, however, the Army settled on maintaining soldier privacy because, in the words of Secretary of the Army Pete Geren, who ultimately made the decision, "You've got to choose the way that more people would get help."[149]

Other changes generated controversy because leaders were committed to their protocols or protective of their turf, because they countered Army tradition, or because they offended some leaders' moral sensibility. "People were hugely upset, because we're used to doing things the way we're used to doing things," Shahbaz remembers, "and we were breaking a lot of eggs."[150] This may have been nowhere more evident than in the Spring 2010 debate over whether chaplains should hold memorial services for soldiers who took their own lives. Historically, that question had been left to the commanders' discretion.[151] For example, in January of that year, when Army Staff Sgt. Thaddeus S. Montgomery, Jr., shot himself in Afghanistan's Korengal Valley, his commander decided that a memorial service was appropriate for the three-tour veteran.[152] Fort Carson's commander Mark Graham had made a similar decision in 2006.[153] Likewise, when then Maj. Gen. William Troy became commander of U.S. Army Alaska in 2009, he insisted that suicide victims be given memorial services identical to those of other soldiers who died on active duty. "When you do a memorial service in a different way [for a suicide victim]," he told the *Anchorage Daily News*, "I think that you're adding to the stigmatization of a soldier who has a behavioral health problem. You don't mean to, but what you're doing is, you're making it look like it's his fault."[154]

Not all general officers, however, shared these commanders' compassionate perspective. Chiarelli sought the opinion of the Army's two- and three-star generals via email and through in-person meetings, and some of them relayed to the Vice Chief that in their view memorial services were solely for soldiers who had died in combat and that soldiers who had taken their own lives weren't deserving of the honor.[155] Some leaders, the *Washington Post* reported, "argued that suicide was dishonorable" or that "it was wrong to salute troops who had shown a lack of resolve."[156] "People believed it somehow negated the sacrifice" of others, remembers Colonel Chris Philbrick, who was Chief of Staff of the Suicide Prevention Council, adding, "Is the individual not worthy of remembrance and recognition of the sacrifices they had made prior to their suicide? That's what General Chiarelli got us to look more broadly at, a soldier's life prior to their suicide."[157]

Indeed, the counterargument didn't sway Chiarelli, who saw the services as a means of reducing stigma and who regretted his own treatment of soldiers who had taken their lives under his command.[158] After weeks of debate in which he invited generals to offer their competing perspectives, he held a video teleconference and told his junior flag officers that memorial services would be held for soldiers who took their own lives. The message, Shahbaz remembers, was "I've heard your concerns, now here's what we are going to do. . . . And the time for debate, discussion and dissention is over. And everybody said 'Yes, sir,' and they started doing it."[159]

If some line officers were upset that the Vice Chief was changing their traditions in a way that they imagined undermined the warrior ethos, others were not thrilled with being called upon to account for why suicides had taken place in their commands. In 2009, Chiarelli instituted monthly meetings of the Suicide Prevention Council that featured briefings from the commander of any unit that had had a death that the Armed Forces medical examiner ruled a suicide.

Ostensibly, these meetings were opportunities to aggregate information in order to better understand the larger suicide problem.[160] "I decided that what we were going to do was when we had a suicide [that] commanders were going to brief me, not so I could beat them up, but so we could all learn," Chiarelli explains. In his view, "If you want to get anything done in the Army, you have to have commanders involved and leadership involved" and "listen to the reports so we could apply lessons learned."[161] For the Vice Chief, they were "the most gut-wrenching meeting[s] I go to," and reporters noted that he was sometimes "visibly angry" by their conclusion.[162] He was not alone in these sentiments; describing the conferences, David Finkel has written that "no one would prefer to be here. . . . It is a brutal, depressing meeting. At the end, people always walk out looking stunned."[163] Some of the attendees affirm this. Raymond Palumbo, who in 2010 had just taken command of U.S. Army Alaska as a major general, and who recalls that he "had the displeasure of doing that twice," remembers that "you're talking about something that you don't even want to talk about. You wish it hadn't happened."[164] The unpleasantness of the task, however, didn't dissuade Palumbo of its necessity because "the leaders that are listening are going to take those lessons learned and go back to their little piece of the Army and do their best to fix it. And then you can listen to General Chiarelli, a very passionate and caring human being. You could feel the leadership of the Army was embracing this."[165]

Palumbo's sentiment was not, however, universally shared. Like every other aspect of the Army's approach to suicide prevention, and mental health more generally, the Vice Chief's suicide briefings were controversial. Most generals would probably second Palumbo's comment that they'd have preferred to be elsewhere. "Some were probably scared out of their minds," Philbrick, who sat to Chiarelli's left at these meetings, argues. "Nobody wants to be called to the principal's office."[166] Rebecca

Porter, who as chief of behavioral health was always sitting along the wall to the Vice Chief's right during these briefings, agrees. "I think they dreaded it," she says of the generals tasked with briefing the council. "Those meetings and those [video teleconferences] were painful for just about everybody involved."[167]

Perhaps unsurprisingly, the task force staff and the AMEDD have quite different views of Chiarelli's treatment of the commanders briefing him. "General Chiarelli had a good command relationship," McGuire recalls. "He wasn't out for blood. He was out to educate leaders."[168] Philbrick likewise maintains that the Vice Chief "would never stick his finger in the chest and say 'You screwed up'" to one of the generals briefing him.[169] Long-form journalism that discusses these briefings has largely echoed Palumbo's sentiments, portraying them as excruciating efforts to uncover the keys to preventing suicide and evidence of Chiarelli's commitment on this issue.[170]

While no one doubted Chiarelli's commitment, some members of AMEDD questioned his approach and recall that the meetings were sometimes tense and had a punitive dimension. From their perspective, Chiarelli's passion for the issue sometimes left him impatient with commanders and the seeming intractability of the problem. "General Chiarelli, as a very successful line commander coming up through the Army . . . was accustomed to, [he] direct[s] something to happen, it happens. [He] direct[s] something not to happen, and it doesn't happen," Porter recalls. And while he understood that suicide was "more complex than that," in her view, he took "kind of an old school approach to it," demanding information and action from other generals "as he would if it were a division or a brigade or a battalion commander," a relationship in which the sentiment was "you know this can't happen . . . [and] this is not going to go well for you if it happens again."[171]

In Porter's estimation, the task force became "very adversarial and not collaborative, the way I think it was originally intended to be. . . . It was more about who could deflect responsibility for the suicide."[172] She was not alone in her feelings. "Many of us disagreed with the approach that was being taken by General Chiarelli and his group," Eric Schoomaker remembers. In his view, the power dynamic inherent in the meetings made them disciplinary rather than educational:

> Whether it's intended or not, when a four-star general calls everyone together and goes down through the command chain and identifies the commander of the unit where they've had a suicide and they try to do a walk-back of all of the factors that led to the suicide, that becomes a form of punishment. It isolates that commander, even when there's nothing that he could have done about it. . . . When that's looked at with almost forensic specificity, that becomes a punitive exercise, too often. And I don't think we spent enough time trying to figure out what worked, what solutions were out there."[173]

The result was that, as much as everyone in the room was committed to figuring out how to prevent suicides, the different constituencies—Chiarelli, the task force, the Medical Command, and the unit commanders—were sometimes at odds with one another, with the various groups trying to cast themselves as having done their best and another group as having fallen down on the job. The Medical Department, and particularly the Behavioral Health Department, felt that it became a convenient scapegoat. In one instance when a general noted that a soldier who had had a long history of trouble on post—a ticket for driving well over two times the posted speed limit, a positive urinalysis, other issues—had never seen a psychologist, Porter, in her usual seat to Chiarelli's right, suppressed a wry laugh, thinking Behavioral Health was again going to be cast as having failed.[174] Chiarelli noticed and asked her what she was laughing about, and the lieutenant colonel replied that she felt like it was another instance where Behavioral Health was being unjustly blamed. "To General Chiarelli's credit," Porter recalls, "he looked back at that General . . . [and] said, 'Did you tell anybody else about any of these other things that were going on with the soldier? Did you call Behavioral Health and ask them, 'Hey if you are seeing this guy you need to know this stuff?'"'[175] Of course, he hadn't.

There are multiple reasons why a commander might try to shift blame to Behavioral Health for a soldier's suicide. Commanders called before Vice Chief probably sought to protect their units and reputations. Many, having come up in an Army where it had been sidelined, probably viewed mental health as outside of their purview and expertise and that responsibility for it lay with those in AMEDD. Either way, in

Porter's estimation, assigning fault for a suicide seem not only unproductive but also not in keeping with good science.[176]

The task force's relationship with AMEDD was also somewhat fraught. From McGuire's perspective, "it was a delicate dance," and information was not always effectively shared. "We would spend months trying to research something," she recalls, "and then we'd find out that there is a beautiful study that's sitting on a shelf somewhere in AMEDD. We found so much information that could have been useful in suicide prevention that wasn't being operationalized."[177] For her part, Porter felt as though her perspective and her office's expertise were sometimes left out when they could have been helpful. Because the Suicide Council had been explicitly designed to operate under Chiarelli's authority and because the Surgeon General didn't have the necessary authority to aggregate information from across the Army, information that came from one area of the Army sometimes wasn't passed to another until the council meetings.[178] "I wouldn't see the full report until that meeting when General Chiarelli started talking about it," she recalls, "and so it was hard to go to those meetings [and] be prepared to address a question."[179] As it was for any other officer called on to answer questions at those meetings, hearing for the first time that Behavioral Health was being held responsible for failing a soldier who died by suicide when she was "sitting there in front of the Vice Chief of staff of the Army" was unpleasant and exasperating.[180] "My staff and I felt the frustration of not being able to get information, not being able to prepare a background so that we could contribute in a meaningful way in the meetings," she recalls.[181] While McGuire asserts that "there was friction [between the task force and the AMEDD], but it was healthy friction," Porter has a quite different view: "I would say we were pretty impotent. From General Schoomaker on down, we were pretty impotent when it came to dealing with that task force."[182]

* * *

If there was tension inside the Pentagon's Gardner Room, where the council met, the public perception of the army's efforts were initially very positive. In June 2010, for example, the Vice Chief received a warm welcome before the Senate Armed Services Committee in a hearing dedicated to military suicide. Chiarelli assured the panel that, "over the past 12 months, the Army's commitment to health promotion, risk reduction, and suicide prevention has changed Army policy, structure, and processes."[183] In his prepared testimony, he offered some good news—suicides appeared to be decreasing across the active component of the force—which he suggested "would seem to indicate that the refocused efforts by our Army are beginning to work."[184] He also test drove several findings that would emerge as central to the Red Book—an "alarming" increase in "high-risk behavior" among the troops and the compressed lifespan of a soldier, in which she endured in a few years of Army life the number of stressors that a civilian generally encountered in a lifetime.[185]

Unlike the contentious hearings on veteran suicide a few years earlier, Chiarelli was universally praised for his efforts. Sen. James Inhofe (R-OK) pointed out that Chiarelli had "really made a study" of suicide, and Sen. Mark Udall (D-CO) began his questioning by lauding Chiarelli for "the attention you've paid to these important issues."[186] When Chiarelli vehemently disagreed with how the Army's efforts to diagnose mTBI had been treated in the National Public Radio story, Udall responded by telling the general, "I respect the passion in your response."[187] Sen. Mark Begich (D-AK) made a point of thanking Chiarelli and calling him "definitely passionate about trying to resolve this issue."[188] In contrast to the VA suicide hearings of 2007 and 2008, which featured the Bowmans, Omvigs, and Luceys describing how the system had failed their children, no family members of soldiers, sailors, airmen, or Marines who had taken their own lives testified that the military had similarly not done enough.

Several factors explain the relative kid gloves with which Congress handled the military's senior leaders that morning. One is that going after those responsible for suicide prevention hammer and tongs no longer made sense politically in 2010, as it had in 2007 and 2008, when critiquing the VA had been a convenient proxy for attacking the Bush administration as a whole. A Democratic Congress was, quite simply, unlikely to condemn failures occurring under a Democratic administration. Another is that, if the VA's reputation as an intractable bureaucracy made it a soft target, the military's privileged place in American culture rendered it a poor object of critique. A third is that the Army had indeed made significant strides toward understanding and addressing the issue. Either way, the Senate Armed Services Committee largely gave the generals a pass, accepting both their argument that the issue was complex and their claim that the services were doing all that they could to mitigate it.

That the Army could not quickly resolve the problem, however, eventually wore out this sympathetic view of the Army. As 2010 turned into 2011 and the suicide rate rose, newspaper coverage presented the Army as flummoxed and ineffective. In October, the *New York Times* pointed out that "nearly 20 months after the Army began strengthening its suicide prevention program and working to remove the stigma attached to seeking psychological counseling, the suicide rate among active service members remains high and shows little sign of improvement."[189] The article profiled several soldiers who had committed suicide at Fort Hood, pointing out the problems that ensued from repeated deployments and multiple prescriptions for anti-depressants. Three days earlier, the *New York Daily News* had pointed to a disturbing pattern of Army leaders belittling suicidal soldiers.[190] In January, when the Army reported that there had been 156 suicides in the active-component Army during the previous year—about the same number as the two preceding years—a doubling of suicides in the National Guard, and a 25 percent increase in the Reserves, newspapers described the Army as "puzzled" by the persistently high rate.[191] In July, when the Army's suicide figures passed their previous record, set in June 2010, the *Washington Post* presented the Army as nearly hapless. Detailing the work of the task force, the Army STARRS program, and Chiarelli's personal interest in the subject, Greg Jaffe wrote that "so far, the efforts have not resulted in a significant change in the suicide rate in the Army"; even the increased dwell time that Casey and others had long sought, he wrote, "does not appear to have had a significant impact."[192]

Criticism of the Army's suicide prevention efforts frustrated Chiarelli. In his view, suicide was not an Army issue, it was a public health issue that affected the entire nation, and the Army had been more aggressive than anyone in attempting to address it.[193] "We started talking about the tragedy of suicide in the United States military, and I got beat up as I worked this problem," he recalls. "I wanted to work this problem. But I got beat up and raked over the coals for 160 and 180 suicides in the United States Army, [the] active-component Army, every single year, when there are 41,000 in the current civilian population. This is a huge problem for all of us. . . . These numbers, these are not military problems, these are society's problems."[194] After years of increases with few solutions, however, the popular perception remained that the Army had failed to appropriately serve its soldiers.

* * *

Faced with a suicide rate that for the first time in history had crept above the civilian rate, the Army sought to learn why soldiers were killing themselves and what could be done about it. This effort, however, is less a medical story than a cultural one. On multiple levels, the Army's grappling with suicide was also a way of grappling with larger questions about what kind of organization the Army was and had become during the wars. Although media coverage emphasized links between combat trauma and suicide, Army research framed the problem as a form of slow violence, an indirect result of the strain that the war's demand for repeated deployments, short dwell times, more soldiers, and an emphasis on combat readiness had placed on leaders, soldiers, and families. Throughout its suicide prevention literature, the Army framed suicide as a problem of failed leadership, arguing that the demands of the wars prevented junior leaders from developing the skills to ensure their soldiers' well-being. If these claims were in effect a subtle anti-war discourse and a jeremiad, they also engendered an increasingly punitive approach in which leaders were encouraged to be more firm disciplinarians and purge soldiers the sake of institutional effectiveness. Internally as well, the work of Peter Chiarelli's Suicide Prevention Council also became increasingly controversial, with both unit leaders and Medical Department representatives feeling pressed to explain why suicides had happened on their watch in ways that did not always seem productive.

The story of the Army's efforts to address suicide, then, was a story of an organization struggling to adapt to new challenges under tremendous stress that threatened it at an existential level. That these efforts were only marginally successful reveals the extent to which the wars' demands threatened the Army's capacity not only to care for its soldiers but to maintain its position as an essential, highly respected institution in American culture.

8

"The Challenge to the VA Is Execution and Implementation"

VA Suicide Prevention in a Moment of Mistrust

A month after Rep. Leonard Boswell (D-IA) reintroduced the Joshua Omvig Veterans Suicide Prevention Act, and nine months before George W. Bush signed it, the Veterans' Health Administration began implementing its mandates. In response to the bill's call for "the designation of a suicide prevention counselor at each Department of Veterans Affairs medical facility," the VA's deputy undersecretary for health operations and management announced that each VA facility would receive funding for a full-time suicide prevention coordinator.[1] That the memo stipulated that staff hired into the position must be solely dedicated to suicide prevention and that the positions should be filled immediately, lest the funding evaporate, betrayed a sense of urgency.[2]

Three months later, though, the inspector general relayed more bad news in a report that found critical shortcomings in the VA's suicide prevention efforts. The VA had yet to develop a twenty-four-hour hotline for veterans in crisis.[3] Only half of VA facilities were effectively screening for suicide, even fewer had special programs in place for high-risk veterans, and only two-thirds provided cognitive behavioral therapy "specifically focused on suicide prevention."[4] In nearly half of VA facilities, severely depressed patients waited more than a week for a mental health evaluation.[5] The majority of front-line staff had not been taught how to intervene with suicidal veterans, and just over 10 percent of VA facilities mandated training that specifically addressed the needs of OEF (Operation Enduring Freedom)/OIF (Operation Iraqi Freedom) veterans.[6] Unsurprisingly, the report concluded that "more work remains."[7]

As chapter 3 shows, veteran suicide became a primary piece of evidence for critics of the Iraq War and the Bush administration. This chapter explores how the VA addressed veteran suicide amid these critiques.

After 2007, the VA met with increased scrutiny and pressure from a skeptical Congress and frustrated veterans' groups, and an array of cultural products cast the agency as lethargic and apathetic. In response, the VA mounted a vigorous effort to understand the magnitude of veteran suicide and develop protocols to better address it. By 2014, the VA, which at the wars' onset had been woefully unprepared to address veteran suicide, had built a robust program of surveillance, outreach, and intervention. Staff members knew how many veterans took their lives each year and which veterans were most at risk, and they had developed protocols for engaging them in effective treatment.

On a local level, however, a lack of resources and a set of unrealistic requirements left many hospitals and clinics struggling to provide care to a growing number of veterans whom the VA's outreach campaigns had inspired to seek help. Stories of veterans who took their lives after they failed to receive proper care thus remained at the forefront of Americans' understandings of veteran suicide and overshadowed the significant strides that the VA had made. As the wars wound toward their unsatisfying conclusions, the VA remained a symbol of a nation indifferent to the suffering of the men and women sent to fight two misguided wars.

* * *

In the summer of 2007, the VA opened its Center of Excellence in Suicide Prevention at the Canandaigua, New York, VA hospital. The center was run by Jan Kemp, a VA nurse who had written her doctoral dissertation on women service members' readjustment after Operation Desert Storm and who had previously been director of education at the Mental Illness Research, Education, and Clinical Center (MIRECC) in Salt Lake City and Denver. There she, along with the suicidologists Larry Adler and Morton Silverman, had conducted more than forty suicide prevention workshops for VA staff between 2005 and 2007.[8] The Center of Excellence was an effort to address what the VA understood as a significant failure in its suicide prevention efforts. Recognizing that "most Veterans who die by suicide are not in mental health treatment," the VA sought to develop what it termed a "public health approach," one that would "intervene with the entire population of at-risk Veterans."[9] This policy was in keeping with the most current research in the field, but it raised a host of questions for an organization that had underestimated and been unprepared for returning veterans' mental health needs and now faced increasing public and congressional scrutiny: Which veterans were, in fact, "high risk"? How could the VA get them into treatment? And how could the VA ensure that they would receive quality care?

None of these questions were easily answered. The VA's January 2009 *Strategies for Suicide Prevention in Veterans*, for example, found that only three studies had examined suicidal behavior among veterans and that efforts to address high-risk individuals had proven discouraging at best.[10] The document explained that, while intervention for high-risk patients "typically . . . involves tracking individuals after emergency room visits for suicide attempts, providing close mental health follow-up, therapy, or case management," there was no statistical difference in the suicide reattempt rates of those who were the beneficiaries of these programs and those who were not.[11] Perhaps more troubling, given where the VA would focus its efforts, was that the authors "found no studies that assessed the specific effectiveness of any hotlines, outreach programs . . . peer counseling, treatment coordination programs, and new counseling programs."[12] The report's only bright spot was that research confirmed something that seemed transparently obvious: When individuals had less access to firearms and drugs, the number of gunshot and overdose suicides decreased.[13]

Based on these findings, the authors concluded that the VA should establish its suicide prevention plan around five principles first enumerated at a 2004 conference in Salzburg, Austria: identifying those at high-risk; outreach to the potentially suicidal, their families, and their physicians; better engagement with the media; providing quality, evidence-based treatment; and removing "access to lethal means," that is, items that would make a suicide attempt more likely to succeed.[14] These recommendations were not new to Kemp and her staff, who had been working on them since 2007.

The centerpiece of the VA's efforts was its Suicide Prevention Hotline, which launched a month after the Center of Excellence opened in partnership with the Substance Abuse Mental Health Services Administration, which runs 1-800-273-TALK, the national suicide prevention hotline. It was immediately popular; 5,000 veterans called in the first three months, and more than 300 actively suicidal veterans were rescued.[15]

Nonetheless, the hotline proved controversial. When in September 2008 the House Committee on Veterans' Affairs held a subcommittee hearing devoted entirely to its effectiveness, nearly all of the witnesses complained that hotline operators offered insufficient care, that callers didn't receive timely or effective follow-up contacts, and that there was no way to know if veterans with the highest risk were being brought into care.

Some of these critiques seemed peevish. Henry Reese Butler II, the founder of the 1-800-SUICIDE network, argued that "now we have too many" hotlines and that the VA was competing with his program in ways that duplicated efforts but not results.[16] Butler complained that there was no way to tell whether veterans, were, in fact, the ones calling and that by having police or emergency medical services perform rescues, the VA was alienating veterans who needed help. They "have been rescued against their will," Butler claimed. "Now, they do not trust the system," he asserted, and they were likely to "go away, and . . . shoot themselves."[17] These were critiques that Jan Kemp met with commonsense responses—"If they tell us they are a veteran, we acknowledge the fact that they are a veteran"—and whatever the merit of Butler's critiques, they were likely motivated at least in part by financial concerns. Because of the shift of federal funding to 1-800-273-TALK, Butler was now "struggling to keep 1-800-SUICIDE afloat."[18]

Other complaints, though, raised important questions. Vietnam Veterans of America's Thomas J. Berger asked, "What is the definition of 'rescue'?" pointing out that "sixteen hundred rescues represents only .048 percent of the calls." He then asked, "What is the status of the rest of the calls? Is there a follow-up or tracking procedure? For 1 month, 3 months, 6 months?"[19] Kemp's answer was again commonsensical. "It means someone was in crisis," she told the panel. "There was a clear and imminent danger of suicide. And emergency or medical personnel were directed to the right location in time to save someone's life."[20] But on the question of whether veterans who weren't in crisis received effective follow-up, her response was more equivocal. When Tyrone Ballesteros of the National Veterans Foundation testified that test callers were most frequently told to go to a VA hospital and "to hang on and be patient until the facility can contact them"—a response that Rep. Phil Hare (D-IL) called "absolutely unbelievable"—Kemp explained that "we have done a fair amount of looking to see what happens to these people who call us," but she offered no statistics and reflected that "we also very much honor their desire to remain anonymous."[21]

Critics also argued that veterans' experiences were so exceptional that the VA's civilian operators could not appreciate them, much less help veterans deal with them. Butler, whose hotlines were staffed by veterans and police officers, argued that, "unless you have been in a situation where you had to use a gun to both defend your fellow comrades or yourself, and face a gun, you cannot ever really say, 'I know how that feels.' It is like a guy telling a woman on a rape crisis hotline, 'I know how it feels to be raped.' . . . There is no credibility."[22]

Butler's comment is rife with false assumptions—that most suicidal veterans have been in combat and are struggling with those experiences, for example, or that men cannot be victims of sexual violence. However, he was not alone in insisting that only veterans could help each other in crisis. Ballesteros similarly complained that "combat veterans are the only ones that can really understand what another combat veteran has been through."[23]

These claims represent a demand for what the historian Jay Winter has called an "essentialist silence" in which "only those who have been there . . . can claim the authority of direct experience required to speak about these matters" and that is particularly salient among veterans.[24] While Kemp sought to answer these critiques by referencing the VA's own expertise and maintaining that "we have, in the VA, a long history of being able to identify those particular veterans' needs," they reiterated that for many veterans the VA seemed another component of a culture that purported to support them but in fact all but ignored and could never understand their experiences.[25]

Claims of veteran exceptionalism also intersected with critiques of the VA's outreach efforts. Tom Tarantino of Iraq and Afghanistan Veterans of America critiqued new public service advertisements that the VA had deployed throughout the Washington, DC, public transportation system as ineffective because "the silhouette employed in the ad is clearly not of a modern soldier," and he called on the VA to "formulate a message that modern veterans will respond to."[26] Tarantino's critique was at odds with Kemp's testimony that calls from the National Capitol Region had doubled while the advertisements ran, but it pointed to another problem that the VA faced: Although OIF and OEF veterans had been positioned by the war's supporters as "the next greatest generation," many resisted this narrative, particularly as they found their experience at odds with historical recollections of the Second World War.[27] For them, the VA seemed an antiquated institution that offered them little.

Other VA outreach efforts lent credence to his point. A December 17, 2007, letter sent to veterans enrolled in its system under the signature of Michael J. Kussman, the VA's undersecretary for health, was, for example, in every way an impersonal form letter that reeked of bureaucracy, from the "Dear Veteran" salutation to the "In Reply Refer To" header. Its enumeration of the "warning signs" included such numbingly obvious items as "looking for ways to kill yourself" and "feeling that there is no reason for living."[28] It's hard to imagine this letter improving anyone's opinion of the VA; it's not difficult to imagine many of these letters being immediately discarded.

While the VA clearly struggled with outreach in 2007 and 2008, it was having better success monitoring high-risk veterans who were already in its system. In April 2008, Gerald Cross, the VA's principal deputy undersecretary for health, mandated that suicide prevention coordinators "maintain a list of patients at high-risk for suicide" and that they be "kept on the list for a period of at least 3 months after discharge" and "evaluated at least weekly during the first 30 days after discharge."[29] He also required that facilities better coordinate mental health and primary care, an expansion of the successful Primary Care–Mental Health Integration program that had begun in 2007 as a means of reducing stigma with mental health screenings, treatment, and referral during annual wellness visits.[30] As a result of this mandate, veterans who did not keep their appointments received rapid follow-up, and providers helped veterans create an action plan that would help them navigate stressful times.[31] By February 2010, there were more than 9,000 veterans on the high-risk list, though only about 1,200 had safety plans, and by 2012, more than 1,000 veterans were being added to the list each month.[32] In November of that year, the VA hospital in San Juan, Puerto Rico, alone added 95 veterans, and while this is a statistical outlier, that same month VA facilities in nine other cities each added more than 20 veterans to the rolls.[33]

The VA also got better at identifying veterans at high risk for suicide. A September 2009 memorandum encouraged clinicians to focus on veterans diagnosed with "depression, bipolar disorder, substance use disorders, anxiety disorders, and schizophrenia."[34] The memo emphasized that veterans with major depression were nearly ten times more likely to take their lives than those without that diagnosis, pointing out that the suicide rate among OEF/OIF veterans with that diagnosis was nearly three times that of similarly diagnosed veterans of earlier conflicts.[35] A week after this memorandum appeared, Kemp announced a program though which high-risk veterans would be more closely followed and receive an "advanced care package" of weekly screenings and a "written safety plan" that veterans and VA staff could implement during a crisis.[36]

Alongside efforts to better treat suicidal veterans, Kemp instituted an initiative to more accurately track suicide attempts and completions among veterans receiving VA care. Mandated suicide event reporting began at all VA facilities in 2008. In August 2009, Kemp began sending a monthly email cajoling suicide prevention coordinators to complete a spreadsheet tabulating suicide attempts and completions among patients receiving care at their facility as well as those veterans' demographics, including the conflicts in which they had served and whether they were receiving VA health care, their mental health diagnoses, and their history of suicidal behavior.[37] A second spreadsheet asked how many veterans had been deemed high risk and whether a care plan had been constructed.[38]

Within a year, the Center of Excellence had tabulated these reports into epidemiological data that revealed much about how many—and which—veterans in VA care were attempting and completing suicide. Between October 2009 and September 2010, there had been 14,831 suicide attempts among veterans already enrolled in Veterans Health Administration (VHA) care and 762 completions—slightly more than 2 per day.[39] Veterans in the mid-Atlantic region completed the fewest suicides—5—while veterans in Oklahoma, Arkansas, Mississippi, Louisiana, and parts of Texas and Missouri completed the most suicides at 83.[40] Nationally, the attempt rate was 722 per 100,000 individuals and a completion rate of 33.6 per 100,000 individuals—well above the national average for that year of 12.1 per 100,000.[41]

This was not entirely an unfavorable news story, however. This number was marginally lower than the proceeding nine calendar years, in which the average had been 36.4/100,000. In fact, as the VA pointed out in numerous internal documents, "The rate for 2008 continues a longer term trend characterized by decreases from 2001 through 2003 and stable rates since 2003."[42] Contrary to what media coverage had suggested, the suicide rate for veterans receiving VA treatment had remained steady throughout the twenty-first century, beginning at 35.7/100,000 in 2003 and ending at 35.9/100,000 in 2009.[43] Also surprising was how weakly PTSD correlated with suicidality. In fact, every other category of mental health diagnosis—depression, schizophrenia, anxiety, substance abuse disorder, and especially bipolar disorder—was more highly correlated.[44] Moreover, of the 18 veterans who completed suicide each day, only 5 were enrolled in VHA care.[45] Thus the notion that the veteran suicide rate was rising was incorrect, at least among veterans who received VA care, and the VA could rightly claim that the veterans it treated were less likely to contemplate suicide.

Veterans Administration appointees also highlighted their efforts. The VA had requested increases of $319 million and $288 million for mental health treatment in fiscal years 2009 and 2010, respectively, and at a March 2009 budget hearing, VA Secretary Eric Shinseki declared VA mental health fully funded and touted the hiring of "18,000 full-time equivalent staff [and] $4 billion going to mental health programs."[46] The VA's efforts and public claims, however, paid few dividends in a culture that viewed it as obstinate, unprepared, and lethargic. The year 2009 opened with former war correspondent and columnist Joe Galloway complaining that the Bush administration's cuts to the VA budget had created the conditions that led some veterans to commit suicide while they waited to receive care.[47] Most stories, though, were of the struggles that veterans faced in getting treatment at their local VA hospital. In Charleston, West Virginia, it was Tom Vande Burgt, who was diagnosed with PTSD, prescribed a handful of psychotropic drugs, and "shuffled through different counselors and psychiatrists" but who "never [saw] the same doctor more than twice" at a VA clinic that "has not had a psychiatrist on staff for at least six months."[48] In Spokane, the news was of twenty-six-year-old Lucas Senescall, who in 2008 had hanged himself "a few hours after he sought psychiatric help at the Spokane VA." In addition to noting that the local VA didn't even know how many of its patients had taken their own lives that year, the *Spokesman Review* quoted Senescall's bewildered father: "I want to know why, when he was rocking back and forth in his chair with his hands over his mouth to keep from crying, [the psychiatrist] sent him home."[49]

This was one of many stories that pointed out the reality that many facilities struggled to fill vacant mental health positions.[50] In Oregon, the manager of a VA clinic complained that "we're very lacking in mental health capacity here" and "we have to scrap and fight for money and services."[51] In Spokane, the VA mental health staff declared themselves overwhelmed by demand and protested by "refus[ing] to accept new patients," including, presumably, the 134 that the Suicide Prevention Hotline referred to them over the next fiscal year.[52] On Veterans' Day, a Florida newspaper lamented that "the gap between a need for mental-health services and availability of such services is so large that wait times from lack of available health providers discourages soldiers from seeking treatment," complained that the Department of Defense and the Veterans Administration "have the resources, if not yet the priorities, to increase the number of health providers," and closed by insisting that "honoring veterans isn't enough when those who may need help most are abandoned to their own traumas."[53]

Such claims were also increasingly part of popular representations of the wars. In August 2009, the same month that Kemp began requiring clinics to tabulate suicide attempts and completions, the *Christian Science Monitor* reported that "even with a heavy infusion of funding . . . the VA has been hard-pressed to meet veterans' needs" and profiled former Marine Clint Van Winkle, who "didn't find a warm welcome when he sought help at the VA office near his home in Phoenix in 2004," a complaint that Van Winkle more fully expounded in his 2009 memoir *Soft Spots: A Marine's Memoir of Combat and Post-traumatic Stress Disorder*, which presented the VA as enmeshed in a culture that neither understands nor cares to engage veterans' experiences.[54]

Soft Spots is structured around imagined encounters in which fellow Marines appear in Phoenix and demand that Van Winkle reveal the traumas that he endured in Iraq—killing a young girl by destroying her home with machine-gun fire, leaving a dead Marine's body on the battlefield, and standing in the remains of a Marine inside a destroyed amphibious assault vehicle. On his return from Iraq, Van Winkle realizes that no one in the United States understands his combat experiences and that the VA is ill equipped to address his needs. Standing on a motel balcony on his first night back from Iraq, he realizes that "nobody within miles had a clue; nobody even knew where Nasiriyah or Al Kut was located."[55]

Indifference also characterizes Van Winkle's encounters with the VA. At the Phoenix VA hospital, which he describes as "look[ing] more like a homeless shelter than a hospital," he finds an "outdated magazine"—a sign of the VA's obsolescence—and staff members who "seemed as in they had been specifically hand-picked to dole out sub-par service."[56] A nurse and doctor barely make eye contact with him before dispensing medications, like a "drive-through window service."[57] "I wouldn't have known anything about PTSD if it weren't for the Internet," he complains. "An explanation would've been nice. Hell, even a brochure. Something."[58] The VA thus appears chronically uninterested in helping troubled veterans, a place where "a fresh batch of mentally impaired vets had taken their place on the bureaucratic merry-go-round, waiting to be herded into holding cells for their five-minute interrogations and rations of feel-better pills."[59]

Moreover, Van Winkle's experience receiving Eye Movement Desensitization and Reprocessing (EMDR) therapy furthers the exceptionalist argument that had been made before Congress a year earlier. It held that the VA was ill equipped to treat returning Iraq and Afghanistan veterans because practitioners had not experienced combat. Eye Movement Desensitization and Reprocessing therapy was among the evidence-based treatments that the VA had approved early in the Iraq War.[60] Yet while Van Winkle positively describes his EMDR therapy, he premises its effectiveness on his clinician's veteran status:

> I have unconditional trust in this man, which is why I decided to let him perform EMDR on me even though many psychologists question the therapy's validity. But what do psychologists really know about war trauma unless they've experienced it? I'd been to a slew of "mind doctors," and look where that got me. Joe is different, knows more about combat veterans than all of the skeptical psychologists combined. He's mentioned some of his experiences as an LRRP [Long-Range Reconnaissance Patrol ranger] and a Ranger Special Ops team leader in Vietnam. . . . Joe is a bona fide war hero and has more medals for valor than anybody I've ever met. . . . The knowledge Joe has about war can't be learned in college

> courses. There isn't anything I can say that he won't understand. . . . We're connected by war.[61]

In this tour de force expression of veteran exceptionalism, Van Winkle denies the effectiveness of VA medical care by emphasizing his counselor's veteran status as the most important factor in his treatment. *Soft Spots* thus dismisses the efficacy of evidence-based treatments, but it also dramatizes the struggles that many veterans faced and which Americans routinely encountered in newspapers. And indeed, in spite of the VA's proactive efforts at gathering statistics, reaching out to veterans, creating the hotline, and implementing evidence-based therapies, national efforts had translated unevenly into quality care at the local level.

* * *

In 2010 and 2011, VA administrators again came under media fire and congressional scrutiny. Of particular concern was the suicide hotline, which VA officials had trumpeted as a centerpiece of reform. In a March 2010 hearing, Richard Burr (NC-R) suggested that whatever good the hotline was doing, the goal should be to ultimately render it unnecessary: "The earlier we can get them into treatment, the longer we can keep them there, the less likely we are to get a phone call . . . and I think the goal should be to make sure that we don't need the functions [of the hotline]."[62] Burr's comment summarized many legislators' and experts' ambivalence. While it was clearly saving lives, there was a sense that the VA's emphasis on the hotline left many veterans unable to get help until the situation was dire.

When Caitlin Thompson, one of the VA's clinical care coordinators, explained that veterans who called the hotline would receive a follow-up call within twenty-four hours, Burr acknowledged that that call "would initiate a very proactive effort on the part of the VA." He also speculated that for veterans who did not call the hotline but came to their local VA, the results would be the opposite.[63] His evidence for this was the testimony of Daniel Hanson, an OIF veteran who moments earlier had complained that the VA had ignored his mental health issues. "We had a lot of conversations where I just said, I don't know what to do anymore," Hanson had explained. When he had told his VA counselors that he was considering suicide, the response was, "You call this number if you are going to do that."[64] Hanson's experience suggested that the VA's reliance on the hotline had delayed earlier intervention.

Four months later, a House hearing revealed similar sentiments. Retired sailor Melvin Cintron, who had himself struggled with mental health issues during the 1990s, made the point more emphatically. The hotline "is too much of a last alternative," he complained, adding, "Either you don't have enough of a problem and you can wait, sometimes for weeks for an appointment, or you are suicidal."[65] He suggested, instead, that the VA needed "a readily or easily accessible intermediate or non-suicide hotline" and called on the VA to better advertise the resources available for veterans who did not consider themselves suicidal.[66]

Cintron was followed by Linda Bean, whose son Coleman had taken his life in 2008, after returning from Iraq. Explaining that when her son died she tried to get his platoon-mates into VA care, she criticized the care offered by local VA facilities as uneven. One VA representative offered to pick up men who needed help, but another told her, "If they don't walk through the door, we can't help them."[67] Bean's conclusion, like Cintron's, was that the VA had invested heavily in the hotline and saving actively suicidal veterans but that "we need to make sure that people understand there are places to go before you hit the suicide hotline."[68] The problem remained that it was difficult for families and veterans to access mental health care.[69]

As at earlier hearings, the question of whether veterans felt comfortable contacting the VA, or whether they felt that their experiences were too exceptional to be understood by those who hadn't served, loomed large. Cintron explained that a veteran "cannot just go talk to anyone."[70] Similarly, Timothy S. Embree from Iraq and Afghanistan Veterans of America told the committee that "young vets coming back from Iraq and Afghanistan right now can speak to each other" but that "the VA for too long has been dealing with issues that have affected past generations, and they haven't recognized that those issues are affecting this new generation of veterans just in different ways."[71]

Embree was especially critical of a VA public service announcement that featuring actor and veterans' advocate Gary Sinise, who had played Lieutenant Dan in the 1994 movie *Forrest Gump*. In that public service announcement (PSA), which featured clips from the film, Sinise asserted that the "VA cares about you and is reaching out" and asked veterans to "reach back."[72] To Embree, the PSA lacked resonance: "There was a bunch of World War II memorabilia around, and Lieutenant Dan from Vietnam with *Forrest Gump* was talking to me. That didn't speak to me, it didn't make any sense to me. But when I see two young veterans walking up to each other and shaking hands . . . I understood what was going on with those guys."[73] He suggested that the VA connect with veterans on social media and take advantage of Facebook and Twitter.[74] Once more, the specter of a generational disconnect fueled by an antiquated VA and unresponsive to contemporary veterans created another barrier to care.

This was a critique that Kemp evidently took to heart. In the March hearing before the Senate, she had admitted that "we didn't always get the symbols right. We didn't get the right uniforms on the right people asking the right questions. So we are quickly trying to work that out."[75] In fact, though, it was more than getting "the right uniforms on the right people." Beginning in 2010, the VA sought to turn the narrative of veteran exceptionalism to its advantage by casting itself as the only place that understood contemporary veterans' experiences. In a series of public service announcements, the VA acknowledged and validated veterans' feelings of isolation and alienation but encouraged them to seek support. "Perspectives," which the VA premiered in 2011, established this theme, highlighting veterans as exceptional and suggesting that in asking for help veterans would enter a community that, unlike the rest of the nation, understood their experiences. In the ad's logic, accessing mental health services was a way for veterans to remain connected to their military comrades.

The advertisement opens in an elevator, where the African American protagonist, Washington, stands with a woman in a camouflage battle dress uniform, an indicator that VA services do not represent a sharp departure from military service. In the next frame, a directional sign lists "Mental Health Services" alongside "Radiology," "Orthopedics," and "Rehabilitation." A hallway is lined with hospital rooms with families visiting patients with presumably physical and psychological injuries, asserting no divide among the combat wounded. This understanding heightens when Washington runs into his old platoon-mate, Fox, who is white and on crutches. Drawing on the trope of the buddy film, in which men of different backgrounds bond through shared experiences, Fox complains, "Physical therapy makes Kandahar look like summer camp," while Washington says about his counseling, "It's like you said: sessions are tough, but I've got to get back on track."

Most important, though, is the recognition of veterans' unique experiences and the VA as a place where those experiences are affirmed. This is evident in the duo's bantering about Jackson, a fellow soldier who was apparently a constant complainer in Afghanistan, and in Washington's response when Fox asks if therapy is working: "Yeah, man, I've been sleeping like a baby. I even went easier on the Class Six"—a slang term for alcohol that only soldiers recognize.[76] That Washington and Fox's jokes are based on military experiences constructs the VA as a legitimate support source where they will be understood. When Washington remarks, "I'm a veteran, and these services are for us," followed by a concluding montage in which other veterans explain, "We've earned" these benefits, "Treatment works," and "Calling the confidential crisis line can help. I know," the PSA rejects the top-down and impersonal logic of the form letter and relies on veteran solidarity to establish help seeking as not only appropriate but shared.

The VA public service announcements developed between 2012 and 2014 followed suit. The 2012 PSA "Common Journey," for example, establishes difficult challenges and caretaking as central components that separate military service from civilian experiences before positing the VA as the organization that fits this unique role. The commercial begins by establishing the "common journey" that service members take, regardless of their branch of service. In the opening frames, a stocky white recruit shakes the hand of a marine recruiter, and an African American woman determinedly does push-ups while an Army drill sergeant bellows in her ear. In each scene, the camera zooms out and pans left, allowing a bright light that identifies the military as a space of promise and opportunity. As the voice-over announces, "We wanted to . . . stand for something bigger than ourselves and protect the things we love," the frame shifts to a deploying soldier hugging his wife before boarding a bus and then to images of the soldier as protector. Over the image of a 1st Infantry Division soldier holding a dark-skinned child, a voice explains the mission as "help[ing] those in need and mak[ing] a difference." The light then reemerges, establishing that the soldier's deployment has been worthwhile.

As the ad shifts from deployment to return, however, the light that at first fills the screen after another family embrace recedes in the next frame to reveal a landscape of browns and grays and a couple sitting at opposite ends of a long table. The husband, who earlier was happily reunited with his family, now stares off into the distance. "In the service, we had each other's backs," the voice-over explains, "but as veterans it can sometimes feel like we're all alone, even when surrounded by our loved ones." As the wife shows her husband the website for the veterans crisis line, a voice explains, "At the Veterans' Crisis Line, we understand what you're going through." The vantage point then shifts to show the couple from behind, with their arms around each other and the sun again rising before them, signifying the bright future ahead. The PSA's message is unequivocal: Civilians do not understand veterans' experiences, but the VA does.

That the VA is the space in which veterans can find an empathetic community amid a culture that has denied the wars and the veterans' experiences reaches its apotheosis in "The Power of 1," the PSA that accompanied the 2014 campaign. The PSA is set in a diner, where the opening image of a hand-lettered "Welcome Home Troops" sign belies what is happening inside. Daniel, a young veteran, sits alone, pensive and wearing a bracelet suggestive of a dead comrade, as life continues around him. Children play, men talk, and an active-duty soldier is greeted by a waitress, highlighting the veteran's invisibility. The PSA's turning point comes, however, when an aging Vietnam veteran pays for Daniel's breakfast, leaving with it a note reading, "If you ever need to talk, I'm here for you." The message, of course, is that a veteran knows instinctively when and how to reach out to a comrade. The VA is their shared space, their new community.

The VA also took to heart the critique that many veterans might view the Suicide Prevention Hotline as a last resort. In March 2011, the VA rebranded it as the Veterans Crisis Line. The change of words mattered. Framing the line as something that was uniquely for veterans responded in part to their growing sense that most Americans, having made no sacrifices, were oblivious to their experiences and that their gestures of support were hollow. Changing the emphasis from "suicide" to "crisis" encouraged veterans to call sooner and hasten intervention. If veterans and family members could be persuaded to call earlier, then treatment could begin before crisis became tragedy.

The critiques made by members of Congress, veterans and their families, and representatives of veterans' organizations thus played an important role in encouraging the VA to make changes that brought more veterans into the VA's care. Prior to the hotline's name change, about 60 percent of callers had been veterans; afterward, the figure rose to 80 percent.[77] More important, the Veterans Crisis Line not only saved lives when veterans were in imminent danger of harming themselves but also facilitated their entrance into treatment programs. In fiscal year 2010—the first year of the rebranding—operators referred nearly 20,000 veterans, or about 55 per day, to VA mental health care.[78] Between October 2010 and April 2011, the crisis line provided 3,658 immediate rescues, arranged for nearly 1,800 veterans to be admitted to their local VA hospital, and saw 1,102 veterans enroll in VHA health care.[79] In 12,206 other cases—or about 81 per day—suicide prevention coordinators "contacted [the] veteran and arranged for appropriate care."[80] Clearly, as the VA improved its outreach efforts, veterans eagerly took advantage.

* * *

These initiatives, however, did not resolve the VA's problems. Its image as an unresponsive and uncaring bureaucracy endured as media accounts continued to emphasize that, even as the VA set progressive national protocols and made significant strides in framing proper care, veterans took their own lives after struggling to get VA assistance. There was Reuben Paul Santos in the San Francisco Bay area, who "was only able to see a therapist once a month"; Timothy Ryan in Wilmington, Massachusetts, who likewise saw a therapist only once a month even though he "did all the things he was supposed to do" and set up regular appointments with the VA; Curt Fike in Ohio, who told his therapist at the VA that "I feel like something horrible is about to happen" a few months before he shot himself in the chest; as well as others.[81] At times, VA responses seemed shockingly callous. A California VA spokesman, responding to claims that a veteran who had killed his mother, his daughter, and himself hadn't received sufficient treatment even after the police and a non-profit had contacted the VA, deflected responsibility by declaring that "the request was mild. 'It wasn't like "this guy is really in need of mental health."'"[82]

These incidents and the accompanying media firestorms frustrated VA officials, who viewed the anecdotes as obscuring their efforts and successes. These anecdotes, however, also pointed to continuing struggles on a national level. In particular, the investigation into Lucas Senescall's treatment by the VA discovered that, while the Spokane VA had reported 9 veteran suicides in the twelve months beginning in July 2007, in fact 22 veterans—15 of whom "had contact with the medical center"—had completed suicide in that period, thereby highlighting that the VA still didn't have an accurate count of how many veterans killed themselves each year.[83] The monthly spreadsheets that Kemp required only counted those veterans who were already enrolled in the VA system. They still didn't know how many veterans nationwide took their own lives.

The VA had been working to better tabulate these numbers since at least 2006, when Ira Katz announced the 18-per-day figure in response to CBS news' reporting.[84] Extrapolating from national statistics, Ira Katz calculated that there were likely approximately 36,000 suicide attempts among veterans each year, although the VA had reported 10,840 or, as Kemp put it in September 2009, "about 30% of attempts in our population."[85] Understanding exactly how many veterans took their own lives was difficult, however. As Kemp explained in a 2011 email, "We will never know about some of the deaths that occur until they show up on the NDI data—because no one is obligated to tell us."[86] The NDI is the National Death Index, which the Centers for Disease Control and Prevention (CDC) uses to compile the death records for each state. Its figures are what Katz had used in 2006 to arrive at the 18 per day figure. The NDI did not, however, deliver data in real time—there was a two-to-three-year lag in the production of those statistics.[87] As a result, in 2010 the VA had national statistics from 2008 but an incomplete set from its own facilities. A year later, it had internal statistics for all of 2009, but it was still waiting for the Centers for Disease Control to release national data.[88]

Based on these data, though, the VA was able to draw some conclusions, not all of which were encouraging. Although the suicide rate among twenty-first-century veterans was in fact lower than that for all veterans—30.8 per 100,000 compared to 35.7 per 100,000—compared to others in their age cohorts, the VA reported in April 2010, "there is evidence of a 21% excess of suicides through 2007 among OEF/OIF veterans."[89] A year later, the VA was able to identify 2,482 non-fatal attempts and provided longitudinal data that identified 2004 as the high-water mark for veteran suicide, when the rate had crept over 1 per 1,000 veterans before declining, although there was a considerable spike in 2008.[90]

The VA claimed that this, in part, was a promising story—once it had begun more vigorously addressing suicide prevention in 2005, the rate had stabilized.[91] And although the suicide rate for the youngest OEF/OIF veterans, those aged 18–24, was 47.1 per 100,000 in 2009, this was a vast improvement over the previous year, when the rate had been 75.4 per 100,000.[92] It could also claim, as early as 2010, that "there is preliminary evidence which suggests that there are decreased suicide rates in Veterans (men and women) aged 18–29 who use VA health care services relative to Veterans in the same age group who do not since 2006. This decrease in rates translates to approximately 250 lives per year."[93] A year later, a memo from William Schoenhard, the deputy undersecretary for health operations and management, reached a similar conclusion. Using newly revised data from 2008, he found that "the incremental risks associated with mental health diagnoses were greater in OEF/OIF Veterans than in other Veterans" but that "the decreasing suicide rate in Veterans under age 30 who utilize VHA services compared to those who do not, suggest that it may be life-saving to help young veterans enroll in VA and to engage in care."[94]

The delay in getting statistics from the NDI, however, proved an enduring problem for the VA. Without knowing exactly how many veterans took their own lives each year, and what their demographics were, the VA could not appropriately target outreach and treatment efforts. To circumvent this delay, Robert M. Bossarte, a Notre Dame–trained sociologist who had completed a post-doc at the CDC in Atlanta before coming to work for Jan Kemp, went directly to the states. Forty-six states indicated veteran status on death certificates, which would provide one metric for revealing how many veterans who were not receiving VHA care completed suicide.[95] On June 16, 2010, the governor of each state received a letter signed by VA Secretary Eric Shinseki that encouraged them to share the data from the death certificates of individuals who had completed suicide.[96] The study wasn't perfect. Some states didn't respond quickly, and the tendency toward underreporting suicide in general meant that some death certificates didn't accurately identify suicide as the cause of death.[97] Moreover, nearly a quarter of the suicide victims' death certificates that Bossarte and his assistant, Heather Shaw, evaluated didn't identify whether the deceased was a veteran. In particular, they found that "younger or unmarried Veterans and those with lower levels of education were also more likely to be missed on the death certificate."[98]

Nonetheless, the review offered considerable insights. By the following April, Bossarte and Shaw had received data from eleven states and had conducted a sample analysis for Washington State which found that nearly 30 percent of the suicides between 1999 and 2008 had been completed by veterans.[99] Two years later, Kemp and Bossarte had data from twenty-one states and calculated that about 22 percent of suicides were completed by veterans.[100] They also found that, while veteran suicide was assumed to be an issue most affecting Iraq and Afghanistan veterans, older veterans took their own lives far more frequently.[101] Most important, though, was their tabulation of a figure that would soon dominate public discussions of veterans' suicide: "If this prevalence estimate is assumed to be constant across all U.S. states," Kemp and Bossarte wrote, "an estimated 22 Veterans will have died from suicide each day in the calendar year 2010."[102]

At first glance, this figure seemed like bad news. After all, it was higher than the figure of 18 that Katz had leaked six years earlier, and a rising number provided ample opportunity for political critique. In fact, the suicide rate had not actually increased. Rather, because the VA had conducted more vigorous surveillance, it understood the parameters of the problem more fully. Kemp had anticipated this problem in 2009, telling her staff that "the better we are about asking the right questions, doing suicide risk assessments and understanding our patients the more attempts we will know about. Reporting should be going up as the program progresses."[103]

Nonetheless, the figure renewed outrage over the war and the VA's treatment of veterans from Congress, from veterans' advocates, and in the popular press. Sen. Bernie Sanders (I-VT) called the figure "an extraordinary tragedy which speaks to the horror of war and the need for us to do a much better job of assisting our soldiers and their families after they return home." Rep. Jeff Miller (FL-R) promised a hearing "to find out 'if the VA's complex system of mental health and suicide prevention services (is) improving the health and wellness of our heroes in need.'"[104] In the *Washington Post*, Elspeth Ritchie, retired from the Army, wondered "if the VA wants to get its arms around this problem, why does it have such a small number of people working on it?"[105] To drive home the point that the VA didn't seem to be devoting many resources to the problem, the *Post* reported that the data had been crunched by "Bossarte and his sole assistant."[106] While Kemp and Bossarte pointed to epidemiological factors that might explain veteran suicide—that the suicide rate was increasing among all Americans, or that veterans tended to have more risk factors than non-veterans, including "being male; living in a rural area, particularly in the West; and having access to firearm"—veterans groups were quick to condemn the high number as "a crisis" and to insist "that for a variety of reasons—including the fact that many veterans have access to health care through the department—the suicide rate for veterans should be much lower than it is."[107]

This well-meaning outrage in some ways misinterpreted the results of Bossarte's study, but it highlights the degree to which, even as the VA struggled to get its arms around the problem, its efforts, resources, and indeed attitude seemed inadequate to the task. Indeed, while Bossarte was crunching numbers, condemnations of the VA continued to come from all directions. Most explicit was the May 2011 ruling on *Veterans For Common Sense v. Shinseki*, the case that had been wending its way through the courts for so long that the defendant's name had changed when the Obama administration took office. The complaint had encapsulated all of Americans' frustrations with the VA—the sense that insufficient care was available and that the agency had little interest in fulfilling its mandate.

Judge Stephen Reinhardt of the U.S. Court of Appeals for the Ninth Circuit agreed. "The VA's unchecked incompetence has gone on long enough," he wrote. "No more veterans should be compelled to agonize or perish while the government fails to perform its obligations. Having chosen to honor and provide or our veterans by guaranteeing them the mental health care and other critical benefits to which they are entitled, the government may not deprive them of that support though unchallengeable and interminable delays."[108] But Reinhardt reserved his most withering condemnation for the inexplicability of the VA's failure: "If there is any justification for the VA's in maintaining the status quo, it has not told us, and we cannot imagine one," he wrote, adding, "In fact, the only governmental interest we can conceive of is the same as Veterans': expediting the provision of mental health care to save the lives of men and women who have fought for our country."[109] The ruling compelled the VA to immediately provide more complete mental health programming and resolve benefits claims in a timelier fashion.[110]

The decision seemed, finally, an acknowledgment of how deeply the VA had failed its constituents, and it promised some relief. Newspapers around the country celebrated the decision and encouraged the Obama administration not to appeal.[111] Nonetheless, the administration appealed in August, claiming that the court did not have jurisdiction over an executive branch agency.[112] On November 16, the court agreed to rehear the case.[113] Responding to the appeal, one *Huffington Post* article dismissed the administration's claims to have vastly improved veterans' care and suicide prevention in familiar terms: "There are countless stories of being denied disability after a 20-minute interview, being given pills rather than meaningful treatment, having to wait months to be assigned a therapist, having disability benefits arbitrarily terminated."[114] Six month later, however, the court sided with the administration. In an opinion written by Jay Bybee—now a federal judge, but best remembered as the Bush administration Justice Department official who had authored memoranda authorizing torture—the nearly unanimous court found that "we lack jurisdiction to afford such relief because Congress, in its discretion, has elected to place judicial review of claims related to the provision of veterans' benefits beyond our reach" and ordered the case dismissed.[115] The ruling contributed to the enduring sense that the VA was not accountable to veterans or, perhaps, to anyone else.

Two weeks earlier, on April 25, 2012, Patty Murray (D-WA) gaveled the Senate Veterans Affairs Committee to order, remarking that "in each of the previous hearings, the Committee heard from the VA how accessible mental health care services were. This was inconsistent with what we heard from veterans and the VA mental health care providers."[116] Over the next hour and forty-five minutes, the VA's shortcomings in providing care, its perceived disingenuousness in reporting how well it was doing, and its officials' defensive protestations were on full display.

The hearing had been occasioned by the release two days earlier of the VA Office of the Inspector General's *Review of Veterans' Access to Mental Health Care*. The results were not, one imagines, what the VA had been hoping for. The executive summary noted, in bold type, that "VHA's Mental Health Performance Data Is Not Accurate or Reliable," and noted other dispiriting findings.[117] In 2010, the VA had mandated that all new patients receive a mental health screening within twenty-four hours and a follow-up evaluation within two weeks.[118] The VA's 2012 report found that although VA facilities claimed that they were almost always meeting this standard, in fact the number of patients screened within the prescribed window was closer to two-thirds.[119] More troubling was that the staff at VA facilities had determined that they could meet the requirement by "us[ing] the next available appointment slot as the desired appointment date for new patients," which led to results indicating that almost every veteran got the appointment that he or she wanted and that there was thus no wait time.[120] Perhaps most troubling, though, was that, because the treatment of high-risk patients was tied to the overall evaluation of the facilities, and many facilities lacked resources to treat them adequately within the period that the VA had mandated, some facilities began keeping patients off the high-risk list in order to protect themselves.[121]

This cooking of the VA's books resulted from the imposition of a requirement that individual facilities lacked the resources to meet. "Despite the increase in mental health care providers," the Office of the Inspector General wrote, "VHA's mental health care service staff still did not believe that they had enough staff to handle the increased workload and consistently see patients within 14 days of the desired dates."[122] In fact, this was more than a perception. In Salisbury, NC, which is an hour north of Charlotte and where the wait time to see a psychiatrist was eighty-six days, "they had lost three psychiatrists to private practice facilities in the past year."[123] Ultimately, however, the VA wasn't doing all that badly, considering its staffing levels: The VA "had the capacity to schedule 71 percent of their patients within 14 days," and it saw 64 percent of patients within that period.[124]

The problem was thus that VA headquarters had chosen a performance metric without considering whether individual facilities in fact had the resources to meet it. As the report explained,

> Meeting a 14-day wait-time performance target for new appointments was simply not attainable given the ongoing challenge of finding sufficient provider slots to accommodate a growing demand for services. Imposing this expectation on the field before ascertaining the resources required and its ensuing broad promulgation represent an organizational leadership failure.[125]

The Office of the Inspector General's primary recommendation was thus for the VHA to determine whether "mental health staff vacancies represent a systemic issue impeding the Veterans Health Administration's ability to meet mental health timeliness goals."[126] This was a problem that the VA faced in other areas as well. Most infamously, the unrealistic mandate that all patients be seen within two weeks, and the tying of accomplishing that goal to performance reviews and employee bonuses, created a culture that facilitated the falsification of records at the Phoenix VA hospital and at other locations. Kemp, seeing this failure, requested that the VA stop using the follow-up visits as a performance metric because "patients will not be flagged when clinically such attention to true risk is needed, and consequently patients will not receive the care and attention needed."[127] This problem pointed to a larger flaw in the VA's suicide prevention efforts: However well meaning and well researched the initiatives of the VA's Center of Excellence in Suicide Prevention were, underresourced local facilities often struggled to implement them.

* * *

These ideas were evident two days after the report's release in the hearing before Sen. Patty Murray's committee. One of the review's authors, John Daigh, testified that he "believe[d] VA provides very high quality health care to its veterans" and "with respect to quality metrics . . . leads the Nation" but that "access to care metrics" told "quite a different story."[128] Daigh was followed by Nicholas Tolentino, a former Navy corpsman and Iraq combat veteran who had become a VA mental health counselor in Manchester, New Hampshire, before quitting in disgust. "The VA's mental health system is deeply flawed," he explained, complaining that "the goal was to see as many veterans as possible, but not necessarily to provide them the treatment they needed."[129] Counselors were encouraged to medicate veterans and to pack group therapy sessions even if that treatment wasn't appropriate.[130] One supervisor had told him to "have contact with as many veterans as we can, even if we aren't able to help them."[131]

Tolentino's testimony resonated with many veterans whose contact with the VA had little to do with directives that came from Kemp or Schoenhard and everything to do with what happened when they showed up at their local clinic. In response to these critiques, the VA looked, predictably, haplessly bureaucratic. When Murray asked Schoenhard whether Tolentino was correct that hospitals were "gaming the system," he remarked only that the VA did not condone that behavior.[132] When she commented that, "despite having heard about this for 7 years now, here we are today," and whether they "should we be more optimistic this time," he promised "a vision of a modern scheduling package that would, among other things, provide patients the ability to make their own appointments."[133]

Schoenhard's testimony did little to challenge the notion that the VA was an out of touch bureaucracy. On Memorial Day of 2012, Mike Scotti complained in the *New York Times* that "too many veterans waging a lonely and emotional struggle to resume a normal life continue to find the [VA] a source of disappointment rather than healing." Scotti had served as a Marine forward artillery observer during the 2003 invasion of Iraq. In the op-ed, he described his own suicidal feelings and those of his fellow Marines and spoke of the stigma that many returning veterans felt, but he complained that veterans "had gone to the V.A. because they had suicidal thoughts, only to receive a preliminary screening, a pat on the back, a prescription for antidepressants—and a follow-up appointment for several months later." The VA, meanwhile, was marked by "institutional indifference," and the VA representative who had met Scotti's brigade when it returned from Iraq had "a cold, unfeeling manner" that "spoke volumes to me of what I might expect at home." Scotti called for a more responsive VA, one that provided appointments more quickly and didn't rely on the Suicide Prevention Hotline, which he called "a temporary salve," but an effective governmental response seemed impossible to imagine. Privatizing veterans' recovery by relying on non-profit entities that focus on "the importance of interpersonal relationships, goal-setting and outdoor, rehabilitative retreats," he argued, constituted a better option.[134]

This perspective also infuses Scotti's 2012 memoir, *The Blue Cascade: A Memoir of Life after War*.[135] The book, like Clint Van Winkle's 2009 *Soft Spots* and Roxana Robinson's 2013 novel *Sparta*, presents veterans who were struggling with post-traumatic stress and suicidal ideations as also faced with an ineffective, antiquated, and uncaring VA more interested in dispensing medication than providing individualized care.[136] Persistently aligning Iraq veterans' experiences with those of older veterans—particularly Vietnam veterans—*The Blue Cascade* and *Sparta* posit these issues as an enduring crisis in American culture. These popular representations thus stand as potent rejoinders to the VA's claims that it has progressed in its ability to help struggling veterans through increased understanding, outreach, and commitment of resources. In the culture of indifference that these texts highlight, there is no help beyond self-help for suicidal veterans.

Scotti and Conrad Farrell, the fictional protagonist of Robinson's *Sparta*, recall the young men whose suicides first focused critical attention on the VA's efforts. Scotti's youth is marked by the same idyllic patriotism described in coverage of Tim Bowman, Jeffrey Lucey, and Chuck Call.[137] "Our tree-lined neighborhood was full of adventure and good friends," he writes, and it was "a happy childhood filled with love and compassion from my family," but "from an early age, I always wanted to go to a war. That was a man's job: to kill the bad guys. To protect my mom and my dad and brothers and friends."[138] The character Conrad Farrell, whose life and motivations are clearly based on the experiences that Nathaniel Fick describes in his memoir *One Bullet Away*, is nearly identical.[139] Like Scotti, Conrad's childhood is nearly idyllic—the book abounds with references to his loving upper-middle-class family, the quaint Westchester County farmhouse that he grew up in, and the traditions in which he was steeped. Like Fick, he explains his decision to join the Marines by explaining that it is a calling beyond the positions that his peers have taken: "I want to do something big. I don't want to go to graduate school and get another degree. I want to do something that has consequences. This is the biggest challenge I know."[140]

Upon returning from Iraq, however, each man is beset by now familiar symptoms of psychological turmoil. Both Scotti and Robinson begin their respective works with fraught homecomings on Camp Pendleton's parade grounds. Even amid these celebrations, Scotti asks, "How many of these happy unions would soon end because hubby got drunk and saw demons or dead Iraqi kids and Marines and then his wife was the enemy and he wanted out of *all of this*."[141]

Conrad Farrell may as well have been one of the men about whom Scotti was thinking, for the character is haunted by intrusive memories that "came crowding in on him, packed and massing, things he kept out during daytime, things that should be gone and over and done."[142] Walking through New York City at Christmastime, he can see only his men being burned to death inside a Humvee.[143] Scotti, who predicted these memories in others, is hardly immune to them himself. On the Fourth of July, his pride at having served gives way to "the silent series of snapshots layered one upon another: dead dogs on the side of the road with tongues hanging out of their mouths, a dead child with black eyes, Marines dying in helicopter crashes, Marines dying in gun battles, Marines . . . destroying other people's homes with artillery, killing other human beings violently."[144] Unsurprisingly, both men are beset with survivor's guilt.[145]

The haunting impact of their own remembrances are compounded by two related assertions about the veterans' experience: that their fellow Americans cannot—and don't care to—understand their experiences and that the faulty justifications for the war and their sacrifices constitute a betrayal. Both veterans thus finds themselves isolated in a world incapable of understanding the realities of war and in which their sense of betrayal is not shared by those who did not share in the suffering. In *Sparta*, Conrad walks with his parents to their rental car, reflecting that

> he was among people for whom there was no dark undertow. Here there were no sudden black boiling clouds, no exploding vehicles. There would be no crack of an AK-47, no smell of burning flesh. . . . No one would round a corner to find something lying on the street, ripped open and gasping, ruby-colored, terrible. These wives and children and parents, with their cameras and rental cars and balloons, were exempt from all that.[146]

Like Van Winkle four years earlier, Robinson's Conrad finds that "none of this—the dim, enfolding clouds of anxiety, the rising choke of panic, the throttling claustrophobia, the straight-out jolts of terror—had anything to do with the bright, calm world where his parents and everyone else lived."[147] Scotti more cynically asks whether wedding guests "*know that we were in the middle of a war?*" and concludes that "someone who had never been through combat could never even remotely understand what I had been through. . . . I was a creature from some other planet."[148]

Each veteran's sense of themselves as an unappreciated alien is compounded by the growing sense that he had been betrayed by the administration that sent them to the war. As Conrad becomes increasingly distraught, he drunkenly disrupts his family's Christmas Eve celebration and counters their entreaties that he behave by arguing that

> I obeyed the rules when I went over there. But they didn't work. I ended up doing things I should never have done, by any rules. . . . I watched my men die. I watched our troops kill civilians. . . . We went over there for no reason, there were no [weapons of mass destruction]. . . . We lost our men for a lie. What is this about rules?[149]

Scotti has a nearly identical reaction, "finally admit[ing] to myself that the reason no one found weapons of mass destruction in Iraq wasn't because we didn't search carefully enough" and "wonder[ing] if it was just all a myth. If I had fought for my country. If I had made innocent American citizens more safe. Or was everything I had ever believed in just complete and utter bullshit?"[150]

These intertwined frustrations of disconnectedness and betrayal lead both Scotti and Conrad Farrell to contemplate suicide. Indeed, Scotti is emphatic that the Bush administration's dissembling is what pushed him to the brink of suicide. In an internal dialogue, he explains that he "want[s] to die" because "there were no fucking weapons. And that was the point of the whole goddamn war," and as a result "my honor is shit. Everything is shit."[151] Wondering whether "all of these things I'd believed in for my whole life *really were* just a myth," he also wonders whether he "should just raise my hand and say, 'Fuck it. I've had enough'" and "make it look like an accident. . . . Go for a run. Right into the path of a bus."[152] Conrad's decision to take his own life—an effort thwarted by his brother—is *Sparta*'s climax. If partly that decision emerges from a desire to "never again have another fucking nightmare," it is also a means of "never again hav[ing] to look at someone and speak to them in one world, theirs, while you were holding that other world, yours, with its black sinkholes, sealed in your head."[153] As in popular media coverage, then, in these two texts the veteran's suicidal ideation indicts a dissembling, war-hungry administration's exploitation of young Americans' idealistic patriotism and a culture that remains fundamentally separate from the military, eager to welcome troops home but not to listen to their stories.

These intertwined narratives of betrayal and indifference also define each protagonist's relationship with the VA. Although each text acknowledges stigma as an issue that veterans must overcome, the government entity charged with helping veterans appears antiquated, indifferent, and incompetent. Scotti, in fact, never makes it to the VA. His only encounter is the "PTSD seminar" at Camp Pendleton that he describes as a "true child of bureaucracy" with an "uninspired speaker."[154] Echoing the sentiments expressed at congressional hearings, he is incredulous that they "expect guys in their early and midtwenties . . . to immediately admit that they might be feeling symptoms of posttraumatic stress" and wonders whether "they have any idea at all who we are."[155]

Similar accounts of ignorance and indifference mark *Sparta*. Just as Van Winkle found an outdated magazine in the lobby, Conrad discovers a calendar for "October, the year before" on the wall of his examination room, a sign that the VA is, literally, out of date.[156] As in *Soft Spots*, *Sparta's* VA is an organization marked by callous staff and bureaucratic indifference. When Conrad, frustrated that he hasn't been given a chance to talk about his experiences, asks the doctor if he "need[s] to know any more[,]" the doctor responds, "'We get a lot of men with these symptoms."[157] Conrad leaves feeling even more distraught because "there would be no discussion of what had happened to him. . . . All this meant nothing to anyone but him. He was trapped with it forever."[158]

That the VA's ineffectiveness and indifference is chronic is heightened by the presence in each text of Vietnam veterans who figure as an indicator of the culture's indifference, the VA's historic failures, and Iraq and Afghanistan veterans' bleak futures. It is a Vietnam veteran who tells Scotti, "I'll tell you one thing that you better get used to. Nobody back home cares about what's going on over there. . . . The quicker you realize that, the quicker you can move on with your life."[159] In *Sparta*, it is a Vietnam veteran who tells Conrad that "it's always three months" to get an appointment, as Conrad reflects, "*Thirty years. . . . He's been coming here for thirty fucking years and he's still no better.*"[160]

Again echoing congressional testimony, each text suggests that veterans' only source of relief comes through conversations with other veterans. This is Conrad Farrell's expressed desire, although his only contact with such men is through emails with his former platoon members.[161] Scotti meets Korea and Vietnam veterans and immediately feels kinship with them, noting that "there was the trust between men who have been under fire, even if it was in wars thirty years apart."[162] Here the older veteran performs a social function opposite that of the older veteran portrayed in the VA's "The Power of 1" public service announcement. If there is shared suffering, the older veteran in Scotti's memoir is hardly a gateway to care; rather, he is evidence that that suffering is interminable and that the system designed to resolve it has failed.

The resonance between Van Winkle's, Scotti's, and Robinson's vision of wartime U.S. culture and the critiques of the VA that dominated media coverage and congressional testimony illustrate how little the public perception of the VA changed during the years that Kemp and her staff worked assiduously to combat veteran suicide. There was, however, a more positive representation of the VA. On Veterans' Day 2013, HBO premiered Dana Heinz Perry's and Ellen Goosenberg Kent's documentary *Crisis Line: Veterans Press 1*. In an email to her staff, Kemp called it a "fantastic" film that "really speaks to the incredible work that the [Veterans Crisis Line] and all of you do."[163] Indeed, it does offer a sharp contrast to the critiques that Van Winkle, Scotti, and Robinson present. Throughout the documentary, the VA appears competent and sympathetic, with dedicated staffers who respect military service—reminding one veteran, for example, that his children's "dad is a Marine. Do you know what that means to them?"—and in which they both encourage veterans to seek psychological care and assist them in doing so.

It is also an organization filled with staffers who, far from being indifferent to veteran suffering, not only spend hours trying to get a single veteran into care but also bear the emotional scars of their work. Indeed, one of the documentary's central themes is the psychological toll on hotline operators. Robert, one of the operators, describes "the full range of emotions" that he feels when a rescue attempt fails. "I start to question whether I did a good enough job," he says, before admitting that his feelings often transition to anger: "How dare you take your life?" In moments like this, the film establishes a sharp contrast with the VA's portrayal in *Soft Spots*, *The Blue Cascade*, and *Sparta*.

* * *

Crisis Line: Veterans Press 1 won an Oscar for best short documentary on February 22, 2015. However, the VA's suicide prevention efforts continued to garner controversy from all corners. From the political right, both FOX News and the conservative *Washington Times* had in the weeks preceding the film's recognition condemned the VA's efforts to provide gun locks to veterans as "raising concerns about a government-run gun registry."[164] Firearms posed a particular issue, as veterans tend to own and know how to use them. In fact, nearly seven in ten veteran suicides are committed with a firearm.[165] To address this issue, on October 30, 2009, Kemp promised her coordinators "more gun locks than you can imagine."[166] The VA had received nearly a million locks, which are designed to keep a user from being able to load a firearm, from the National Shooting Sports Foundation, and within eighteen months it had distributed 242,100 of them at no cost.[167] By 2013, the VA, which in an act of artfulness pitched the program as a way to keep children safe, was distributing them with no questions asked in whatever quantity veterans requested.[168]

The gun-lock program, called Operation ChildSafe, was part of a much larger effort to reduce veterans' access to a means of self-harm that included better securing poisons and eliminating hanging risks in VA facilities by replacing door hinges, cutting insulation into segments less than a foot long, mandating that wireless Internet routers be covered with "a shatterproof cover that is installed and secured with tamper resistance [*sic*] screws," securing "trash can liners" and "bathrobe sashes," and printing the Veterans Crisis Line number on pill bottles.[169] The program, of course, did not remove the means of suicide from veterans' homes—it merely created a temporary barrier to their use. That barrier is important, however, because most people do not contemplate suicide for very long—often for only a few minutes, the amount of time it takes to smoke a few cigarettes—before attempting it.[170]

Although VA staff and hotline operators often encouraged veterans to remove guns from their homes, it is doubtful that a more ambitious program to limit veterans' access to firearms could have succeeded.[171] Programs that aim to remove guns from veterans' homes are destined to fail for the same reasons that the guns are more likely to be there in the first place: Veterans often come from parts of American society—regionally and culturally—that are deeply attached to guns and deeply skeptical of gun control. Moreover, although many Americans favor greater gun control, a powerful and vocal minority views any encroachment on gun possession as an unreasonable infringement of the Second Amendment.

It was this minority that became critical of programs like Operation ChildSafe. First, in 2013, Rep. Steve Stockman (R-TX) introduced the Veterans Second Amendment Protection Act, which stipulates the requirement that a veteran can only be found mentally incompetent—and thus be stripped of his or her weapons—through the "finding of a judge, magistrate, or other judicial authority."[172] In 2015, the *Washington Times* and FOX News captured—and helped stoke—anxiety about veterans' gun rights through a misreading of the form letter that the VA sent out offering the locks, which simply asked veterans to note how many locks they were requesting. It revealed once again that VA suicide prevention efforts had to navigate a complex terrain in which its efforts were always suspect.

It was not only the political right that remained critical of the VA's efforts. From the left came the Clay Hunt Suicide Prevention Bill, which Barack Obama signed into law in 2015. The provisions were familiar. The law required the VA to better advertise its services and providers, covered the educational expenses of doctoral-level mental health providers who worked for the VA, and called for new research on suicide prevention.[173] The bill was named for Clay Hunt, a Marine who had taken his life in 2011 after serving in Iraq during the 2007 troop surge and whose mother, Susan Selke, described her son's experiences, also in familiar language: "Clay consistently voiced concerns about the care he was receiving, both in terms of the challenges he faced with scheduling appointments as well as the treatment he was receiving for PTSD, which consisted primarily of medication."[174]

The VA's argument, however, remained one of considerable progress despite vastly inadequate resources. Harold Kudler, the VA's chief mental health consultant, told the committee that, although the suicide rate nationally had increased, it was decreasing among veterans who received

VA care, that the VA was treating half again as many patients in 2014 as it had been in 2006, that it had nearly doubled the number of mental health staff, and that the Veterans Crisis Line had referred more than 200,000 veterans to VA care and was rescuing more than two dozen suicidal veterans each day.[175] He explained the efficacy of the evidence-based therapies that the VA had embraced and declared that veterans received "warm and personal care."[176] He acknowledged, however, that these efforts had hardly solved the problem. Although Congress had recently appropriated five billion dollars to the VA, the money was only "a great stop gap. We will be able to squeeze a little more juice from the lemon that exists, but . . . there is not enough juice in the lemon to cover all the capacity that is needed."[177]

* * *

The history of the VA's efforts to address veteran suicide and of veterans' experiences with the VA between 2007 and 2014 reveal the accuracy of both Susan Selke's and Harold Kudler's arguments. Officials at VA headquarters and at the Canandaigua, New York, Center of Excellence in Suicide Prevention, having been slow to acknowledge the magnitude of the issue that they faced and subject to increasing congressional and public pressure, worked assiduously to understand a complicated public health problem and save veterans' lives. Between 2008 and 2015, the VA gained a better understanding of how many veterans took their own lives, which veterans were most at risk, and what interventions were most promising, and it consistently sought to present itself as responsive and caring. Through the creation of the crisis line alone, thousands of suffering veterans were kept from harming themselves and were brought into the orbit of VA care. And when veterans did enter the VA system, they generally received quite good care.

Recognizing these efforts, however, tells only part of the story. However well researched and evidence based the interventions that VA national staff developed were, a lack of resources and a set of frustrating protocols frequently led to their failed implementation. Indeed, as one witness before Congress put it, "The challenge to the VA is the execution and implementation of these policies."[178] Mental health providers and veterans complained of insufficient staff and long wait times, and veterans continued to take their lives after seeking, but not finding, VA care. Before Congress, in the popular press, and in literature about veterans' experience, the VA continued to appear antiquated, intractable, and indifferent. Thus, while the commitment of many VA staffers, and the quality of care that they provide, is beyond question, the issue of veteran suicide during and after the Iraq and Afghanistan Wars highlighted the tragedies that inevitably ensue when the nation fails to prepare for the aftermath of the wars that it chooses to wage.

Conclusion

"They Will Start to Bring the . . . Lessons That They Learned Back into Their Communities"

Rhonda Cornum was a urologist by training, but, as she put it, "I did happen to have a pretty compelling resilience story of my own, which made it more likely that I would have credibility."[1] During the 1991 Gulf War, her helicopter had been shot down, and she was gravely injured, captured by the Iraqi army, and held for several days, during which time she was sexually assaulted.[2] Like many of the encounters that defined the Army's approach to mental health during the Iraq and Afghanistan Wars, the incident that led to Cornum being tapped to lead the Army's Comprehensive Soldier Fitness (CSF) program was somewhat serendipitous. The program was an effort to teach soldiers and their families how to face of the demands of military life with more resilience, and it began in 2008 and ran alongside the Army's other efforts at improving behavioral and mental health outcomes. Rhonda Cornum ended up leading it because Eric Schoomaker was in the hospital.

In the summer of 2008, Cornum was a newly promoted brigadier general and the Assistant Surgeon General. That meant that, when she went to meetings with George Casey, she usually sat along the back wall while Eric Schoomaker sat at the table. At one particular meeting, however, Cornum took Schoomaker's place "at yet another suicide prevention briefing."[3] As she remembers it, Casey "turned to me and he said, 'Rhonda, we have got to do something [other] than just find and treat more and more disease.' And I said, 'Yes sir, we need to make people more psychologically fit before they are exposed to some adverse experience.'"[4] Cornum left the meeting thinking that she had the opportunity to do some research that might help the Army embrace resiliency training—something she had attempted to implement on a smaller scale at Fort McPherson, Georgia, a few years earlier—and brief it to Casey, who, she expected, "would say 'Yes, go do that.'" Her faulty assumption, though, was in believing that "he'd tell someone else to go do that."[5]

Casey was eager to embrace such a program. When he became Chief of Staff in 2007, he brought with him a hard-won understanding of the psychological impact that the war was having on the men and women fighting it. In addition to worrying about what the wars had done to American service members, he ruminated over how poorly prepared soldiers had been for what they experienced. When he considered the resilience skills that people came in with, he recalls, he found that "they were precious few."[6] And when he asked to be shown the Army's programs to "identify, treat, and prevent" mental health issues, he noticed that "they were all identify and treat."[7] Something, he thought, had to be done to better prepare soldiers and their families for deployment's rigors. According to Eric Schoomaker, who served as Casey's Surgeon General, Casey "was one of the first that really forced the issue of 'I see what you're doing when people become ill . . . in the way of formal treatment, but what are you doing to keep people to the left of that threshold?'"[8]

The Army's turn to resiliency can thus be traced to Casey's early weeks as Chief of Staff.[9] This shift was rooted in the notion that people can be trained to become better able to withstand stressful encounters. It became the central plank in the larger Comprehensive Soldier Fitness program, which the Army touts as possibly "the largest deliberate psychological intervention in history" and which, according to Schoomaker's successor Lt. Gen. Patricia Horoho, asked "How do you look at prevention and how do you look at the total person? . . . It really was mental, physical, spiritual, emotional, and financial, when we were trying to provide care to our soldiers and really look at resiliency instead of looking at it from a behavioral health aspect."[10]

The turn toward this model represents an important evolution in the Army's thinking about mental health in the twenty-first century, but it also reveals the extent to which approaches to mental and behavioral health problems continue to be thoroughly debated inside the Army. The army embraced resiliency out of genuine concern for the plight of soldiers and their families, but in doing so it embraced a neoliberal model that seeks to create healthy and productive workers while failing to address the immiserating conditions that they face.[11] In this conclusion, I do not aim to provide a comprehensive history of the Army's resiliency programs or how soldiers and families received them.Rather, I want to illustrate how this shift, like every other aspect of mental health care during the Iraq and Afghanistan Wars, was hotly debated inside the Army. Although many senior leaders and Army Medical Department (AMEDD) officers saw it as a necessity, others were concerned that the shift toward teaching resilience took attention away from treatment and downplayed the realities of the psychological damage that war produced. And yet this embrace of resiliency should not result in a condemnation of the Army. As an institution beholden to its civilian masters, as Peter Chiarelli and others pointed out, the Army had little control over the demands placed on the force. Thus critique should perhaps be directed to a culture that, in acquiescing to a permanent state of war, has made resiliency the Army's only viable path.

* * *

The Army's decision to focus on the prevention of mental health issues as well as on treating them was, essentially, the application of the standard model of disease care. In a public health model, the goal is not only to treat those diagnosed so they can be cured or can manage their symptoms in order to retain their standard of living. It is also to help a population not become ill in the first place. In epidemiological terms, wellness in regard to one variable—say, high blood pressure or lung cancer—can be imagined as a spectrum from very healthy to very ill. If this were the *x* axis of a graph, the *y* axis would be the population, with data points indicating which percentage of the population fell at which point on the wellness scale, ultimately forming a bell curve. The goal of public health programs is thus to reduce the incidence of disease, effectively moving the curve to the left by lowering the percent of the population affected. In theory, cutting down on risk factors and encouraging healthy behavior accomplishes this. That's the goal of vaccinations, smoking-cessation programs, seatbelt laws, and programs that place affordable farmer's markets in food deserts—to produce fewer cases of typhoid fever, lung cancer, trauma deaths, type-II diabetes, and so on.

Notably, of course, prevention campaigns are supposed to go hand in hand with better treatment. Thus, a seatbelt is a preventive measure, but there is still the need for ambulances and trauma centers. This model has historically applied to physiological military medicine during wartime, Schoomaker explains: "We have to design better body armor, design better vehicles so that if you hit an [improvised explosive device] you survive it or you don't get injured. At the same time, if you do get injured we have to treat you."[12] Casey's notion was that this model could be applied to mental and behavioral health. As anthropologist Emily Sogn explains, "Resilience marks the boundaries between 'healthy' and 'unhealthy' responses to war," and "the notion that human responses to trauma fall along a 'bell curve' is an important insight that [experts] argue illustrates the importance of studying resilience."[13] Casey believed that there were things that the Army could do to produce a healthier population capable of weathering the rigors of prolonged war. His question, Schoomaker recalls, was, "What are we doing to shift the whole curve to the left?"[14]

Importantly, Comprehensive Soldier Fitness was not strictly a behavioral health program, and it was not under the AMEDD's control. Schoomaker advised Casey to move it to Army Operations because, in his view, "If you make it a medics problem . . . it won't be well received by the force."[15] Casey agreed and placed the program under Army Operations rather than the Medical Command. In his view, "The docs aren't going to change the culture of the Army. The leaders will."[16]

To create Comprehensive Soldier Fitness, the Army adapted a positive psychology program that had been created by Martin Seligman, a University of Pennsylvania psychologist. In its civilian form, the program trained people to cope with adversity in their personal and professional lives through an essentially neoliberal model that aimed to produce more productive workers by enabling them to cope with the dissatisfaction inherent in contemporary capitalism.[17] As Seligman described it in the *Harvard Business Review*, "Failure is a nearly inevitable part of work; and along with dashed romance, it is one of life's common traumas. [Some people] are almost certain to find their careers stymied, and companies full of such employees are doomed in hard times. It is [more resilient] people who rise to the top." He then promised that "we have learned not only how to distinguish those who will grow after failure from those who will collapse, but also how to build the skills of people in the latter category."[18] When Cornum and Casey approached him in 2008, he writes, he offered a vision in which soldiers could be made more effective and productive by becoming more resilient to failure as a means solving the Army's mental health issues. Reiterating Schoomaker's description of the desire to shift the curve to the left, he writes:

> How human beings react to extreme adversity is normally distributed. On one end are the people who fall apart into PTSD, depression, and even suicide. In the middle are most people, who at first react with symptoms of depression and anxiety but within a month or so are, by physical and psychological measures, back where they were before the trauma. That is resilience. . . . I told General Casey that the army could shift its distribution toward the growth end by teaching psychological skills to stop the downward spiral that often follows failure.[19]

Notably, Seligman here casts being traumatized in war as a "failure," a notion that Cornum also embraced. "Frankly," she told me, "I think we ought to stop being euphemistic and calling it 'mental health.' It really is a psychological dysfunction. We don't send people home when they have got a broken leg saying it's a 'bone health' problem."[20] In both views, psychological diagnoses are imagined as deviations from an imagined acceptable normal capacity, a view that seemingly marks a troubling departure from the Army's historic embrace of mental health issues as "normal responses to abnormal events" and which potentially generates stigma around mental health issues.[21]

The Army embraced Seligman's approach. Cornum argues, "These are skills that people need to deal with any circumstance in their life . . . whether it is disappointment from getting turned down for the prom or not getting a good grade or whether it's not getting promoted if you are in the Army or whether it's being away from your family because you went away to college or being away from your family because you are deployed."[22] In her view, the skills that made a person adaptable to the stresses of everyday life would also help that person become a good soldier. "The thought processes that make you successful are not unique to the Army," she maintains.[23] Amy Adler, who worked on integrating the BATTLEMIND program into Comprehensive Soldier Fitness, agrees. Of Seligman's program, she says, "The basic premise is 'how you think about stuff matters.' That's it. That is the brilliant contribution of this program: How you think matters for how you feel and how you experience the world. . . . It's pretty basic stuff. So I did not have any trouble with the concept of integrating positive psychology in what we were doing."[24]

Over the next four years, the Army increasingly invested in resiliency training. Using the same chain-teaching method that Casey rolled out in an effort to destigmatize post-traumatic stress disorder, in 2009 the Army began sending non-commissioned officers to take the course that Seligman and another Penn psychologist, Karen Ryvich, had developed and become "Master Resiliency Trainers," which would allow them to then teach the material to their subordinates.[25] In Casey's view, these trainers were the key to the entire program's success. "The primary stigma reduction tool was master resiliency training," he argues. Anything else the Army did "wasn't going to make a difference until sergeants were talking to soldiers."[26] In that sense, Comprehensive Soldier Fitness embraced the ethos of the army's suicide prevention efforts and the Embedded Behavioral Health program.

By 2011, the Army was touting the program's success. "There is now sound scientific evidence that Comprehensive Soldier Fitness improves the resilience and psychological health of Soldiers," a December report by Army psychologist Paul Lester exclaimed.[27] The units that had an NCO certified as a Master Resiliency Trainer "exhibited significantly higher [resilience/psychological health] scores" after the implementation of Comprehensive Soldier Fitness than groups whose NCOs had not completed the training.[28] Soldiers in those units showed improvements in "emotional fitness," "social fitness," "adaptability," "good coping," "optimism," and "friendship" while scoring "significantly lower on catastrophizing."[29]

Outside the army, some psychologists worried that the emphasis on resilience was training soldiers to have no issues with killing.[30] As well, historian Jennifer Mittelstadt has compellingly argued that resilience places the onus for well-being too much on the soldier, rather than on the Army that puts him or her in harm's way, and geographer Alison Howell argues that it is a means of "wag[ing] war more effectively" while "reduc[ing] healthcare costs."[31] Inside the Army, however, evidence touting the program's effectiveness was further evidence of its value, and the Comprehensive Soldier Fitness program increasingly became the primary avenue through which behavioral health would be provided. As well, Sogn notes, "Many descriptions of CSF attest to a hope that the project will provide the basis for broader research efforts that extend beyond military communities."[32] Rhonda Cornum and Carl Castro, for example, argue that it should be taught in public schools. "One of my biggest frustrations," Castro admits, was that

> I could never get the Chief of Staff of the Army or anyone in the military to go talk to the Department of Education to see what we can do about instilling resilience in children, because that's where resilience needs to be built. Not once we get them. . . . I always thought our strategic approach as a military was flawed, waiting until you join the military before they did anything. You really needed to be looking much, much earlier.[33]

Particularly striking about Castro's comment is that resiliency is important because it will facilitate continued militarism. In his view, "we"—the Army—will eventually "get them," a sentiment that suggests that culture-wide resilience is necessary because more wars are inevitable and a comment that underscores Howell's critique that "CSF2 [Comprehensive Soldier and Family Fitness Program, as the program was renamed in 2012] contributes to [the] expansion of warfare by creating the human resources necessary to carry out this kind of extensive and persistent form of war."[34]

The embrace of resiliency was central to the larger Comprehensive Soldier Fitness program, which also sought to streamline all of the Army's mental health programs in order to make the program more effective for clinicians and more manageable for soldiers. The increased need for care, as well as the increased attention and funding, meant that mental health programs had proliferated in the previous five years—what Vice Chief of Staff Peter Chiarelli described as "a thousand flowers blooming."[35] This led many soldiers to struggle getting appropriate care, even as options proliferated. "Because people didn't know how to navigate the system, we had more programs than people knew how to access," Horoho remembers.[36] There were the various iterations of BATTLEMIND, Embedded Behavioral Health, Strong Bonds, The Army Family Covenant, suicide prevention, RESPECT-Mil and others, totaling 230 programs aimed at soldier and family wellness. What this meant is that soldiers at one post might be getting radically different care than soldiers at another post or that their care would change radically if they were moved to a new location.[37]

The task of ensuring that there was "one kind of flower" fell to Rebecca Porter, who described her three years as head of Army behavioral health as "the hardest job I've ever had in the Army."[38] She spent a year traveling throughout the Army's installations, learning how different hospitals were doing everything from screening soldiers to prescribing drugs before determining which programs were most effective and which should be consolidated or eliminated.[39] Porter and the Surgeon General also worked to create Army-wide standards so soldiers and families who changed posts would be assured that their treatment would not change.[40] In the end, the Army reduced the number of behavioral health programs from 230 to 11.[41]

The narrowing down of programs pushed all behavioral health under the umbrella of Comprehensive Soldier Fitness. This, along with the increased privileging of resilience, frustrated some of the people who had spent the previous few years developing programs and who felt that this new emphasis did not pay enough attention to treatment. Even Chiarelli, who championed the streamlining, admits to being "a reluctant warrior on the resiliency front. I felt that people had gotten the ear of the Chief [and] claimed you could train post-traumatic stress out of someone." And while he is in favor of helping people become more resilient,

> I do not believe that it is the panacea. . . . I believe that there are certain traumatic events that somebody is going to go through, and . . . they could have all the resiliency training in the world, but they are going to come out of it with post-traumatic stress. . . . I mean, you get good colds and bad colds right? Some colds develop into pneumonia.[42]

Others worried that that the pendulum had swung too far toward resiliency at the expense of other meaningful programs. When Chiarelli retired, the after-action reviews of suicides stopped. Horoho wasn't sure that they were beneficial. "What I found over time is [that] they were reflective of the past, [and] I got to a point where I am not sure they were as effective as they could have been," she remembers. Instead, she explains, the Army began "looking at it from what should we be doing to really improve resiliency and readiness."[43]

While this decision likely pleased those in AMEDD who questioned those sessions' utility, emphasizing resilience in suicide prevention didn't set as well with Bruce Shahbaz, one of the primary authors of the 2010 suicide prevention handbook. Although he believed that the "new leaders should have the opportunity to shape what it is they want the organization to be doing" and that the new leadership was "just taking a different tack from what previous folks had done," he was skeptical of the emphasis on resilience.[44] He recalls telling Horoho that resilience wasn't going to help people who already had mental health issues. "Telling someone who has cancer that they should eat better and exercise more to prevent cancer really doesn't help that person with cancer," he explains. "They need chemotherapy."[45] His fear was that "we were swinging perhaps too far in the direction of prevention without acknowledging that we still needed to be able to have the response piece in there" because "You can't necessarily build resilience in someone who is currently depressed. You need to treat them for depression and then work with them on resilience."[46]

Chuck Engel was also somewhat disappointed by the Army's shift to resilience. In one sense, Engel was skeptical of the need to standardize mental health across the entire Army. Although the current vogue was to "come up with a uniform process that creates a uniform product," he explains, he believed that "variation in and of itself isn't bad, unmeasured variation is bad, because variation is actually the source of innovation. But only if you measure it." Instead, he believed, "you [need a system] to assess naturally occurring innovation and decide whether it's something that could actually be put into the program."[47] However, Engel was at the end of his career, and he did not have Horoho's ear as he had had Eric Schoomaker's or Kevin Kiley's. "She brought in her own new team," he recalls, "and so a lot of the ideas, a lot of the lessons learned, I think, fell a little bit to the wayside."[48]

Particularly difficult for Engel was that another of the initiatives central to the Comprehensive Soldier Fitness program, the Patient-Centered Medical Home, undid some of his work on RESPECT-Mil. Essentially one-stop shopping for soldiers and their families, the Patient-Centered Medical Home in particular integrated "internal behavioral health consultants" into primary care facilities at the rate of one clinician "per 3,000 enrolled adult[s]."[49] For Engel, this seemed a shift away from training primary care physicians and that the concept that he had struggled to get the doctors themselves to accept had also grown unpopular with psychologists who felt that their expertise was being supplanted. "When the move to the Patient-Centered Medical Home happened, the psychologist community in particular really elbowed to the table, and I found myself kind of on the outside looking in," he recalls.[50]

For the psychiatrist who had sacrificed so much to get the program off the ground and make it work across the Army, this transition was painful. Partly, it was because he was so invested in the program. "I took such ownership over what we were doing, it was really hard for me to give it up" he admits. And while he remained very proud of what he had accomplished, he struggled to step away.[51]

It was not simply ego. Engel worried that the shift to the Patient-Centered Medical Home was not necessarily in the best interests of the patients. Although some aspects of RESPECT.Mil were integrated into the new program, he was concerned the emphasis was on making sure specialists were available in each clinic. "That's the easy thing to do," Engel believes, "because you just hire somebody, you plop them in the clinic, you say 'All done,' with no integrated mental health in primary care. Nobody monitors what that social worker is doing . . . [and] meanwhile you have a lot of people going up to the turnstyles that never get referred to that person."[52] He worried that the specialist's presence would mean less patient screening and less long-term follow-up.[53]

* * *

The shift to resiliency, the implementation of Comprehensive Soldier Fitness, and the shift to the Patient-Centered Medical Home define the Army's approach to mental health in the putative aftermath of the Iraq War and as the Afghanistan War remains a stubbornly enduring—though perhaps largely invisible to most Americans—conflict. Its evolution and embrace is also a microcosm of the debates over mental health care that have animated the entire history of these wars. Paying attention to these debates illuminates the broader cultural contours of the wars themselves and of the military's place in U.S. culture.

For many Americans outside the military, PTSD, mTBI, and related issues were troubling issues that portended permanent disability and illustrated why the wars were ill conceived and, perhaps, morally wrong.

They raised important questions about the government's attention to soldiers who were imagined as simultaneously heroic and vulnerable as well as about the wisdom and cost of the United States' imperial adventures. Pointing to troops' and veterans' mental health issues thus became a primary means through which opposition to the wars could be expressed. In a climate that equated critique of the war with a failure to support the troops, arguing that the wars were disastrous because they had endangered soldiers' and veterans' psychological well-being, damaged their brains, or prompted them to take their lives facilitated a declaration that the wars were misbegotten and mismanaged disasters while ensuring the accuser's patriotic bona fides.

Popular critiques of the wars' mental health implications often failed to appreciate the challenges that the Army and the Department of Veterans Affairs faced. Indeed, the story of efforts to address the mental health needs of service members and their families in the twenty-first century is not a history of callous or disinterested inaction; or of a few zealous leaders who motivated their recalcitrant subordinates to finally take the issue seriously; or of a linear path from ignorance to knowledge, from inability to capacity, from failure to success. Rather, it was a story of an Army that worked incredibly hard to care for soldiers under the unprecedented strain of the nation's longest wars.

Efforts to address post-traumatic stress, depression, anxiety, traumatic brain injury, and other issues were marked by uncertainty, frustration, and, not infrequently, acrimonious debate. Researchers and clinicians in the Army Medical Department, committed to deliberate and rigorous empirical research, sought to reach conclusions and develop treatments that were scientifically and ethically defensible. As they did so, they endeavored to advance a medical understanding of combat stress as often normal and treatable to a public that feared an epidemic, even as they engaged in vigorous internal debates over the etiology and treatment of post-traumatic stress and traumatic brain injury. Line commanders like George Casey and Peter Chiarelli, eager to implement strategies that could help troops who were at risk in the field or who were already suffering, grew frustrated by the AMEDD's deliberative process, which they felt moved too slowly to meet battlefield needs and refused to yield many definitive answers. At the same time, both groups struggled to create a culture that embraced—or at least did not deride—mental health care and learned that this could only be achieved through organic, bottom-up efforts that built upon positive individual experiences and trusted relationships. As they did so, the entire Army increasingly worried that its burgeoning suicide rate reflected a deterioration of the culture that it had worked so hard to cultivate in the aftermath of the Vietnam War and subtly critiqued the wars for the institutional damage that they had wrought.

The story of mental health during the United States's twenty-first-century wars, then, is not simply a medical story. It is a cultural story—one about the culture of the U.S. Army, and about contemporary U.S. culture as a whole. Both the U.S. Army and U.S. culture as a whole entered the Iraq and Afghanistan Wars predisposed to worry about the troops' mental health but having vastly underestimated how the wars would proceed and with knowledge of behavioral health that was limited in important but unrecognized ways.

The cultural history of military mental health during the Iraq and Afghanistan Wars thus offers important lessons about these wars, about the relationship between military and society, and for the future. The first is that, however acrimonious these intra- and intermural debates about mental health were, every stakeholder in these debates was deeply invested in helping American soldiers and veterans. Bob Filner thought the VA needed to be prodded to acknowledge its failures, and Jan Kemp believed that the suicide rate could be lowered with better treatment. Every time Peter Chiarelli exclaimed that his "kids" were hitting improvised explosive devices that day, Eric Schoomaker worried that an unproven treatment might do more harm than good. As much as Geoff Ling and Leanne Young's research showed that a blast wave alone could in fact damage the brain, Ling respected Charles Hoge's research illustrating the need to carefully differentiate between PTSD and mTBI because "misdiagnosing somebody . . . is a clinical tragedy."[54] If Patricia Horoho believed that an emphasis on resilience and the Patient-Centered Medical Home would help soldiers become more resilient, Bruce Shahbaz and Charles Engel worried about treating those already afflicted.[55] Sometimes these debates were quite heated, marked by distrust and defensiveness. But most of the time, they were debates about evidence and interpretation. Quite simply, popular narratives of inaction and indifference oversimplify this complexity.

This leads to the second lesson. Mental health during wartime was defined by the limitations of medical, military, and popular understandings of the issues that service members and veterans faced, the circumstances in which they occurred, and the tensions that these limits produced. Inside the medical community, whether in the Veterans Administration or the Army, behavioral health issues were medical problems to be solved through rigorous, evidence-based research. That understanding, as well as the pace at which they believed those solutions could be achieved, frequently put clinicians and researchers at odds with soldiers, veterans, leaders, families, and politicians who either understood these issues as permanently disabling or as demanding more immediate solutions. The distance between popular and medical understanding of both the problem and the solution often resulted in criticism and defensiveness rather than in collaboration. This history thus speaks to the need for a better understanding of medicine and science among policy makers, military leaders, and the general public.

Addressing these issues was never as simple as hiring more providers, training them to treat those affected, and making treatment more available. However important each of those was, it was not always clear what maladies service members and their families faced or what treatments would work. At the beginning of the wars, for example, the Army and VA faced multiple problems simultaneously: a dearth of clinicians prepared to treat post-traumatic stress, a lack of evidence-based treatments, and limited understandings of the causes of suicide or the nature of concussive brain injury. As Charles Engel put it, "Those on the mental health side are doing the best they can with limited tools in difficult environments and with constantly evolving challenges."[56] The process of addressing these issues moved at a much slower pace than that of the wars. Soldiers were returning from Iraq and Afghanistan with post-traumatic stress, suicidal ideation, and traumatic brain injury faster than AMEDD and VA scientists could conduct trials on Eye Movement Desensitization and Reprocessing (EMDR) and blast injury.

Addressing the needs of both contemporary and future service members and veterans thus requires a long-term investment in the military's and the VA's clinical and research capacities. In 2018, the United States military remains engaged around the world, with troops dying in combat in Africa, fighting the Islamic State in the Middle East, and cautiously watching escalating tensions on the Korean Peninsula. As George Casey put it, the major lesson from the struggle to understand and treat mental health issues from these wars is that "you can't let it fall off the table again."[57] The military needs to continue the project that began with the Behavioral Health Assessment Teams of the 1990s so it doesn't go into another conflict without a means of identifying the behavioral health issues that the troops are likely to face, and it needs to maintain a full complement of clinicians prepared to use the latest evidence-based treatments to address them. This means that, as the Army resets after these prolonged wars, it is imperative that the hard-won knowledge about post-traumatic stress, blast, and other issues not be forgotten and instead become the stepping stone for more advanced research into the best methods for surveillance, diagnosis, and treatment. This means appreciating—and fully funding—the Walter Reed Army Institute of Research, the Army STARSS program, and collaborative partnerships among military, VA, and civilian researchers. As Pete Chiarelli put it in the final speech of his Army career, if he and many Americans had "come to believe the invisible injuries of post-traumatic stress and traumatic brain injury represent the signature wounds of this war," that knowledge was not enough; rather, "while we have made significant progress in recent years, we must, must, must continue our effort."[58]

Third, Americans as a whole must recognize what the Army learned over the course of these wars. Mental health is as much a cultural issue as a medical one. The capacity to successfully identify and address mental health issues thus depends upon the cultivation of a culture that appreciates them as real and worthy of treatment. As the Army's efforts prior to 2007 showed, this could not be achieved through platitudes or orders; it has to happen from the ground up, through personal relationships that help individuals at all levels appreciate both that mental health issues are inevitable and that they can be addressed. It requires that those who have struggled and improved speak out about their experiences, and it demands that we strike the right balance between promoting resilience and encouraging treatment.

The Army, in a very real way, has led on this. "Because we've raised such a ruckus about the fact that war is hell on our emotional well-being," Gale Pollock says, "hopefully we're going to be the ones, because we're so vocal about it, that are going to force advances in mental health as well." In her vision, returning veterans who have learned in the service that it is okay to speak up about and seek help for mental health issues,

> when they leave the military and they go back to their communities and they are hanging out with their friends and family and buddies and workmates, they are going to be able to say, "Hey, something is not right. What's going on? Talk to me." And they will start to bring the experiences that they had, the lessons that they learned, back into their communities. And then in the communities they will start to say, "Why aren't we doing this more? Why isn't there more discussion?"[59]

This is a noble goal, and one that is within reach, but it cannot be achieved unless the culture as a whole makes a similar commitment to encouraging individuals to be cognizant of their own mental health and supportive of those around them as they seek care. Americans on the whole must become more open to the realities that many people will, at some point, face a mental health issue or a brain injury, and they must be more open to both talking and hearing about those experiences—particularly from veterans who are often lauded, sometimes pitied, and rarely carefully listened to.

The most important lesson, however, is one that the Army has long appreciated but that Americans must more fully embrace. When I interviewed George Casey in the lobby of an Arlington, Virginia, hotel in the spring of 2016, one of his first comments to me was that "the human mind was not made for war. That's the starting point for everything."[60] This was something that David Mangelsdorff and Jim Stokes knew when they trained psychologists to combat battle fatigue in an imagined war against the Soviets and that Amy Adler and Carl Castro knew when they put BATTLEMIND training together. It was something that Tim Bowman's death reflected. Soldiers cannot endure the strain and carnage of war and return unchanged.

In the end, this returns us to Comprehensive Soldier Fitness. In January 2011, *American Psychologist*, the official journal of the American Psychological Association, devoted an entire issue to the program. That October, *American Psychologist* published several pieces critical of the endeavor. One, by the psychologist John Dyckman, complained that, in promoting resiliency, Seligman, Ryvich, Casey, and others had "ignored the obvious fact that the best way to prevent posttraumatic stress disorder is to avoid trauma in the first place. They neglected to consider that fighting unnecessary wars would be a good place to start."[61] In response to Dyckman, Seligman wrote that "we are strongly for peace and for conflict resolution, and that is what advocacy directed to Congress and to the White House is for. The Army, however, carries out the orders of our politicians, and in doing so, deserves the very best that psychology can offer by way of prevention and treatment."[62]

In closing, I want to assert that Dyckman and Seligman are both correct. As its critics point out, CSF2 is an inherently neoliberal program that does little to oppose the conflicts in which the Army found itself in the twenty-first century. However, our critiques should not rest on the Army that has embraced resiliency but on the culture as a whole—a culture that, in acquiescing to permanent war has left the Army with few options but to attempt to produce soldiers capable of withstanding the rigors of multiple deployments.[63]

Despite the fact that every service member is a volunteer, the decision to go to war is not their own. This was a point that several of the people whose stories are part of this larger history made when I spoke with them. "These are people who make enormous sacrifices, and they do it without question," Charles Engel says, "and we don't invent the wars, we are just told to go there by the larger society."[64] Gale Pollock echoed Engel: "Wars are fought because of political agendas of a nation. . . . [Those] of us who have to go do it, we know it's a really bad idea. We are the ones who suffer from it," she says, adding, "We don't sit around and go, 'We're bored. Lets go to war today.' We know the costs."[65] Moreover, Pollock, along with several others, alluded to the disingenuousness with which the Bush administration, and the nation as a whole, approached the Iraq War. "The media and, I would say, the politicians painted it that we are going to be the welcomed, we're going to be treated like returning heroes," she argues. "And you know, we were treated more like invaders."[66]

The result of that rush to war, and the culture's acquiescence to it, has been nearly a quarter century of conflict and more than two million individual years of service fought by a relatively small all-volunteer force.[67] Faced with those demands, the Army in a sense has little choice but to insist that its soldiers become more resilient and capable of withstanding the challenges inherent in their professions because it and they lack the capacity to change the immiserating factor—the war itself. The turn to resilience, then, might best be understood as a coping mechanism of an Army beholden to a larger culture in thrall to perpetual war, and it is perhaps Americans' willingness to so easily acquiesce to military intervention that deserves the most severe critiques as we contemplate the mental health outcomes of the nation's longest wars.

ACKNOWLEDGMENTS

This project was conceived at one institution, begun at another, largely written at a third, and finished at a fourth. It draws primarily upon thousands of pages of documents acquired through the Freedom of Information Act and several dozen interviews. As a result, there are many people who have helped me along the way and who deserve recognition.

First and foremost, Beth Bailey is responsible for setting me on the path to writing this book. In the spring of 2014, Beth invited me to give a talk at Temple University's Center for the Study of Force and Diplomacy. I had just finished my first book, *Forever Vietnam*, and told her that I'd love to come talk about it. She told me that, rather than presenting something that was already done, I ought to present my new research. This invitation led to me writing, presenting, and, of course, getting great feedback on what would eventually become chapter 6 and to thinking that I had the seeds of a longer project. Beth, who is a formidable historian of war and society, is unmatched in her intellectual generosity and commitment to mentoring junior scholars. Everyone should be so lucky as to benefit from her kindness.

Beth also introduced me to Clara Platter at New York University Press. Early on, Clara saw the potential for a book, believed in it even when I didn't, and was endlessly patient as the project evolved in new directions. She also arranged for two very gracious reviewers to read the manuscript and provide feedback that made the prose tighter and the arguments stronger. Martin Coleman stepped in at a key moment to oversee the production. Jennifer Dropkin's copyediting made the book more readable. Derek Gottlieb prepared the index.

I began writing this book as a visiting faculty member at Skidmore College. Tillman Nechtman, Erica Bastress-Dukehart, Eric Morser, Jenny Day, Matt Hockenos, Jordana Dym, Dan Nathan, and Greg Pfitzer welcomed me into a collegial community of serious scholars, innovative teachers, and wonderful people. My good luck continued when I was awarded a writing fellowship at the University of Utah's Obert C. and Grace A. Tanner Humanities Center. Bob Goldberg, who is himself a first-class historian and even better person, and his staff of Beth James, John Boyack, and Susan Anderson have created an intellectual community without equal. There, I had time to think, research, and write and to contemplate the thoughtful criticism of my "fellow fellows," especially Martin Padgett and Eric Hinderaker. Matt Basso, Margrit Schneider, Lauren Tice Bergenheier, Mary Beth Coffman Osoro, and Brian Osoro deserve special mention for making Emma's and my time in Utah so wonderful. My year in Utah was also made memorable by a chance encounter that led to an immediate friendship with Dave Littlefield, truly one of the most remarkable people I've ever met and who left us suddenly and much too soon. That the last time I saw Dave he came to hear me present my work and brought me a newspaper article that he thought was useful for my research tells you some, but not all, of what the world is missing without his charm and infectious optimism.

The time to write and think that the Tanner Fellowship afforded me gave me the opportunity to significantly reimagine this project as one that would describe the internal debates and organizational challenges that attended the Army's and VA's efforts to address mental health issues. I am grateful to a number of current and former civilian and uniformed Army personnel who facilitated my research: the tireless Freedom of Information Act officers who sent me thousands of pages of documents; M. Sgt. Craig Zentkovich, Lt. Col. Daniel Elliott, John Manley, M. Sgt. Jeremy H. Crisp, and Lt. Col. Jason Billington at the U.S. Army Office of the Chief of Public Affairs–Northeast who graciously prepared and renewed the memoranda of understanding that enabled me to interview current and former Army staff; Hannah Kleber at the Uniformed Services University of the Health Sciences; S. Sgt. C. R. Vega and April D. Cunningham at the General Officer Management Office; and especially Marlon Martin and Maria Tolleson at the Office of the Surgeon General/U.S. Army Medical Command Directorate of Public Affairs, who arranged many of the interviews and accompanied me on several. All of these people handled my multiple requests with grace and good humor, and they always made sure that I got the best information that they could provide while never attempting to sway my analysis. Writing a contemporary history would be impossible without those documents and that access, and this book could not have been written without the Army's openness.

I am also grateful to the many Army leaders, clinicians, and researchers who spoke with me at length about their experiences and perspectives. Peter Chiarelli was the first person that I reached out to, and he welcomed me to OneMind's offices in Seattle for a conversation that challenged my assumptions and reframed my project. Eric Schoomaker insisted that, if I wanted to fully understand these issues, I needed to talk to many of the people who worked under him and facilitated those conversations. Bruce Shahbaz arranged for me to spend a day at the Pentagon with the Health Promotion, Risk Reduction, Suicide Prevention Task Force. Several people have answered repeated emails to help me clarify points that were somewhat abstract for a non-scientist. Many people have graciously introduced me to colleagues with whom they thought I should speak. Several shared documents from their personal collections that would otherwise have been unavailable: Eric Schoomaker, James Martin, James Stokes, Paul Bliese, Carl Castro, John Holcomb, and Kristy Kaufmann. Kelsey Glander transcribed many of the interviews.

I also want to recognize the FOIA officers at the Department of Veterans Affairs who filled my first request, resulting in more than three thousand pages of material on suicide prevention efforts. I also want to recognize that a number of the VA researchers and leaders whose stories appear in this book were initially willing to speak with me but that those interviews were canceled when the VA Public Affairs Office informed me that staff are not allowed to contribute to research if the author stands to profit financially from its publication. Like the soldiers and Army civilians whom I interviewed, many of the VA staff were eager to share their experiences and perspectives. It's a pity that the VA wouldn't allow that.

I completed the writing of this book at Washington & Jefferson College, where I am grateful for the friendship of Jennifer Sweatman, Nichole Fifer, Amparo Alpanes, Michael Lewis, Patrick Schmidt and Arlan Hess, and Mike and Christy Shaughnessy. I am also grateful to my departmental colleagues Tom Mainwaring, Victoria List, and Patrick Caffrey for their support, to Amanda Holland-Minkley for her wise counsel, to Charlie Hannon and Jamie March for the fishing trips, and to the Bliss Club for reminding me to occasionally go take a hike.

The George Washington University American Studies network remains strong; Melani McAlister and Jennifer James never stopped giving me good advice after I graduated, and my friendships with Julie Elman, Katie Brian, Kevin Strait, and Jeremy Hill, among others, have endured and helped me write this book. Jennifer Mittelstadt deserves special mention for reading a chapter draft and introducing me to Kristy Kaufmann. Ken MacLeish and I shared a number of conversations, some drafts, and a few good meals as we worked on parallel projects. That I chose to use a flight from DC to Pittsburgh to grade blue book exams led to a chance meeting with Dave Garrow, who in addition to being a craft beer aficionado is a first-rate historian. Our conversations helped me expand my idea of what is possible in this book. Ronald Shectman, George Litsakis, and Beth Wilson helped me stay psychologically healthy while writing this book. Ed Martini, Rebecca A. Adelman, Scott Laderman, and Megan Kate Nelson have sustained me through this process, particularly when the going got tough. Thanks.

Anne Litchfield and Sasha and Eduardo Nunes hosted me on many occasions when I came to DC. Back home, the Allegheny Commons Dog Park crew provided me with good conversation and lots of laughs at the end of each writing day. I am, as ever, grateful to my family, who has not only supported me and asked about how the research was going but also listened graciously as I prattled on about topics that might not always seem appropriate for family gatherings: Mary Anne Kieran; Kathleen Kieran; Kathleen Dolley; Susan Gilmore; Jamie, Elsbeth, Sam, and Megan Iannone; and Carrie, Rob, Marshall, and Nora Wagner. Not all families are related by blood, of course. Thanks to Jon Hoffmann, Leslie Horton, Lisa Keller, Abby Vanim, Ben Wojtyna, and Rachel Vanim for welcoming us into the "Pittsburgh Family." Lisa deserves special mention for checking every footnote in the manuscript.

As always, Emma Gilmore Kieran has been with me every step of the way, and this book couldn't have been written without her love and support. There is not a day that goes by that I do not recognize how amazing you are and all that you have done for me. Thanks for always believing in me. I love the life that we have built together, and I love you.

NOTES

INTRODUCTION

1 "These Unseen Wounds Cut Deep; A Mental Health Crisis Is Emerging, with One in Six Returning Soldiers Afflicted," *Los Angeles Times*, November 14, 2004.

2 Sen. Barack Obama (D-IL), speaking on "Epilepsy and Returning Wounded Soldiers," *Congressional Record*, April 20, 2005, S3996.

3 Gregg Zoroya, "Brain Injuries Range from Loss of Coordination to Loss of Self," *USA Today*, March 4, 2005. On this article's significance in framing the use of this term, see John M. Kinder, "The Embodiment of War: Bodies for, in, and after War," in *At War: The Military and American Culture in the Twentieth Century and Beyond*, ed. David Kieran and Edwin A. Martini (New Brunswick: Rutgers University Press, 2018), 226–227.

4 Dana Priest and Anne Hull, "Soldiers Face Neglect, Frustration at Army's Top Medical Facility," *Washington Post*, February 18, 2007.

5 Michael Abramowitz and Steve Vogel, "Army Secretary Ousted," *Washington Post*, March 3, 2007; and Josh White, "Surgeon General of Army Steps Down," *Washington Post*, March 13, 2007.

6 "The Warriors' Second Front at Home," *New York Times*, April 13, 2007.

7 Emily Sogn, "Internal Frontiers: Health, Emotion, and the Rise of Resilience-Thinking in the U.S. Military" (PhD diss., The New School, 2016), 3.

8 Senate Armed Services Committee, *Nominations before the Senate Armed Services Committee*, 110th Congress, 1st Session (2007), 712.

9 Sogn, "Internal Frontiers," 4.

10 Lou Michel, "Tragic Consequences," *Buffalo News*, November 4, 2007.

11 Sarah J. Hautzinger and Jean Scandlyn, *Beyond Post-traumatic Stress Disorder: Homefront Struggles with the Wars on Terror* (Walnut Creek, CA: Left Coast Press, 2014), 16, quoted in Sogn, "Internal Frontiers," 5; Kenneth T. MacLeish, *Making War at Fort Hood: Life and Uncertainty in a Military Community* (Princeton: Princeton University Press, 2013), 2–3.

12 Eric Schoomaker, interview with the author, December 16, 2016.

13 Ann Jones, *They Were Soldiers: How the Wounded Return from America's Wars—The Untold Story* (Chicago: Haymarket Books, 2013); Aaron Glantz, *The War Comes Home: Washington's Battle with America's Veterans* (Berkeley: University of California Press, 2008); Yochi Dreazen, *The Invisible Front: Love and Loss in an*

Era of Endless War (New York: Crown, 2014); David Finkel, *Thank You for Your Service* (New York: Sarah Crighton Books, 2013); and C. J. Chivers, *The Fighters: Americans in Combat in Afghanistan and Iraq* (New York: Simon & Schuster, 2018).

14 The most comprehensive outline of the Army's efforts, which mentions many issues that I take up in this book, is Charles W. Hoge et al., "Transformation of Mental Health Care for U.S. Soldiers and Families during the Iraq and Afghanistan Wars: Where Science and Politics Intersect," *American Journal of Psychiatry* 173, no. 4 (2016): 334–343.

15 Beth Linker, *War's Waste: Rehabilitation in World War I America* (Chicago: University of Chicago Press, 2011); and John M. Kiner, *Paying with Their Bodies: American War and the Problem of the Disabled Veteran* (Chicago: University of Chicago Press, 2015).

16 See, e.g., Peter Leese, *Shell Shock: Traumatic Neurosis and the British Soldiers of the First World War* (New York: Palgrave McMillan, 2002); Tyler Downing, *Breakdown: The Crisis of Shell Shock in the Somme* (London: Little, Brown, 2016); Stefanie Linden, *They Called It Shell Shock: Combat Stress in the First World War* (Solihull: Helion & Co., 2016); and Peter Hodgkinson, *Glum Heroes: Hardship, Fear and Death—Resilience and Coping in the British Army on the Western Front, 1914–1918* (Solihull: Helion & Co., 2016).

17 See, e.g., Thomas Childers, *Soldier from the War Returning: The Greatest Generation's Troubled Homecoming from World War II* (Boston: Mariner Books, 2009); Patrick Hagopian, *The Vietnam War in American Memory: Veterans, Memorials, and the Politics of Healing* (Amherst: University of Massachusetts Press, 2011), 49–78; and Wilbur J. Scott, *Vietnam Veterans since the War: The Politics of PTSD, Agent Orange, and the National Memorial* (Norman: University of Oklahoma Press, 2004).

18 Jennifer Terry, *Attachments to War: Biomedical Logics and Violence in Twenty-First Century America* (Durham, NC: Duke University Press, 2017), 6.

19 Alison Howell, "Resilience, War, and Austerity: The Ethics of Military Human Enhancement and the Politics of Data," *Security Dialogue* 46, no. 1 (2015): 15–31.

20 Rebecca A. Adelman, *Figuring Violence: Affective Investments in Perpetual War* (New York: Fordham University Press, 2019), 4.

21 Ibid., 3.

22 Zoe H. Wool, *After War: The Weight of Life at Walter Reed* (Durham, NC: Duke University Press, 2015); Erin M. Finley, *Fields of Combat: Understanding PTSD among Veterans of Iraq and Afghanistan* (Ithaca, NY: Cornell University Press, 2011); Sogn, "Internal Frontiers"; Hautzinger and Scandlyn, *Beyond Post-traumatic Stress*; and MacLeish, *Making War at Fort Hood.*

23 Chivers, *The Fighters*, xviii.

24 MacLeish, *Making War at Ford Hood*, 17.

25 On the intersection of masculinity and mental health stigma broadly, see Finley, *Fields of Combat*, 73–88.

26 Dave Baiochi, *Measuring Army Deployments to Iraq and Afghanistan* (Santa Monica, CA: RAND Corp., 2012), 1; U.S. Army, *2012 Army Posture Statement* (Arlington, VA: Department of the Army, 2012), www.army.mil; Hoge et al., "Transformation of Mental Health Care," 334.

27 Pete Geren, interview with the author, June 6, 2016.

28 That said, this book does not offer a general history of PTSD, TBI, or suicide, nor does it detail the debate over PTSD's legitimacy. On the history and debate, see Jerry Lembcke, *PTSD: Diagnosis and Identity in Post-empire America* (Lanham, MD: Lexington Books, 2013); Alan Young, *The Harmony of Illusions: Inventing Post-traumatic Stress Disorder* (Princeton, NJ: Princeton University Press, 1995); David Morris, *The Evil Hours: A Biography of Post-traumatic Stress Disorder* (New York: Houghton-Mifflin, 2015); Patrick Hagopian, *The Vietnam War in American Memory: Veterans, Memorials, and the Politics of Healing* (Amherst: University of Massachusetts Press, 2009), 49–78; Judith Herman, *Trauma and Recovery: The Aftermath of Violence—From Domestic Abuse to Physical Terror* (New York: Basic Books, 2015), 7–32; and Jonathan Shay, *Achilles in Vietnam: Combat Trauma and the Undoing of Character* (New York: Simon & Schuster, 1995). For one example of how historians have retroactively applied the category of post-traumatic stress disorder to earlier conflicts, see Eric T. Dean, *Shook over Hell: Post-traumatic Stress, Vietnam, and the Civil War* (Cambridge, MA: Harvard University Press, 1997).

29 Lembcke, *PTSD*, 166.

30 Kinder, "The Embodiment of War," 227.

31 Ibid. Also see Jerry Lembcke, "'Shell Shock' in the American Imagination: World War I's Most Enduring Legacy," *Peace and Change* 41, no. 1 (2016): 79–80.

32 There is some debate over the material reality of each of these conditions. See, e.g., Lembcke, *PTSD*, xi. See also Young, *The Harmony of Illusions*, 5.

33 Kinder, "The Embodiment of War," 227.

34 Nancy J. Hirschmann and Beth Linker, "Disability, Citizenship, and Belonging: A Critical Introduction," in *Civil Disabilities: Citizenship, Membership, and Belonging*, ed. Nancy J. Hirschmann and Beth Linker (Philadelphia: University of Pennsylvania Press, 2015), 1–21; quoted at 4. See also Kinder, *Paying with Their Bodies*, 7 and Michel Foucault, *Madness and Civilization: A History of Insanity in the Age of Reason*, trans. Richard Howard (New York: Vintage, 1988), 24–25.

35 Kinder, "The Embodiment of War," 227.

36 For a critique of how this can be used perniciously, see ibid.; and Lembcke, *PTSD*, 139–171.

37 Megan Kate Nelson, *Ruin Nation: Destruction and the American Civil War* (Athens: University of Georgia Press, 2012), 161, 175; and Kinder, *Paying with Their Bodies*, 6, 3.

38 Kinder's contends that "the disabled soldier remains a popular symbol among critics of American militarism and government indifference"; see Kinder, *Paying with Their Bodies*, 292.

39 Among the best examples are Beth Bailey, *America's Army: Making the All-Volunteer Force* (Cambridge, MA: Harvard University Press, 2009); Jennifer Mittelstadt, *The Rise of the Military Welfare State* (Cambridge, MA: Harvard University Press, 2015); and Aaron B. O'Connell, *Underdogs: The Making of the Modern Marine Corps* (Cambridge, MA: Harvard University Press, 2013).

40 Charles Engel, interview with the author, January 18, 2016.

41 Here, I am following Claire Potter's call to "look for, honor, and interrogate [sources] for conflict and indecision . . . as well as for decision making that pushed . . . forward." See Claire Potter, "When Radical Feminism Talks Back: Taking an Ethnographic Turn in the Living Past," in *Doing Recent History: On Privacy, Copyright, Video Games, Institutional Review Boards, Activist Scholarship, and History That Talks Back*, ed. Claire Bond Potter and Renee C. Romano (Athens: University of Georgia Press, 2012), 174.

42 Regarding my interviews, I am guided by Claire Potter's point that "few historians have been trained to handle heightened scrutiny by their own subjects" and her call to be attentive "to how our published work responds to the web of obligations and disagreements that we create in our wake" (ibid., 159–160). I have been gratified by the openness of everyone who has spoken with me, and I hope I honor their generosity by portraying their stories accurately.

43 On Army STARRS, see National Institute of Mental Health, "Army Study to Assess Risk and Resilience in Servicemembers (Army STARRS): A Partnership between NIMH and the U.S. Army" (Washington, DC: National Institutes of Health), www.nih.gov.

44 Kirby Dick, director, *The Invisible War*, documentary (Los Angeles: Cinedigm Corp., 2012).

45 Finley, *Fields of Combat*, 69.

CHAPTER 1. "AT THE TIME PEOPLE HADN'T BEEN ASKING THOSE SORTS OF QUESTIONS"

1 Until 1974 the Army had often commissioned psychologists as first lieutenants and captains, and not, as was most common for newly commissioned officers, as second lieutenants. See E. R. Worthington, "Combat Psychology: The Role of the Division Psychologist in Time of War," in *Army Medical Department Psychology Symposium, 13–17 November 1978*, ed. A. David Mangelsdorff and Gerald L. Bryan (El Paso, TX: William Beaumont Army Medical Center, 1978), 8.

2 Ben Shephard, *A War of Nerves: Soldiers and Psychiatrists in the Twentieth Century* (Cambridge, MA: Harvard University Press, 2000), 343; E. R. Worthington, "AMEDD Psychology in the Seventies," in Mangelsdorff and Bryan, *Army Medical Department Psychology Symposium*, 8; John Prados, *Vietnam: The History of an Unwinnable War* (Lawrence: University Press of Kansas, 2009), 275–276; Beth Bailey, *America's Army: Making the All-Volunteer Force* (Cambridge, MA: Harvard University Press, 2009), 37.

3 American Psychiatric Association, *Diagnostic and Statistical Manual* of Mental Disorders, 3rd ed. (*DSM-III*; Washington, DC: American Psychiatric Association, 1980). On the controversial history of post-traumatic stress disorder, see Allan Young, *The Harmony of Illusions: Inventing Post-traumatic Stress Disorder* (Princeton, NJ: Princeton University Press, 1995); David J. Morris, *The Evil Hours: A Biography of Post-traumatic Stress Disorder* (Boston: Houghton-Mifflin, 2015); Gerald Nicosia, *Home to War: A History of the Vietnam Veterans Movement* (New York: Carroll & Graf, 2001); and Wilbur J. Scott, *Vietnam Veterans since the War: The Politics of PTSD, Agent Orange, and the National Memorial* (Norman: University of Oklahoma Press, 1995).

4 Bailey, *America's Army*, 88–129, esp. 123. See also Andrew Bacevich, The New American Militarism: How Americans Are Seduced By War (New York: Oxford University Press, 2005), 108-109.

5 Richard Halloran, "Panel Appointed to Study Military Manpower," *New York Times*, July 9, 1981; and Drew Middleton, "Army Overseas Cites Problems of Female G.I.'s," *New York Times*, June 28, 1981; quote at Middleton.

6 Bailey, *America's Army*, 79–82.

7 Drew Middleton, "U.S. 7th Army: Ready to Fight; Spirit in Germany High despite Bad Conditions," *New York Times*, May 17, 1981.

8 Ibid.

9 John F. Kennedy, "Remarks of Senator John F. Kennedy in the United States Senate, National Defense, Monday, February 29, 1960," John F. Kennedy Presidential Library and Museum, www.jfklibrary.org; Andrew Bacevich, *Breach of Trust: How Americans Failed Their Country and Their Soldiers* (New York: Metropolitan Books, 2013), 82, and *The New American Militarism*, 42. See also Drew Middleton, "What If Land War Came to Europe?" *New York Times*, October 12, 1983.

10 Ronald Reagan, "June 9, 1982: Address to the Bundestag in West Germany," Miller Center, University of Virginia, http://millercenter.org. For an overview of Reagan's remilitarization, see Bacevich, *The New American Militarism*, 105–111.

11 Bacevich, *Breach of Trust*, 83. For an overview of this strategy, see also Bacevich, *The New American Militarism*, 45–47; and Bailey, *America's Army*, 174.

12 Greg Schneider and Renae Merle, "Reagan's Defense Buildup Bridged Military Eras; Huge Budgets Brought Life back to Industry," *Washington Post*, June 9, 2004; and Bailey, *America's Army*, 191–195.

13 Drew Middleton, "U.S. Seventh Army Undergoes a Renovation," *New York Times*, October 16, 1983.

14 Ibid.; Drew Middleton, "War Game Said to Show Allied Readiness," *New York Times*, September 24, 1984; George C. Wilson, "The Military; Fighting Back; Army Has Recovered from Wounds of War," *Washington Post*, April 16, 1985.

15 Raymond Palumbo, interview with the author, April 24, 2016. See also John Barbanel, "War Exercise Transforms Fort Drum into a Dutch Plain 'Battle,'" *New York Times*, July 20, 1981; David Cloud and Greg Jaffe, *The Fourth Star: Four Gener-*

als and Their Epic Struggle for the Future of the United States Army (New York: Crown, 2011), 68–77.

16 Carl Castro, interview with the author, April 15, 2016; Benedict Carey, "David H. Marlowe Dies at 83; Helped Army Nurture Bands of Brothers," *New York Times*, January 9, 2015.

17 James Martin, interview with the author, July 21, 2017.

18 Ibid.

19 Robert Shisko, *The European Conventional Balance: A Primer* (Santa Monica: RAND Corp., 1981), 6.

20 Ibid., 18.

21 Ibid., 9.

22 Ibid., 10.

23 Raymond J. Keller, "Threat," in *Proceedings: Users' Workshop on Combat Stress, 1981*, by A. David Mangelsdorff and T. Paul Furukawa (Fort Sam Houston, TX: Health Care Studies, Academy of Health Sciences, 1983), 10.

24 William H. Thornton, "Threat," in *Proceedings: Second Users' Workshop on Combat Stress*, 1982, by A. David Mangelsdorff and T. Paul Furukawa (Fort Sam Houston, TX: Health Care Studies, Academy of Health Sciences, n.d.), 19.

25 Gregory Lucas Belenky, Paul Newhouse, and Franklin Del Jones, "Prevention and Treatment of Psychiatric Casualties in the Event of a War in Europe," in Mangelsdorff and Furukawa, *Proceedings: Users' Workshop on Combat Stress*, 1981, 93–98, quoted at 94.

26 William F. Schultheis, "Combat Stress in Perspective," in Mangelsdorff and Furukawa, *Proceedings: Users' Workshop on Combat Stress*, 1981, 7–9. Notably, the 1973 war was the most frequent touchstone for U.S. planners in the 1980s.

27 James W. Stokes, "Standardization Agreement on Combat Stress Terminology," in *Proceedings: Third Users' Workshop on Combat Stress*, 1983, by A. David Mangelsdorff, James M. King, and Donald E. O'Brien (Fort Sam Houston, TX: Health Services Command, 1983), 214–219, esp. 215.

28 Ibid.

29 Bailey, *America's Army*, 174–177; James Stokes, interview with the author, March 28, 2017; and James Martin, interview with the author, July 21, 2017.

30 Maj. Gen. Joseph J. Skaff, "Soldiers: They Deserve Good Leadership," *Military Review: The Professional Journal of the US Army* 65, no. 12 (1985): 40–43; and Lt. Col. Jeffrey L. House, "Leadership Challenges on the Nuclear Battlefield," *Military Review: The Professional Journal of the US Army* 65, no. 3 (1985): 60–69.

31 House, "Leadership Challenges on the Nuclear Battlefield," 61.

32 Henry L. Thompson, "Stress-Train: Training for High Performance," *Military Review: The Professional Journal of the US Army* 65, no. 2 (1985): 58.

33 Henry L. Thomspon, "Sleep Loss and Its Effect in Combat," *Military Review: The Professional Journal of the US Army* 65, no. 2 (1985): 19.

34 See, e.g., Federico M. V. Tamayo, "Treatment and Prevention of Battlefield Psychiatric Casualties: Implications for the Next War," in Mangelsdorff and Furukawa, *Proceedings: Users' Workshop on Combat Stress*, 1981, 61–90, esp. 64–65; Jerry Melcher, "First Cavalry Combat Stress Course," in *Proceedings: Fourth Users' Workshop on Combat Stress: Lessons Learned in Recent Operational Exercises*, by A. David Mangelsdorff, James M. King, and Donald E. O'Brien (Fort Sam Houston, TX: U.S. Army Health Services Command, 1985), 9–25, esp. 14; and James W. Stokes, "Standardization of Training Packages for Army-Wide Training about Battle Fatigue," in *Proceedings: Fifth Users' Workshop on Combat Stress*, by A. David Mangelsdorff, James M. King, and Donald E. O'Brien (Fort Sam Houston, TX: U.S. Army Health Services Command, 1986), 98–126, esp. 109–110, 125–126.

35 Schultheis, "Combat Stress in Perspective," 7; and Federico M. V. Tamayo, "Training Personnel in the Prevention and Management of Battlefield Psychiatric Casualties: Considerations for Training Based on the Next War," in Mangelsdorff and Furukawa, *Proceedings: Users' Workshop on Combat Stress*, 1981, 49–60, esp. 50.

36 James Stokes, "Fact Sheet: Management of Combat Stress and Battle Fatigue," November 21, 1983, 218–219, in Stokes, "Standardization Agreement on Combat Stress Terminology."

37 James Stokes, interview with the author, March 28, 2017. See also Shephard, *A War of Nerves*, 184, 218, 220, 339–340, 348–349; Erin Finley, *Fields of Combat: Understanding PTSD among Veterans of Iraq and Afghanistan* (Ithaca, NY: Cornell University Press, 2011), 93.

38 Melcher, "First Cavalry Combat Stress Course," 18.

39 Ibid., 9.

40 Stokes, "Standardization of Training Packages," 98–99; Bailey, *America's Army*, 174–177.

41 Stokes, "Standardization of Training Packages," 109.

42 Ibid., 114, 128.

43 Ibid., 99.

44 KIA: killed in action; WIA: wounded in action. See James Stokes, "Stress Anxiety Instruction #310-4 (Anxiety, Used Up Luck)"; courtesy of James Stokes, copy in author's possession.

45 James Stokes, "Stress Anxiety Instruction #310-7 (Anxiety, Stomach Pain)"; courtesy of James Stokes, copy in author's possession.

46 Ibid.

47 Stokes, "Fact Sheet: Management of Combat Stress and Battle Fatigue," 217.

48 Ibid., 219.

49 Schultheis, "Combat Stress in Perspective," 9.

50 Robert Stretch, "Post Traumatic Stress Disorder among Army Nurses," 33–45, and Gary Palmer, "Toma VAMC Post Traumatic Stress Treatment Program," 56–58, both in Mangelsdorff, King, and O'Brien, *Proceedings: Fifth Users' Workshop on Combat Stress.*

51 Belenky, Newhouse, and Jones, "Prevention and Treatment of Psychiatric Casualties," 96. Also see G. L. Belenky, "Combat Psychiatry: Nigeria, Israel, Federal Republic of Germany, United Kingdom, 13 June–19 July, 1978," trip report (Washington, DC: Walter Reed Army Institute of Research, 1978), 103.

52 James W. Stokes, "Simulation of Combat Stress Casualties by Live Role Players and 'Constructive (Paper) Casualties," in Mangelsdorff and Furukawa, *Proceedings: Second Users' Workshop on Combat Stress*, 1982, 144. See also 125–193, esp. 129 and 132, and "Standardization of Training Packages," 109.

53 See, e.g., Bruce Boman, "Combat Stress, Post-traumatic Stress Disorder, and Associated Psychiatric Disturbance," *Psychosomatics* 27, no. 8 (1986): 567–573; Kathy A. Pearce et al., "A Study of Post Traumatic Stress Disorder in Vietnam Veterans," *Journal of Clinical Psychology* 41, no. 1 (1985): 9–14; Johanna Gallers et al., "Post-traumatic Stress Disorder in Vietnam Combat Veterans: Effects of Traumatic Violence Exposure and Military Adjustment," *Journal of Traumatic Stress* 1, no. 2 (1988): 181–192; David W. Foy, "Combat Related Post-traumatic Stress Disorder Etiology: Replicated Findings in a National Sample of Vietnam-Era Men," *Journal of Clinical Psychology* 43, no. 1 (1987): 28–31; Robert H. Stretch, "Incidence and Etiology of Post-traumatic Stress Disorder among Active Duty Army Personnel," *Journal of Applied Psychology* 16, no. 6 (1986): 464–481; Zahava Solomon, "A 3-Year Prospective Study of Post-traumatic Stress Disorder in Israeli Combat Veterans," *Journal of Traumatic Stress* 2, no. 1 (1989): 59–73; Steven Michael Silver and C. U. Iacono, "Factor-Analytic Support for *DSM-III*'s Post-traumatic Stress Disorder for Vietnam Veterans," *Journal of Clinical Psychology* 40, no. 1 (1985): 5–14; Lawrence C. Kolb, "The Post-traumatic Stress Disorders of Combat: A Subgroup with a Conditioned Response," *Military Medicine* 149, no. 3 (1984): 237–243; Duane Sherwin, "The Psychodynamic Origin of Post-traumatic Stress Disorder in American Vietnam Veterans—An Ongoing Saga," *Dynamische Psychiatrie* 21 (1988): 348–358; and Stephen L. O'Brien, "Symptoms of Post-traumatic Stress Disorder in Falklands Veterans Five Years after the Conflict," *British Journal of Psychiatry* 159 (1991): 135–141.

54 See, e.g., Douglas F. Zatzick et al., "Posttraumatic Stress Disorder and Functioning and Quality of Life Issues in a Nationally Representative Sample of Male Vietnam Veterans," *American Journal of Psychiatry* 154, no. 12 (1997): 1690–1695; and Thomas C. Neylan et al., "Sleep Disturbances in the Vietnam Generation: Findings from a Nationally Representative Sample of Male Vietnam Veterans," *American Journal of Psychiatry* 155, no. 7 (1998): 929–933.

55 James Stokes, interview with the author, March 28, 2017.

56 Ibid.

57 Ibid.

58 James Martin, email to the author, October 30, 2018.

59 Ibid.

60 James Martin, interview with the author, July 21, 2017.

61 James Stokes, interview with the author, March 28, 2017.

62 Jesse Harris, "Soldier Stress and Operation Urgent Fury," 162–168, esp. 167–168, and Jesse Harris et al., "Panel on Grenada," 147–161, esp. 152, both in Mangelsdorff, King, and O'Brien, *Proceedings: Fifth Users' Workshop on Combat Stress.*

63 Douglas Porch, *Counterinsurgency: Exposing the Myths of the New Way of War* (New York: Cambridge University Press, 2013), 291; and Bacevich, *The New American Militarism*, 35, 52.

64 David H. Marlowe, J. Martin, and R. Gifford, "Observations and Initial Findings of the WRAIR Stress Evaluation Team, Operation Desert Shield: 22 Sep–6 Oct 90" (Washington, DC: Walter Reed Army Institute of Research, November 15, 1990), n.pag.; courtesy of James A. Martin, copy in author's possession.

65 Ibid., 6–8.

66 Ibid., 11.

67 Ibid., 8.

68 Ibid., 14.

69 Ibid.

70 Frederick J. Manning to Chief of Staff, "Human Issues in Troop Return after Operation Desert Storm" (February 12, 1991); courtesy of James A. Martin, copy in author's possession.

71 David H. Marlowe, K. Wright, and R. Gifford, "Some Considerations on the Human Issues in Troop Return after Operation Desert Storm," memorandum (February 8, 1991), 2; courtesy of James A. Martin, copy in author's possession.

72 Ibid., 3–4, 6, 10.

73 Ibid., 8.

74 That Operation Desert Storm, and in particular the inclusion of Vietnam veterans in post-war parades and celebrations, was part of the larger healing after Vietnam has been well documented. See, among others, Bacevich, *The New American Militarism*, 35–37; Jerry Lembcke, *The Spitting Image: Myth, Memory, and the Legacy of Vietnam* (New York: New York University Press, 1998), 2; Patrick Hagopian, *The Vietnam War in American Memory: Veterans, Memorials, and the Politics of Healing* (Amherst: University of Massachusetts Press, 2011), 522; and George Lipsitz, *The Possessive Investment in Whiteness: How White People Profit from Identity Politics* (Philadelphia: Temple University Press, 2009), 82–83.

75 Marlowe, Wright, and Gifford, "Some Considerations on the Human Issues in Troop Return," 3–4.

76 Ibid., 5–6.

77 Ibid., 9. On the creation of the Combat Action Badge, see Lisa Burgess, "Army Announces New Badge to Replace CCB in Honoring Non–Front Line Troops," *Stars and Stripes*, May 6, 2005, www.stripes.com.

78 Marlowe, Wright, and Gifford, "Some Considerations on the Human Issues in Troop Return," 3.

79 Ibid., 8.

80 See, e.g., William H. Reno to CDR, ACENT, "Psychological Preparation for Combat—Technical Guidance for Commanders" (January 14, 1991); and M. A.

Fischl, "Combat Stress Decompression" (March 7, 1991); courtesy of James A. Martin, copies in author's possession.

81 James Stokes, interview with the author, March 28, 2017.

82 See ibid.; and James Martin, interview with the author, July 21, 2017.

83 James Martin, interview with the author, July 21, 2017.

84 James A. Martin, "Combat Psychiatry: Lessons Learned from the War in Southwest Asia," in *Proceedings: Eighth Users' Stress Workshop*, ed. A. David Mangelsdorff (Fort Sam Houston, TX: U.S. Army Health Services Command, 1992), 137–141.

85 James Martin, interview with the author, July 21, 2017.

86 James A. Martin, interview with the author, February 23, 2017.

87 Ibid.

88 U.S. Army Medical Research Unit Europe, "Mental Health Issues: Lessons from Operation Desert Shield/Storm" (n.d.), 1; courtesy of James A. Martin, copy in author's possession.

89 Ibid., 13.

90 Ibid., 1.

91 Ibid., 9, 7.

92 Ibid., 7.

93 Martin, "Combat Psychiatry: Lessons Learned," 137.

94 Ibid., 141.

95 James Martin et al., "Psychological Well-Being among U.S. Soldiers Deployed from Germany to the Gulf War," *Journal of the U.S. Army Medical Department* (September/October 1992): 33.

96 Robert Rosenheck et al., *Returning Persian Gulf Troops: First Year Findings* (Washington, DC: Department of Veterans Affairs, 1992), i.

97 Rosenheck et al., *Returning Persian Gulf Troops*, i.

98 Steven Southwick, *Psychological and Neurobiological Consequences of the Gulf War Experience* (Frederick, MD: U.S. Army Medical Research and Materiel Command, 1997), 6.

99 Otto F. Wahl, "Mental Health Consumers' Experience of Stigma," *Schizophrenia Bulletin* 25, no. 3 (1999): 470–471.

100 On this anxiety, see Charles C. Engel, Kurt Kroenke, and Wayne J. Katon, "Mental Health Services in Army Primary Care: The Need for a Collaborative Health Care Agenda," *Military Medicine* 159, no. 3 (1994): 203–209; and Finley, *Fields of Combat*, 83. Intriguingly, the only study on stigma in the military that appeared in the 1990s found that most commanders did not hold negative views of subordinates who had sought mental health care; see T. L. Porter and W. B. Johnson, "Psychiatric Stigma in the Military," *Military Medicine* 159, no. 9 (1994): 602–605.

101 George Casey, interview with the author, April 6, 2016.

102 Michael S. Tucker, interview with the author, January 12, 2016.

103 Ibid.

104 Rebecca Porter, interview with the author, December 17, 2015.

105 Ibid.

106 John Sloan Brown, *Kevlar Legions: The Transformation of the U.S. Army, 1989–2005* (Washington, D.C.: United States Army Center for Military History, 2011), chap. 3, quoted at 1, www.history.army.mil.

107 Carl Castro, interview with the author, April 15, 2016.

108 Ibid.

109 Elspeth Cameron Ritchie, interview with the author, December 15, 2015.

110 Charles Engel, interview with the author, January 18, 2016.

111 Carl Castro, interview with the author, April 15, 2016.

112 Rebecca Porter, interview with the author, December 17, 2015.

113 Charles Engel, interview with the author, January 18, 2016.

114 Carl Castro, interview with the author, April 15, 2016.

115 Chuck Sudetic, "U.S. Opens a MASH with a Difference in Croatia," *New York Times*, November 16, 1992.

116 Amy Adler, interview with the author, April 21, 2016.

117 Ibid.

118 Ibid. Also see P. T. Bartone, M. A. Vaitkus, and A. B. Adler, "Psychological Issues in Peacekeeping Contingency Operations," Report no. WRAIR/TR-94-0022 (Washington, DC: Walter Reed Army Institute of Research, 1994), 3, 6; Spencer J. Campbell et al., "Operation Joint Guard (SFOR) Bosnia: Assessment of Operational Stress and Adaptive Coping Mechanisms of Soldiers" (Washington, DC: Walter Reed Army Institute of Research, 1998), 5; and Robert K. Gifford, James N. Jackson, and Kathleen B. DeShazo, "Report of the Human Dimensions Research Team Operation Restore Hope, 26 January–5 March 1993" (Washington, DC: Walter Reed Army Institute of Research, 1993), 5.

119 Carl Castro, interview with the author, April 15, 2016.

120 Amy Adler, interview with the author, April 21, 2016.

121 Ibid. See, e.g., J. Morano, "The Relationship of Workplace Social Support to Perceived Work-Related Stress among Staff Nurses," *Journal of Post-anesthesia Nursing* 8, no. 6 (1993): 395–402; T. Wynne, "Increasing Concern Is Expressed over Stress in the Workplace. How Far Should Stress Management be a Responsibility of Individuals or Organizations?" *Journal of Nurse Management* 1, no. 6 (1993): 293–296; S. L. Crawford, "Job Stress and Occupational Health Nursing: Modeling Health Affirming Choices," *AAOHN Journal* 41, no. 11 (1993): 522–528; G. Thomas, "Working Can Be Hazardous to Your Health," *Cancer Nurse* 89, no. 6 (1993): 35–38; A. I. Meleis et al., "Stress, Satisfaction, and Coping: A Study of Women Clerical Workers," *Health Care Women International* 10, no. 4 (1989): 319–334; P. Carayon, "Job Design and Job Stress in Office Workers," *Ergonomics* 36, no. 5 (1993): 463–477; and S. Enzoe et al., "Work Stress in Japanese Computer Engineers: Effects of Computer Work or Bioeducational Factors," *Environmental Research* 63, no. 1 (1993): 148–156.

122 Elizabeth M. Fowler, "More Stress Found in Workplace," *New York Times*, September 12, 1989.

123 "Job Stress: Rating Your Workplace," *New York Times*, July 10, 1991.

124 Amy Adler, interview with the author, April 1, 2016.

125 Ibid.

126 Ibid.

127 Gifford et al., "Report of the Human Dimensions Research Team Operation Restore Hope, 26 January–5 March 1993," 3, 9.

128 Campbell et al., "Operation Joint Guard (SFOR) Bosnia," 17, 19.

129 Amy Adler, interview with the author, April 1, 2016.

130 Bartone, Vaitkus, and Adler, "Psychological Issues in Peacekeeping Contingency Operations," 9–10; Bradley F. Powers, Mark A. Vaitkus, and James A. Martin, "Observations from a U.S. Army Medical Unit Deployed to Support the U.N. Protection Force in Croatia" (Frederick, MD: U.S. Army Medical Research, Development, Acquisition, and Logistics Command [Provisional], 1993), 3.

131 Powers, Vaitkus, and Martin, "Observations from a U.S. Army Medical Unit," 4, 5–6.

132 Gifford, Jackson, and DeShazo, "Report of the Human Dimensions Research Team Operation Restore Hope," 2. See also Campbell et al., "Operation Joint Guard (SFOR) Bosnia," 16.

133 Campbell et al., "Operation Joint Guard (SFOR) Bosnia," 16–17.

134 Powers, et. al., "Observations from a U.S. Army Medical Unit Deployed to Support the U.N. Protection Force in Croatia,", 6.

135 Amy Adler, interview with the author, April 1, 2016.

136 Ibid.

137 Ibid.

138 Ibid.

139 Paul Bliese, interview with the author, December 3, 2015.

140 Ibid.

141 Ibid.

142 Rebecca Porter, interview with the author, December 17, 2015.

143 Paul Bliese, interview with the author, December 3, 2015.

144 Carl Castro, interview with the author, April 15, 2016.

145 Headquarters, Department of the Army, *Leaders' Manual for Combat Stress Control*, Field Manual no. 22-51(Washington, DC: Headquarters, Department of the Army, September 29, 1994), chap. 7, "Stress Issues in Army Operations," 7-3-4, appendix A, "Leader Actions to Offset Battle Fatigue Risk Factors," 11-1-2, and chap. 2, "Stress and Combat Performance," section 2, "Combat Performance and Combat Stress Behaviors," 6-1-9. Also see James Stokes, "Stress Dimensions in Operations Other than War," booklet, March 1996, 3–17 (booklet extracted from James Stokes, "Combat Stress Control in Operations Other than War" [March 1994]), courtesy of James W. Stokes, copies in author's possession.

146 Headquarters, Department of the Army, *Leaders' Manual for Combat Stress Control*, chap. 2, "Combat Stress Behaviors," 2-11a, and chap. 5, "Battle Fatigue," 5-5d.

147 David Brown, "Army Takes New Tack in Fighting the Unseen Enemy," *Washington Post*, November 23, 1998.

148 For a listing of some of the most commonly cited potential causes, see Robert W. Haley, T. L. Kurt, and J. Hom, "Is There a Gulf War Syndrome? Searching for Syndromes by Factor Analysis of Symptoms," *Journal of the American Medical Association* 227, no. 3 (1997): 215–222; David Brown, "Army Takes New Tack in Fighting the Unseen Enemy"; and A. David, S. Ferry, and S. Wessely, "Gulf War Illness: New American Research Provides Leads but No Firm Conclusions," *British Medical Journal* 314 (January 25, 1997): 239–240.

149 David Brown, "Army Takes New Tack in Fighting the Unseen Enemy."

150 Ibid.

151 Paul D. Bliese et. al., "Post-deployment Mental Health Screening Instruments: How Good Are They?" Proceedings for the Army Science Conference (24th), November 29–December 2, 2005, Orlando, Florida, 2; E. Jones, K. C. Hyams, and S. Wessely, "Screening for Vulnerability to Psychological Disorders in the Military: An Historical Survey," *Journal of Medical Screening* 10, no. 1 (2003): 44–46.

152 Amy Adler, interview with the author, April 1, 2016.

153 Ibid.

154 William Zung, "A Self-Rating Depression Scale," *Archives of General Psychiatry* 12, no. 1 (1965): 63–65.

155 Amy Adler, interview with the author, April 21, 2016. Also see Ann H. Huffman et al., "The Impact of Deployment Length and Deployment Experience on the Well-Being of Male and Female Military Personnel" (Fort Detrick, MD: U.S. Army Medical Research and Materiel Command, 1999), 8–9.

156 Amy Adler, interview with the author, April 21, 2016.

157 Huffman et al., "The Impact of Deployment Length," 8.

158 Ibid., 10, 14, 20.

159 Ibid., 20.

160 Amy Adler, interview with the author, April 21, 2016; Jones, Hyams, and Wessely, "Screening for Vulnerability," 44.

161 Amy Adler, interview with the author, April 21, 2016.

162 Indeed, Bacevich argues that military leaders "expected future American wars to replicate Desert Storm—large-scale conventional conflicts" featuring "overwhelming force . . . [and] a quick exit" (*The New American Militarism*, 54).

CHAPTER 2. "THE PSYCHIATRIC COST OF SENDING YOUNG MEN AND WOMEN TO WAR"

1 David Maraniss and Anne Hull, "Combat's Bitter Revelations," *Washington Post*, March 9, 2003.

2 Erica Goode, "Learning from the Last Time," *New York Times*, March 25, 2003. See also Antonio Regaldo, "The War inside the Mind," *Wall Street Journal*, April 9, 2003; Joseph B. Verrengia, "For Some Iraq Veterans, Forgetting Hardest Part," *Lowell (MA) Sun*, April 21, 2003.

3 Erica Goode, "Learning from the Last Time"; Ron Martz, "GIs Get Combat Stress Counsel," *Atlanta Journal-Constitution*, July 30, 2003.

4 Erin Finley, *Fields of Combat: Understanding PTSD among Iraq and Afghanistan Veterans* (Ithaca, NY: Cornell University Press, 2011), 103. See also Vernon Loeb, "Military Cites Elevated Risk of Suicides in Iraq," *Washington Post*, January 15, 2004; and "Troop Numbers; Foreign Soldiers in Iraq," *Al Jazeera*, December 14, 2011.

5 Loeb, "Military Cites Elevated Risk."

6 "OIF Tips—Helping a Soldier in Distress; Leader's Hip Pocket Training Guide," March 2004; and "OIF Tips—How to Face the Injured and the Dead," courtesy James B. Stokes; copies in author's possession.

7 Finley, *Fields of Combat*, 103. Also see Loeb, "Military Cites Elevated Risk."

8 U.S. Army Surgeon General and HQDA G-1, *Operation Iraqi Freedom (OIF) Mental Health Assessment Team (MHAT) Report, 16 December 2003* (Arlington, VA: Department of the Army, 2003), 32, posted at Global Security, www.globalsecurity.org.

9 Carl Castro, interview with the author, April 15, 2016.

10 Ibid.

11 U.S. Army Surgeon General and HQDA G-1, *Operation Iraqi Freedom (OIF) Mental Health Assessment Team (MHAT) Report*, "Annex A: WRAIR Report of Soldier Health and Well-Being Assessment," A-8; A-3.

12 Ibid., "Executive Summary," 5.

13 Ibid., "Introduction," 11.

14 Ibid., "Findings," 12; and Finley, *Fields of Combat*, 103. Finley also offers an overview of some of the issues (ibid.).

15 U.S. Army Surgeon General and HQDA G-1, *Operation Iraqi Freedom (OIF) Mental Health Assessment Team (MHAT) Report*, "Findings," 16. "Combat Operational Stress Control" is a term that the military began using in the 2000s; while the 1994 Field Manual 22-51 is titled "Leaders' Manual for Combat Stress Control," the 2006 Field Manual 4-02.51 is titled "Combat and Operational Stress Control."

16 Ibid., "Executive Summary," 5. Also see Finley, *Fields of Combat*, 103.

17 U.S. Army Surgeon General and HQDA G-1, *Operation Iraqi Freedom (OIF) Mental Health Assessment Team (MHAT) Report*, "Recommendations," 22, 25.

18 Carl Castro, interview with the author, April 15, 2016.

19 Ibid.

20 George Casey, interview with the author, October 13, 2017.

21 Eric Schmitt, "U.S. Army Finds Its Suicide Rate in Iraq Higher than for Other G.I.'s," *New York Times*, March 26, 2004. See also Drew Brown, "Army Acts to Prevent Suicides," *Philadelphia Inquirer*, March 26, 2004.

22 See also Charles W. Hoge et al., "Combat Duty in Iraq and Afghanistan, Mental Health Problems, and Barriers to Care," *New England Journal of Medicine* 351, no. 1 (2004): 13–22, quoted at 17. Also see Charles W. Hoge et al., "Walter Reed Army Institute of Research Contributions during Operations Iraqi Freedom and Endur-

ing Freedom: From Research to Public Health Policy," chapter 5 of *Combat and Operational Behavioral Health*, senior ed. Elspeth Cameron Ritchie (Falls Church, VA: Office of the U.S. Surgeon General; Fort Detrick, MD: Borden Institute, 2011), 78, posted at American Ukrainian Medical Foundation, http://aumf.net.

23 Hoge et al., "Walter Reed Army Institute of Research Contributions during Operations Iraqi Freedom and Enduring Freedom: From Research to Public Health Policy," 77.

24 Charles Hoge, interview with the author, December 31, 2015.

25 Ibid.; Kevin Kiley, interview with the author, December 11, 2015.

26 Hoge et al., "Combat Duty in Iraq and Afghanistan, Mental Health Problems, and Barriers to Care," 15.

27 Ibid., 17.

28 Ibid., 17, 13.

29 Ibid., 13, 20–21.

30 Ibid., 20.

31 Matthew J. Friedman, "Acknowledging the Psychiatric Cost of War," *New England Journal of Medicine* 351 (2004): 75–77, quoted at 75.

32 Friedman, cited in Anahad O'Connor, "The Reach of War: The Soldiers; 1 in 6 Iraq Veterans Is Found to Have Stress-Related Disorder," *New York Times*, July 1, 2004.

33 Friedman, cited in "Iraq Vets Reporting High Rates of Stress," *St. Petersburg (FL) Times*, July 1, 2004.

34 Lisa M. Krieger and Esther Landhuis, "Iraq, Afghanistan Leave Invisible Wounds on Soldiers," *San Jose (CA) Mercury News*, July 1, 2004.

35 Jane Eisner, "Coping with War—Body and Soul," *Philadelphia Inquirer*, July 8, 2004.

36 Neil Amato, "Like Anybody with a Son Serving in Iraq, the Past Year Has Been Agonizing for the Family of UNC Assistant Coach Joe Holladay," *Durham Herald-Sun*, February 10, 2004; "Paratroopers Form Northern U.S. War Front: Coalition Destroys Iraqi Convoy," *Calgary Herald*, March 27, 2003; Paul D. Bliese et al., "Timing of Postcombat Mental Health Assessments," *Psychological Services* 4, no. 3 (2007): 141–148, esp. 143; Paul Bliese, interview with the author, December 3, 2015; and Amy Adler, interview with the author, April 21, 2016.

37 Paul Bliese, interview with the author, December 3, 2015.

38 Amy Adler, interview with the author, April 21, 2016.

39 Hoge et al., "Combat Duty in Iraq and Afghanistan, Mental Health Problems, and Barriers to Care," 21.

40 Finley, *Fields of Combat*, 105; Kenneth T. MacLeish, *Marking War at Fort Hood: Life and Uncertainty in a Military Community* (Princeton: Princeton University Press, 2013), 122.

41 Rebecca Porter, interview with the author, December 16, 2016.

42 Paul Bliese, interview with the author, December 3, 2015.

43 Bliese et al., "Timing of Postcombat Mental Health Assessments," 142.

44 Hoge et al., "Combat Duty in Iraq and Afghanistan, Mental Health Problems, and Barriers to Care"; and Amy Adler, interview with the author, April 21, 2016.

45 Paul Bliese, interview with the author, December 3, 2015.

46 Bliese et al., "Timing of Postcombat Mental Health Assessments," 144.

47 Ibid.

48 Ibid., 147.

49 Ibid., 141–148.

50 Paul Bliese, interview with the author, December 3, 2015; Amy Adler, interview with the author, April 21, 2016; Joshua L. Wick, "Casper Native Retires as Commanding General," *Casper (WY) Journal*, July 28, 2011.

51 William Winkenwerder to Assistant Secretary of the Army (M&RA), Assistant Secretary of the Navy (M&RA), and Assistant Secretary of the Air Force (M&RA), "Post-Deployment Health Reassessment," March 10, 2005, 1, posted at Military Health System, https://health.mil.

52 Ibid., 2.

53 Amy Adler, interview with the author, April 21, 2016.

54 Paul Bliese, interview with the author, December 3, 2015; "MCMR/DA Memo 25–02: (MCMR-UWX) STO-W FY05 Milestone," n.d.; courtesy of Paul D. Bliese, copy in author's possession.

55 House Committee on Veterans' Affairs, *The Department of Defense and Department of Veterans Affairs: The Continuum of Care for Post Traumatic Stress Disorder*, 109th Congress, 1st Session (2005), 3. For a longer discussion of the Pelkey's struggles, see Yochi Dreazen, *The Invisible Front: Love and Loss in an Era of Endless War* (New York: Crown, 2014), 158–168.

56 House Committee on Veterans' Affairs, *The Department of Defense and Department of Veterans Affairs: The Continuum of Care for Post Traumatic Stress Disorder*, 3.

57 Thomas Ricks, "Missed the War, in Place for the Peace," *Washington Post*, May 17, 2003.

58 Kevin Sullivan, "Troops Ready for a Change in Guard," *Washington Post*, July 22, 2003; Thomas E. Ricks, "U.S. Alters Tactics in Baghdad Occupation," *Washington Post*, May 25, 2003; Neela Bannerjee, "After the War, Firearms," *New York Times*, June 19, 2003; Peter Slevin and Vernon Loeb, "Plan to Secure Postwar Iraq Faulted," *Washington Post*, May 19, 2003; Thomas E. Ricks, "On Patrol in Sweltering Baghdad, A U.S. Patrol Turns Up the Heat," *Washington Post*, June 15, 2003; Daniel Williams, "Attacks in Iraq Traced to Network," *Washington Post*, June 22, 2003; Rajiv Chandrasekaran and Peter Finn, "U.S. Cracks Down on Iraqi Resistance," *Washington Post*, July 1, 2003.

59 Thomas E. Ricks, "Comfort Rare in Iraq, Even for U.S. Troops," *Washington Post*, May 26, 2003.

60 Ibid.

61 "Pvt. Robert L. Frantz, 19, San Antonio, TX," *Washington Post*, July 6, 2003; Thomas E. Ricks and Rajiv Chandrasekaran, "In Postwar Iraq, the Battle Widens,"

Washington Post, July 7, 2003; Molly Moore, "3 U.S. Soldiers Killed in Baghdad," *Washington Post*, July 7, 2003.

62 Rajiv Chandrasekaran and Molly Moore, "Urban Combat Frustrates Army," *Washington Post*, July 8, 2003.

63 House Committee on Veterans' Affairs, *The Department of Defense and Department of Veterans Affairs: The Continuum of Care for Post Traumatic Stress Disorder*, 4.

64 Ibid.

65 Ibid.

66 Ibid., 4–5.

67 Ibid.

68 Ibid., 5.

69 Ibid.

70 Ibid.

71 Ibid., 9; see also 4, 5.

72 Ibid., 5–6.

73 "Captain Michael J. Pelkey," *Memory Of*, May 31, 2006, www.memory-of.com.

74 House Committee on Veterans' Affairs, *The Department of Defense and Department of Veterans Affairs: The Continuum of Care for Post Traumatic Stress Disorder*, 6.

75 Ibid., 3.

76 Ibid., 6.

77 Ibid., 7.

78 Bob Filner, interview with the author, April 11, 2017.

79 Ibid.

80 House Committee on Veterans' Affairs, *The Department of Defense and Department of Veterans Affairs: The Continuum of Care for Post Traumatic Stress Disorder*, 10.

81 Ibid.

82 Ibid., 11. Also see House Committee on Veterans' Affairs, *The Department of Defense and Department of Veterans Affairs: The Continuum of Care for Post Traumatic Stress Disorder*, 11. The National Vietnam Veterans Readjustment Study concluded in 1988 that nearly 90 percent of veterans, and more than 85 percent of Vietnam veterans with a PTSD diagnosis, had never been homeless or vagrant. See Richard A. Kulka et al., *Contractual Report of Findings from the National Vietnam Veterans Readjustment Study*, vol. 2, *Tables and Findings* (Research Triangle Park, NC: Research Triangle Institute, 1988), VII-18-1, posted at PTSD: National Center for PTSD, www.ptsd.va.gov.

83 House Committee on Veterans' Affairs, *The Department of Defense and Department of Veterans Affairs: The Continuum of Care for Post Traumatic Stress Disorder*, 11.

84 Ibid., 16.

85 Ibid.

86 Ibid.

87 Ibid., 21–22.

88 Marilyn Elias, "Mental Disorders Are on the Rise among Afghanistan, Iraq Veterans," *USA Today*, March 31, 2005; Kulka et al., *Contractual Report of Findings*, IV-1-1; VI-2-1.

89 Tom Udall, "Commentary: Troops Need Mental-Health Support Now," *Santa Fe New Mexican*, May 15, 2005.

90 Charlie Patton, "Mental Health Forum to Concentrate on Issues Related to Military Service," *Florida Times-Union*, May 24, 2005.

91 Elias, "Mental Disorders Are on the Rise"; Jonathan Finer, "Battle-Hard G.I.'s Learn to Release Their Pain," *Washington Post*, June 14, 2005.

92 U.S. Army Surgeon General, *Operation Iraqi Freedom (OIF-II) Mental Health Advisory Team (MHAT-II) Report, 30 January 2005* (Falls Church, VA: Office of the U.S. Army Surgeon General), 3.

93 Ibid., 10.

94 Ibid., 11, 3.

95 Ibid., 12.

96 House Committee on Veterans' Affairs, *The Department of Defense and Department of Veterans Affairs: The Continuum of Care for Post Traumatic Stress Disorder*, 25–26.

97 Ibid., 25, 27.

98 Ibid., 27–28.

99 Ibid., 30.

100 Ibid., 59.

101 Finley, *Fields of Combat*, 94.

102 House Committee on Veterans Affairs, *PTSD and TBI as Emerging Issues in Force and Veterans' Health*, 109th Congress, 2d Session (2006), 82.

103 Ibid., 77.

104 Verrengia, "Some Iraq Vets Find Forgetting the Hardest Part about Killing." On how popular understandings of PTSD differ from scientific understandings of the condition, see Jerry Lembcke, *PTSD: Diagnosis and Identity in Post-empire America* (Lanham, MD: Lexington Books, 2013), 11; Allan Young, *The Harmony of Illusions: Inventing Post-traumatic Stress Disorder* (Princeton, NJ: Princeton University Press, 1995), 5, 10; and David J. Morris, *The Evil Hours: A Biography of Post-traumatic Stress Disorder* (New York: Houghton, Mifflin, Harcourt, 2015), 162.

105 Ben Shephard, *A War of Nerves: Soldiers and Psychiatrists in the Twentieth Century* (Cambridge, MA: Harvard University Press, 2000), 391.

106 American Psychiatric Association, *Diagnostic and Statistical Manual of Mental Disorders*, 4th ed. (*DSM-IV*; Washington, DC: American Psychiatric Association, 1994).

107 Allison G. Harvey and Richard A. Bryant, "Acute Stress Disorder: A Synthesis and Critique," *Psychological Bulletin* 128, no. 6 (2002): 886–902; and Dan Koren,

I. Arnon, and E. Klein, "Long-Term Course of Chronic Posttraumatic Stress Disorder in Traffic Accident Victims: A Three-Year Prospective Follow-Up Study," *Behaviour Research and Therapy* 39, no. 12 (2001): 1449–1458, esp. 1450.

108 Edward B. Blanchard et al., "One-Year Prospective Follow-Up of Motor Vehicle Accident Victims," *Behavior Research and Therapy* 34, no. 10 (1996): 775–786, esp. 784.

109 Here I am paraphrasing a point made by Eric B. Schoomaker in an interview with the author, December 16, 2015.

110 Bob Coalson, "Nightmare Help: Treatment of Trauma Survivors with PTSD," *Psychotherapy: Theory, Research, Practice, Training* 32, no. 3 (1995): 381–388, esp. 384–385; Candice M. Monson et al., "Cognitive Processing Therapy for Veterans with Military-Related Posttraumatic Stress Disorder," *Journal of Consulting and Clinical Psychology* 74, no. 5 (2006): 898–907, esp. 901; and John G. Carlson et al., "Eye Movement Desensitization and Reprocessing Treatment for Combat PTSD," *Psychotherapy: Theory, Research, Practice, Training* 33, no. 1 (1996): 104–113, esp. 104. Also see Alexander C. McFarlane, "Posttraumatic Stress Disorder: A Model of the Longitudinal Course and the Role of Risk Factors," *Journal of Clinical Psychiatry* 61, suppl. 5 (2000): 15–20, esp. 17.

111 House Committee on Veterans' Affairs, *The Department of Defense and Department of Veterans Affairs: The Continuum of Care for Post Traumatic Stress Disorder*, 30. Also see Robert Jay Lifton, *Home from The War: Vietnam Veterans, Neither Victims nor Executioners* (New York: Simon & Schuster, 1973); and Gerald Nicosia, *Home to War: A History of the Vietnam Veterans Movement* (New York: Carroll & Graf, 2001), 162–165, 194–195, and 517–520.

112 Laura A. Pratt, D. J. Brody, and Q. Gu, "Anti-Depressant Use in Persons Aged 12 and Over: United States, 2005–2008," *NCHS Data Brief* 76 (Oct 2011): 1–8, esp. 1, posted at Centers for Disease Control and Prevention, www.cdc.gov; and David Morris, *The Evil Hours: A Biography of Post-traumatic Stress Disorder* (New York: Houghton-Mifflin, 2015), 161.

113 In his study of post-traumatic stress disorder during the Iraq and Afghanistan Wars, Jerry Lembcke argues that the dominant media image has been of the criminalized veteran. This conclusion overlooks the far greater number of stories that portray veterans as merely struggling to readjust. See Lembcke, *PTSD*, 1–8.

114 Joe Fahy, "35% of Returning Troops Sought Mental Health Aid," *Pittsburgh Post-Gazette*, March 1, 2006; Edward Colimore, "Posttraumatic Stress Comes Home," *Philadelphia Inquirer*, April 9, 2006; and Bill Hendrick, "Vets Return Fraught with Psychological Hurdles, Study Finds," *Atlanta Journal-Constitution*, August 2, 2006.

115 Jeffrey Gettleman, "When Soldiers Come Marching Home, Trauma Often Sets In," *New York Times*, January 15, 2006.

116 Ibid.

117 Bob Kerr, "Major Sees an Epidemic on the Way," *Providence (RI) Journal*, August 6, 2006.

118 Jim Carney, "Disorder Familiar among Soldiers," *Akron (OH) Beacon Journal*, January 2, 2006.

119 Charles W. Hoge, Jennifer L. Auchterlonie, and Charles S. Milliken, "Mental Health Problems, Use of Mental Health Services, and Attrition from Military Service after Returning from Deployment to Iraq or Afghanistan," *Journal of the American Medical Association* 295, no. 9 (2006), 1023–1032, esp. 1024, 1027.

120 Ibid., 1028.

121 Ibid., 1030.

122 Benedict Cary, "Iraq Veterans Using Clinics to Assist Mental Health," *New York Times*, March 1, 2006.

123 Shankar Vedantam, "Veterans Report Mental Distress," *Washington Post*, March 1, 2006.

124 Jordan Lite, "Emotional Toll Huge on Soldiers in Iraq," *New York Daily News*, March 1, 2006.

125 Cary, "Iraq Veterans Using Clinics." See also Vedantam, "Veterans Report Mental Distress."

126 Amy Adler, interview with the author, April 21, 2016.

127 Frederick J. Manning to Chief of Staff, Department of the Army, "Human Issues in Troop Return after Operation Desert Storm," February 12, 1991, 4–6; courtesy of James Martin, copy in author's possession.

128 On this controversy, see George S. Everly, Jr., and Jeffrey T. Mitchell, "The Debriefing 'Controversy' and Crisis Intervention: A Review of Lexical and Substantive Issues," *International Journal of Emergency Health* 2, no. 4 (2000): 211–225; and Bryan E. Bledsoe, "Critical Incident Stress Management (CISM): Benefit or Risk for Emergency Services?" *Prehospital Emergency Care* 7, no. 2 (2003): 272–279.

129 Amy Adler, interview with the author, April 21, 2016.

130 Ibid.

131 Carl Castro, interview with the author, April 15, 2016.

132 For a brief overview of the program, see Finley, *Fields of Combat*, 105.

133 Carl Castro, interview with the author, April 15, 2016.

134 Charles Hoge, interview with the author, December 31, 2015.

135 Ibid.

136 Amy Adler, interview with the author, April 21, 2016.

137 Charles Hoge, interview with the author, December 31, 2015; and Amy Adler, interview with the author, April 21, 2016.

138 Amy Adler, interview with the author, April 21, 2016; Carl A. Castro, et al., "Battlemind Training: Transitioning Home from Combat," report (Silver Spring, MD: Walter Reed Army Institute of Research, November 2, 2016), 18.

139 Amy Adler, interview with the author, April 21, 2016; Castro, et al., "Battlemind Training," 21.

140 Castro et al., "Battlemind Training," 1.

141 Castro, et al., "Battlemind Training," 19.

142 Castro et al., "Battlemind Training," 19; Finley, *Fields of Combat*, 105.

143 Amy Adler, interview with the author, April 21, 2016.

144 Ibid.

145 Castro et al., "Battlemind Training," 1.

146 Finley, *Fields of Combat*, 106.

147 Amy Adler, interview with the author, April 21, 2016.

148 Castro et. al., "Battlemind Training," 1.

149 Amy Adler, interview with the author, April 21, 2016.

150 Ibid.; and Castro et. al., "Battlemind Training," 1.

151 Amy Adler, interview with the author, April 21, 2016.

152 House Committee on Veterans' Affairs, *Post-Traumatic Stress Disorder and Traumatic Brain Injury as Emerging Trends in Force and Veterans Health*, 109th Congress, 2d Session (2006), 17; Senate Committee on Armed Services, *The State of the United States Army*, 110th Congress, 1st Session (2007), 84.

153 Carl A. Castro, Charles W. Hoge, and Anthony L. Cox, "Battlemind Training: Building Soldier Resiliency," in *Human Dimensions in Military Operations—Military Leaders' Strategies for Addressing Stress and Psychological Support*, Meeting Proceedings RTO-MP-HFM-134, Paper 42 (Neuilly-sur-Seine: RTO, 2006), 42-1 to 42-6.

154 WRAIR Land Combat Study Team, "BattleMind Training: Preparing for War: What Soldiers Should Know and Do" (Washington, DC: Walter Reed Army Institute of Research, September 11, 2006).

155 Ibid.

156 Elaine Scarry, *The Body in Pain: Making and Unmaking the World* (New York: Oxford University Press, 1985).

157 WRAIR Land Combat Study Team, "Pre-Deployment Battlemind for Leaders" (Washington, DC: Walter Reed Army Institute of Research, n.d.).

158 General Accountability Office, "Post-traumatic Stress Disorder: DoD Needs to Identify the Factors Its Providers Use to Make Mental Health Evaluation Referrals for Servicemembers" (Washington, DC: U.S. General Accountability Office, 2006), 1, www.gao.gov.

159 Ibid., 20.

160 Ibid.

161 Ibid., 30–31.

162 Gregg Zoroya, "Many War Vets' Stress Disorders Go Untreated," *USA Today*, May 11, 2006.

163 Lolita C. Baldor, "Study Shows Combat Stress Aid Falls Short," *Washington Post*, May 11, 2006; Shankar Vedantam, "GAO: Few Troops Are Treated for Disorder; Post-traumatic Stress Risk Gauged," *Washington Post*, May 11, 2006; and Amy Clark, "Study: Stressed Vets Not Getting Help," *CBS News*, May 11, 2006.

164 Clark, "Study: Stressed Vets Not Getting Help"; and Shankar Vedantam, "Pentagon Faults Report Questioning Veterans' Mental Health Care," *Washington Post*, May 13, 2006.

165 Vedantam, "GAO: Few Troops Are Treated for Disorder"; "Defense Should Move Quickly on PTSD," *St. Paul (MN) Pioneer Press*, May 18, 2006.

166 Zoroya, "Many War Vets' Stress Disorders Go Untreated." See also Rep. Mike Michaud (D-ME), speaking on "GAO PTSD Report Release," *Congressional Record*, May 11, 2006, H2506.

167 Barbara Boxer to Kevin Kiley, Press Release of Senator Boxer: "Boxer Calls Defense Department's Failure to Properly Treat Soldiers with PTSD Inexcusable," May 11, 2006, available through the *Wayback Machine* at https://web.archive.org.

168 Lisa Chedekel and Matthew Kauffman, "Special Report: Mentally Unfit, Forced to Fight," *Hartford (CT) Courant*, May 14, 2006. As early as May 2004, newspapers in Canada and the United Kingdom expressed concern that soldiers with mental health issues were being redeployed to combat zones and that the suicide rate in Iraq was evidence that "the military is totally unprepared for the onslaught of post-traumatic stress disorders coming in the months ahead"; see Lynda Hurst, "Troops in Iraq on Suicide Watch," *Toronto Star*, April 11, 2004. Also see Andrew Gumbel, "Is America Sending Battle-Weary, Clinically Stressed Soldiers back into the Heat of Iraq?" *Independent*, April 3, 2003; and Suzanne Goldenberg, "Broken US Troops Face Bigger Enemy at Home," *Guardian*, April 3, 2004.

169 Chedekel and Kauffman, "Special Report: Mentally Unfit, Forced to Fight."

170 Ibid.

171 Ibid.

172 Charles Hoge, interview with the author, January 31, 2015.

173 Christopher Ivany, interview with the author, December 16, 2015.

174 Ibid.

175 Elspeth Cameron Ritchie, interview with the author, December 15, 2015.

176 Elspeth C. Ritchie, "Psychiatric Medication for Deployment," *Military Medicine* 159, no. 10 (1994): 647–649.

177 Elspeth Cameron Ritchie, interview with the author, December 15, 2015.

178 Ibid.

179 Ibid.

180 Ibid.

181 Kevin Kiley, interview with author, December 11, 2015.

182 Elspeth Cameron Ritchie, interview with the author, December 15, 2015.

183 Ibid.

184 Ibid.

185 Kevin Kiley, interview with author, December 11, 2015.

186 Ibid.

187 Ibid.

188 Charles Hoge, interview with the author, January 31, 2015.

189 Kevin Kiley, interview with author, December 11, 2015.

190 On the racial and economic dimensions of contemporary military service, see, among others, Hector Amaya, "Latino Immigrants in the American Discourses of Citizenship and Nationalism during the Iraqi War," *Critical Discourse Studies* 4, no. 3 (2007): 237–256; Irene Garza, "Advertising Patriotism: The 'Yo Soy El Army' Campaign and the Politics of Visibility for Latina/o Youth," *Latino Studies* 13, no. 2

(2015), 245–268; Gina M. Perez, *Citizen, Soldier, Student: Latina/o Youth, JROTC, and the American Dream* (New York: New York University Press, 2015); and Beth Bailey, *America's Army: Making the All-Volunteer Force* (Cambridge, MA: Belknap Press of Harvard University Press, 2009), 255–258. On the historical question of whether military service was comparable to other careers, see Jennifer Mittelstadt, *The Rise of the Military Welfare State* (Cambridge, MA: Harvard University Press, 2015), 46–72.

191 Office of the Surgeon, Multinational Force–Iraq, and Office of the Surgeon General, United States Army Medical Command, *Mental Health Advisory Team (MHAT-III) Operation Iraqi Freedom 04–06 Report, 29 May 2006* (Falls Church, VA: Office of the U.S. Surgeon General, 2006), 7, 32.

192 Ibid., 6, 7, 41.

193 Paul Bliese, interview with the author, December 3, 2015.

194 Office of the Surgeon, Multinational Force–Iraq, and Office of the Surgeon General, United States Army Medical Command, *Mental Health Advisory Team (MHAT-III) Operation Iraqi Freedom 04–06 Report*, 6, 55.

195 Ibid., 53.

196 Office of the Surgeon, Multinational Force–Iraq, and Office of the Surgeon General, United States Army Medical Command, *Mental Health Advisory Team (MHAT) IV: Operation Iraqi Freedom 05-07, Final Report* (Falls Church, VA: Office of the U.S. Surgeon General, November 17, 2006), 3, posted at National Technical Reports Library, https://ntrl.ntis.gov.

197 Carl Castro, interview with the author, April 15, 2016.

198 Office of the Surgeon, Multinational Force–Iraq, and Office of the Surgeon General, United States Army Medical Command, *Mental Health Advisory Team (MHAT) IV*, 14.

199 Ibid., 19–21, 71.

200 Amy B. Adler et al., "The Impact of Deployment Length and Experience on the Well-Being of Male and Female Soldiers," *Journal of Occupational Health Psychology* 10, no. 2 (2005): 131.

201 See, e.g., William D. S. Kilgore et al., "The Effects of Prior Combat Experience on the Expression of Somatic and Affective Symptoms in Deploying Soldiers," *Journal of Psychosomatic Research* 60, no. 4 (2006): 379–385.

202 Office of the Surgeon, Multinational Force–Iraq, and Office of the Surgeon General, United States Army Medical Command, *Mental Health Advisory Team (MHAT) IV*, 77.

203 Ibid., 76.

204 On the amenities available to U.S. troops in Iraq, see, e.g., Meredith Lair, *Armed with Abundance: Consumerism and Soldiering in the Vietnam War* (Chapel Hill: University of North Carolina Press, 2011), 222–224.

205 Office of the Surgeon, Multinational Force–Iraq, and Office of the Surgeon General, United States Army Medical Command, *Mental Health Advisory Team (MHAT) IV*, 18.

206 Ibid., 23.

207 Carl Castro, interview with the author, April 15, 2016.

208 Ibid.

209 Office of the Surgeon, Multinational Force–Iraq, and Office of the Surgeon General, United States Army Medical Command, *Mental Health Advisory Team (MHAT) IV*, 21, 18–19.

210 Carl Castro, interview with the author, April 15, 2016.

211 Ibid.

212 Shephard, *A War of Nerves*, 348.

213 Office of the Surgeon, Multinational Force–Iraq, and Office of the Surgeon General, United States Army Medical Command, *Mental Health Advisory Team (MHAT) IV*, 76.

214 Ibid.

215 Ibid. See also Aaron Glantz, *The War Comes Home: Washington's Battle With America's Veterans* (Berkeley: University of California Press, 2009), 30.

216 Office of the Surgeon, *Multinational Force-Iraq, Mental Health Advisory Team (MHAT) IV*, 62.

217 See Hoge, Auchterlonie, and Milliken, "Mental Health Problems, Use of Mental Health Services, and Attrition"; General Accountability Office, "Post-traumatic Stress Disorder"; Chedekel and Kauffman, "Special Report: Mentally Unfit, Forced to Fight"; and Office of the Surgeon, Multinational Force–Iraq, and Office of the Surgeon General, United States Army Medical Command, *Mental Health Advisory Team (MHAT) IV*.

218 Rep. James McDermott (D-WA), speaking on "Iraq Week in the House of Representatives," *Congressional Record*, June 13, 2006, H3893.

219 Sen. Bernie Sanders (I-VT), speaking on "U.S. Strategy in Iraq," *Congressional Record*, February 5, 2007, S1569.

220 Sen. Jim Bunning (R-KY), speaking on "Expressing the Sense of Congress on Iraq—Motion to Proceed," *Congressional Record*, February 17, 2007, S2186; Carl Hulse and Jeff Zeleny, "Senate Rejects Renewed Effort to Debate Iraq," *New York Times*, February 18, 2007. The text of Gregg's resolution in fact stipulated that "Congress should not take any action that will endanger United States military forces in the field"; see Sen. Judd Gregg (R-NH), speaking on "Expressing the Sense of Congress on Iraq—Motion to Proceed," *Congressional Record*, February 17, 2007, S2191.

221 Sen. Robert Byrd (D-WV), speaking on "Expressing the Sense of Congress on Iraq—Motion to Proceed," *Congressional Record*, February 17, 2007, S2196.

222 Sen. Barbara Boxer (D-CA), speaking on "Amendment No. 762," *Congressional Record*, March 28, 2007, S4014.

223 Public Law 109-163 (2006), 3349, 2301.

224 Sen. Barbara Boxer (D-CA), "SA 4466," *Congressional Record*, June 21, 2006, S6317.

225 Assistant Secretary of Defense for Health Affairs, "Policy Guidance for Deployment-Limiting Psychiatric Conditions and Medications," November 7, 2006, 1, posted at Military Health System, https://health.mil.

226 Ibid., 3–5.
227 Ibid., 3.
228 Ibid., 3, 7.
229 Ibid., 4.
230 Ibid., 5–6.

CHAPTER 3. "CALLOUS DISREGARD OF VETERANS' RIGHTS IS OF A PIECE WITH THE ADMINISTRATION'S ENTIRE APPROACH TO WAR"

1 George W. Bush, "The President's Radio Address, March 19, 2005," *Public Papers of the Presidents of the United States: George W. Bush, 2005, Book 1, January 1–June 30, 2005* (Washington, DC: Government Printing Office, 2005), 481.
2 Editorial, "Two Years Later," *New York Times*, March 18, 2005. Also see Jaff Warzer and Robert F. Worth, "Blast Kills 122 at Iraqi Clinic in Attack on Security Recruits," *New York Times*, March 1, 2005; Robert F. Worth, "More than a Dozen Killed by Rebels in Central Iraq," *New York Times*, March 8, 2005, "Across Iraq, Fresh Mass Graves and Fatal Bomb Attacks," *New York Times*, March 10, 2005, and "Mine Kills Marine near Baghdad in Day of Deadly Insurgent Attacks," *New York Times*, March 31, 2005; and James Barron and Kirk Semple, "2 Soldiers, Friends from Queens, Die on a 'Routine' Patrol in Iraq," *New York Times*, March 5, 2005.
3 Robert C. McFadden, "Hundreds of Rallies Held across U.S. to Protest Iraq War," *New York Times*, March 20, 2005; Patty Allen-Jones, "Hundreds Rally in Sarasota against the War in Iraq," *Sarasota (FL) Herald-Tribune*, March 20, 2005; and "Antiwar Protests Continue for Second Day in U.S. Cities, *Providence (RI) Journal*, March 21, 2005.
4 See testimony in House Committee on Veterans' Affairs, *Stopping Suicides: Mental Health Challenges within the U.S. Department of Veterans Affairs*, 110th Congress, 1st Session (2007), 7; and "Possible Delay of Deployment May Mean More Family Time for Troops," *Kewanee (IL) Star Courier*, March 1, 2004.
5 House Committee on Veterans' Affairs, *Stopping Suicides*, 7.
6 Edward Wong and Jason Horowitz, "Italian Hostage, Released in Iraq, Is Shot by G.I.'s," *New York Times*, March 5, 2005. Also see John F. Burns, "On Way to Baghdad, Death Stalks Main Road," *New York Times*, May 29, 2005.
7 On the improved communication between deployed troops and those at home, see Susan Carruthers, "Communications Media, the U.S. Military, and the War Brought Home," in *At War: The Military and American Culture in the Twentieth Century and Beyond*, ed. David Kieran and Edwin A. Martini (New Brunswick, NJ: Rutgers University Press, 2018), 258–278.
8 "Vets' Suicide Rate 'Stunning,'" *CBS News*, November 13, 2007. Also see Christine Lagorio, "The Full Story: Veterans and Suicide," *CBS News*, March 20, 2008.
9 "House Committee on Veterans' Affairs, *Stopping Suicides*, 7.
10 Ibid., 10.
11 Chris Adams, "VA System Ill-Equipped to Treat Mental Anguish of War," McClatchy DC Bureau, February 5, 2007.

12 Ibid.

13 Lisa Black and Stacy St. Clair, "Wars' Hidden Toll," *Chicago Tribune*, January 17, 2010.

14 Lagorio, "The Full Story."

15 Ibid.

16 Ibid.

17 Adams, "VA System Ill-Equipped to Treat Mental Anguish of War.

18 On the social role of death, see Thomas W. Lacquer, *The Work of the Dead: A Cultural History of Mortal Remains* (Princeton, NJ: Princeton University Press, 2015).

19 Richard Bell, *We Shall Be No More: Suicide and Self-Government in the Newly United States* (Cambridge, MA: Harvard University Press, 2012), 9. See also Terri L. Snyder, *The Power to Die: Slavery and Suicide in British North America* (Chicago: University of Chicago Press, 2015) 11–12.

20 Erin Finley, *Fields of Combat: Understanding PTSD among Iraq and Afghanistan Veterans* (Ithaca, NY: Cornell University Press, 2011), 120.

21 House Committee on Veterans' Affairs, *Stopping Suicides*, 7.

22 Tom Infield, "Rumsfeld Hints at Much Broader Military Mission," *Twin Cities (MN) Pioneer Press*, January 17, 2002.

23 Melissa T. Brown, *Enlisting Masculinity: The Construction of Gender in US Military Recruiting Advertising during the All-Volunteer Force* (New York: Oxford University Press, 2012), 161.

24 Beth Bailey, *America's Army: Building the All-Volunteer Force* (Cambridge, MA: Harvard University Press, 2009), 250.

25 Drew Brown, "Military: Mixed Bag for Army; Re-enlistment Figures Rise, but Recruitment Lags," *Grand Forks (ND) Herald*, January 19, 2006; Ruth Pawder, "Fewer High School Seniors Opt for Military," *Bergen County (NJ) Record*, February 2, 2006; Carrie Anne Platt, "It Takes a Family to Raise a Soldier: Nurturing Recruitment through the Army Reserve's 'Help Them Find Their Strength' Campaign," paper presented at the annual meeting of the American Studies Association, Oakland, CA, October 2006, 1, copy in author's possession.

26 Platt, "It Takes a Family to Raise a Soldier," 1, 9. As the *Dallas Morning News* explained, these were advertisements "aimed at 'influencers'—parents, coaches, teachers and other adults who can be an obstacle to recruiting"; see Richard Whittle, "Aggressive Ad Efforts Tackle Army Recruiting Woes from Many Angles," *Dallas Morning News*, June 2, 2005. As Beth Bailey explains, a new "focus on parents" emerged because "struggling recruiters learned that, for this cohort of youth, parents' opinions really did matter" (*America's Army*, 251, 229). See also Brown, *Enlisting Masculinity*, 162, 164–165.

27 Brown, *Enlisting Masculinity*, 165; Donna Miles, "Army Recruiting Campaign Focuses on Prospects, Influencers," *DOD News*, August 30, 2005, www.defense.gov; and Ann Scott Tyson, "Army Debuts New Slogan in Recruiting Commercials," *Washington Post*, November 22, 2006. One columnist derisively referred to this advertising campaign as one in which the "Army has at last been repositioned as

a finishing school"; see Seth Stevenson, "Uncle Sam Wants Your Parents . . . to Let You Enlist," *Slate*, August 22, 2005, www.slate.com. On the campaign in general, see Bailey, *America's Army*, 251; and Platt, "It Takes a Family to Raise a Soldier."

28 Platt, "It Takes a Family to Raise a Soldier," 2. See also Brown, *Enlisting Masculinity*, 165.

29 RandomJunkGuy, "Army Strong Commercial 4," November 10, 2008, YouTube, www.youtube.com/watch?v=2TUbnGXqI1s. See also Brown, *Enlisting Masculinity*, 166.

30 Stevenson, "Uncle Sam Wants Your Parents"; and Platt, "It Takes a Family to Raise a Soldier," 12.

31 Platt, "It Takes a Family to Raise a Soldier," 13; and Brown, *Enlisting Masculinity*, 165.

32 Rise Advertising, "Two Things," YouTube, September 9, 2012, www.youtube.com/watch?v=HGiiVq9mV20. As Brown explains, these ads "contain a masculine subtext, a subtle suggestion that the Army will make a boy into a man" (Melissa Brown, *Enlisting Masculinity*, 165). As Platt explains, this series of commercials focuses on "a white father-son dyad" and offers "a concrete visualization of military masculinity" (Platt, "It Takes a Family to Raise a Soldier," 12).

33 Platt, "It Takes a Family to Raise a Soldier," 12, 2.

34 Ibid., 12; see also Brown, *Enlisting Masculinity*, 165.

35 Robert Burns, "Army Launching 'Army Strong' Ad Campaign," *Associated Press*, October 9, 2006. On the challenges of selling the Army while acknowledging the risks inherent in military life, see Bailey, *America's Army*, 245–247.

36 Stuart Elliot, "Army's New Battle Cry Aims at Potential Recruits," *New York Times*, November 9, 2006. On the silence on this matter, see Melissa Brown, *Enlisting Masculinity*, 165.

37 Stevenson, "Uncle Sam Wants Your Parents." The failure to acknowledge these realities amounts to what Jay Winter has termed a "political silence," one "chosen to suspend or truncate open conflict over the meaning and/or justification of violence"; see Jay Winter, "Thinking about Silence," in *Shadows of War: A Social History of Silence in the Twentieth Century*, edited by Efrat Ben-Ze'ev, Ruth Ginio, and Jay Winter (Cambridge: Cambridge University Press, 2010), 5.

38 "Changing the Course," *Augusta (GA) Chronicle*, November 9, 2006.

39 Lyndsey Layton, "The Story behind the Iraq Study Group," *Washington Post*, November 21, 2006.

40 James A. Baker III and Lee H. Hamilton, co-chairs, et al., *The Iraq Study Group Report* (New York: Vintage Books, 2006), xii–xiv.

41 Eugene Robinson, "A 'Surge' in Wasted Sacrifice," *Washington Post*, December 19, 2006.

42 John Kerry, "When Resolve Turns Reckless," *Washington Post*, December 24, 2006.

43 Adam Gorlick, "Marine Returns from Iraq to Emotional Ruin, Suicide," *Seacoast Online*, October 17, 2004. It is striking as well how much this sentence parallels

the final scene of Ron Kovic's memoir of the Vietnam War and its aftermath, *Born on the Fourth of July* (New York: McGraw-Hill, 1976).

44 Bob Herbert, "Death of a Marine," *New York Times*, March 19, 2007.

45 Gorlick, "Marine Returns from Iraq to Emotional Ruin, Suicide." See also Irene Sege, "'Something Happened to Jeff'; Jeffrey Lucey Returned from Iraq a Changed Man; Then He Killed Himself," *Boston Globe*, March 1, 2005.

46 Dennis Magee, "Parents Push for Soldier's Story to Continue beyond Suicide," *Waterloo-Cedar Falls (IA) Courier*, May 12, 2006, in Senate Committee on Veterans' Affairs, *Hearing on Mental Health Issues*, 110th Congress, 1st Session (2007), 18.

47 Ibid.

48 Kevin Giles, "VA Medical Records Don't Mention That Veteran Felt Suicidal," *Minneapolis Star Tribune*, January 27, 2007.

49 Kevin Giles, "A Marine's War at Home," *Minneapolis Star-Tribune*, February 5, 2007.

50 Tara Tuckwiller, "'The Closest Thing to Hell': Mother Copes with Son's Suicide," *Charleston (WV) Gazette*, August 6, 2006.

51 Ibid.

52 David Shaffer, "Haunted by War, Unwilling to Seek Help," *Minneapolis Star-Tribune*, February 18, 2007.

53 Chris Adams, "VA System Ill-Equipped to Treat Mental Anguish of War."

54 House Committee on Veterans' Affairs, *Stopping Suicides*, 7.

55 Magee, "Parents Push for Soldier's Story to Continue beyond Suicide."

56 Herbert, "Death of a Marine."

57 Beth Walton, "Soldier Suicides: Veterans Are Killing Themselves in Record Numbers," *Minneapolis Citypages*, March 26, 2008.

58 House Committee on Veterans' Affairs, *Stopping Suicides*, 68.

59 Jennifer Jacobs, "Pair Help Iraq Veterans 'Survive Peace,'" *Des Moines (IA) Register*, May 12, 2006, in Senate Committee on Veterans' Affairs, *Hearing on Mental Health Issues*, 16; and Magee, "Parents Push for Soldiers Story to Continue beyond Suicide."

60 Jacobs, "Pair Help Iraq Veterans 'Survive Peace'"; Adams, "VA System Ill-Equipped to Treat Mental Anguish of War"; and Herbert, "Death of a Marine." See also Sege, "Something Happened to Jeff."

61 Adams, "VA System Ill-Equipped to Treat Mental Anguish of War."

62 Platt points to another "Help Them Find Their Strength" commercial in which a father and son discuss enlisting over a game of pool; see Platt, "It Takes a Family to Raise a Soldier," 10. As Andrew Bacevich explains, both Budweiser and Miller based advertising campaigns on returning veterans; see Andrew Bacevich, *Breach of Trust: How Americans Failed Their Soldiers and Their Country* (New York: Metropolitan Books, 2013), 33–34.

63 Adams, "VA System Ill-Equipped to Treat Mental Anguish of War."

64 Sege, "Something Happened to Jeff."

65 Amy Goodman, "The Uncounted Casualties of War," *Salt Lake (UT) Tribune*, August 2, 2007.

66 Drew Brown, "U.S. Military Investigating Deaths of Iraqis Killed in Firefight," Knight Ridder Washington Bureau, March 16, 2006.

67 George W. Bush, "The President's News Conference, June 14, 2006," *Public Papers of the Presidents of the United States: George W. Bush, 2006, Book 1, January 1–June 30, 2006* (Washington, DC: Government Printing Office, 2006), 1127.

68 "Kick Blame Upstairs," *Brattleboro (VT) Reformer*, June 6, 2006. On the larger debate, see "Just a Few Bad Apples?" *Economist*, January 20, 2005; and Susan Sontag, "Regarding the Torture of Others," *New York Times*, May 23, 2004.

69 Thomas E. Ricks and Ann Scott Tyson, "Troops at Odds with Ethics Standards," *Washington Post*, May 5, 2007.

70 "Rules of Engagement," *New York Times*, Match 8, 2005. See also Kristi Ceccarossi, "Towns Mull Use of Guard in War Zones," *Brattleboro (VT) Reformer*, February 23, 2005.

71 Ibid.

72 Andrew Bacevich, "What's an Iraqi Life Worth?" *Washington Post*, July 9, 2006.

73 Ibid.

74 Ibid.

75 Ibid.

76 Here, I am following Judith Butler's argument that "war or, rather, the current wars rely upon and perpetuate a way of dividing lives into those that are worth defending, valuing, and grieving when they are lost, and those that are not quite lives, not quite valuable, recognizable, or, indeed, mournable"; see Judith Butler, *Frames of War: When Is Life Grievable?* (London: Verso, 2009), 42–43.

77 Brett T. Litz et al., "Moral Injury and Moral Repair in War Veterans: A Preliminary Model and Intervention Strategy," *Clinical Psychology Review* 289, no. 8 (2009): 700. See also Ann Jones, *They Were Soldiers: How the Wounded Return from America's Wars—The Untold Story* (Chicago: Haymarket Books, 2013), 102–103.

78 Ann Jones, *They Were Soldiers*, 102.

79 Senate Subcommittee of the Committee on Appropriations, *Department of Defense Appropriations for Fiscal Year 2003*, 107th Congress, 2d Session (2002), 71.

80 Instead, it explains that "The National Guard is best situated to coordinate the efforts of local, state, and federal agencies so as to provide the nation with the best possible disaster response capabilities" (ibid., 185).

81 Dave Moniz, "Rate of Guard Deaths Higher," *USA Today*, December 13, 2004.

82 Army National Guard, "Annual Financial Report, Fiscal Year 2005," 23, 16, U.S. Army website, www.army.mil; Lawrence J. Korb, "All-Volunteer Army Shows Signs of Wear," *Atlanta Journal-Constitution*, February 27, 2005; and Josh White and Ann Scott Tyson, "Military Has Lost 2,000 in Iraq," *Washington Post*, October 26, 2005.

83 Senate Subcommittee of the Committee on Appropriations, *Department of Defense Appropriations for Fiscal Year 2008*, 110th Congress, 1st Session (2007), available through the Government Printing Office, www.gpo.gov.

84 Ann Scott Tyson, "Possible Iraq Deployments Would Stretch Reserve Force," *Washington Post*, November 6, 2006; Bailey, *America's Army*, 246–247; David Von Drehle, "Wrestling with History," *Washington Post*, November 13, 2005; Eric Schmidt, "General Says the Current Plan Is to Maintain 120,000 Soldiers in Iraq through 2006," *New York Times*, January 25, 2005; Korb, All-Volunteer Army Shows Signs of Wear"; White and Tyson, "Military Has Lost 2,000 in Iraq," *Washington Post*, October 26, 2005; Ron Harris, "Changing Role of National Guard Takes Toll on Citizen Soldiers," *Port St. Lucie (FL) Tribune in Fort Pierce*, August 1, 2005; Harry Levins, "Extended Duty Sways Debate over Reservists," *St. Louis Post-Dispatch*, March 21, 2005; and James Janega, "Morale Woes Rattle Guard," *Chicago Tribune*, August 7, 2005.

85 See, e.g., Marla Jo Fisher and Michael Coronado, "Hardships in the Guard," *Orange County (CA) Register*, May 17, 2006; Dana Hull, "Post-Traumatic Stress: A New Generation," *San Jose Mercury News*, October 8, 2006; and Thom Shanker and Michael Gordon, "Strained, Army Looks to Guard for More Relief," *New York Times*, September 22, 2006.

86 Magee, "Parents Push for Soldiers Story to Continue beyond Suicide."

87 House Committee on Veterans' Affairs, *Stopping Suicides*, 7.

88 Ibid.

89 "Editorial: The Nation Is Failing Its Mentally Wounded," *Minneapolis Star-Tribune*, January 31, 2007.

90 Ibid.

91 Ibid.

92 "Readers Write," *Minneapolis Star-Tribune*, February 1, 2007.

93 Ibid.

94 Ibid.

95 "Balancing Budget on Veterans' Backs," *Pittsfield (MA) Berkshire Eagle*, February 20, 2007.

96 Ibid.

97 Mary L. Dudziak, *War Time: An Idea, Its History, Its Consequences* (Oxford: Oxford University Press, 2012), 8.

98 Bacevich, *Breach of Trust*, 35.

99 "Medical Scandal Goes beyond Walter Reed; Stories Tell of Poor Care in Many Places and for a Long Time," *Minneapolis Star-Tribune*, March 6, 2007.

100 Larry Evans, "Mental Wounds of War," *Sarasota (FL) Herald-Tribune*, March 6, 2007.

101 Ibid.

102 Rep. Bob Filner of California, speaking on H.R. 327, *Congressional Record*, March 21, 2007, H2773.

103 Sen. Harkin of Iowa, speaking in favor of S. 479, *Congressional Record*, February 1, 2007, S1529.

104 Ibid.

105 Senate Committee on Veterans' Affairs, *Hearing on Mental Health Issues*, 110th Congress, 1st Session (2007), 13.

106 House Committee on Veterans' Affairs, *Stopping Suicides*, 7.

107 Ibid.

108 Ibid., 7, 9.

109 Ibid., 14.

110 Ibid., 28.

111 Sen. Amy Klobuchar of Minnesota, speaking on "Veterans Health Care," *Congressional Record*, March 7, 2007, S2753–S2754.

112 Ibid., S2753.

113 Ibid.

114 Ibid., S2754.

115 Ibid.

116 Ibid.

117 Ibid.

118 Senate Committee on Veterans' Affairs, *Hearing on Mental Health Issues*, 110th Congress, 1st Session (2007), 4, 6.

119 Ibid., 5. This was a point that Rockefeller returned to later in the hearing (ibid., 46). It was part of a broader critique that Democrats leveled over the Bush administration's willingness to increase the national debt in order to fight the war. See, e.g., , Rep. Allyson Schwartz of Pennsylvania, "Democrats are Prepared to Lead the Country in a New Direction," *Congressional Record*, September 19, 2006, H6713; Sen. Byron Dorgan of North Dakota, "The 109th Congress," *Congressional Record*, September 26, 2006, S10113; Rep. Chris Murphy of Connecticut, "Impact of the President's Budget on America," *Congressional Record*, February 5, 2007, H1197; and Rep. Lynn Woolsey of California, "20th Special Order on the War in Iraq," *Congressional Record*, April 17, 2007, H3449.

120 On the VA's budget during the early years of the Iraq War in general, see Aaron Glantz, *The War Comes Home: Washington's Battle against America's Veterans* (Berkeley: University of California Press, 2009), 118–128.

121 See "Statement of Michael J. Kussman, MD, MS, MACP, Acting Under Secretary of Health, Department of Veterans Affairs, Before the House Committee on Appropriations Subcommittee on Military Construction, Veterans' Affairs and Related Agencies" (Washington, DC: Department of Veterans Affairs, March 8, 2007), www.va.gov.

122 "VA Announces Action Plan," *Psychiatric News*, November 5, 2004; and President's New Freedom Commission on Mental Health, *Achieving the Promise: Transforming Mental Health Care in America: Final Report* (Rockville, MD: Department of Health and Human Services, 2003), available through the University of North Texas Government Documents Department Cybercemetery at https://govinfo.library.unt.edu.

123 Antonette M. Zeiss and Bradley E. Karlin, "The Role of Psychology in Emerging Federal Health-Care Plans," in the *Oxford Handbook of Clinical Psychology*, 2nd ed., ed. David H. Barlow (Oxford: Oxford University Press, 2014), 171.

124 Ibid., 172.

125 Finley, *Fields of Combat*, 118.

126 Zeiss and Karlin, "The Role of Psychology," 171.

127 Ibid., 172–173.

128 Glantz, *The War Comes Home*, 119–120, 123.

129 General Accountability Office, "VHA Health Care: Spending for Mental Health Strategic Plan Initiatives Was Substantially Less than Planned" (Washington, DC: General Accountability Office, 2006), 1, www.gao.gov.

130 Ibid., 14.

131 Ibid., 10, 17.

132 Ibid.

133 Ibid., 21.

134 Ibid., 24.

135 Ibid., 25.

136 Sen. Harkin of Iowa, speaking in favor of S. 479, *Congressional Record*, February 1, 2007, S1530.

137 See Rep. Tim Walz of Minnesota speaking in favor of HR 327, *Congressional Record*, March 21, 2007, H2776; Rep. Bruce Braley of Iowa, speaking in favor of HR 327, *Congressional Record*, March 21, 2007, H2777; and Rep. Dave Loebsack of Iowa, speaking in favor of HR 327, *Congressional Record*, October 23, 2007, H11872.

138 Rep. McNerney of California, speaking in favor of HR 327, *Congressional Record*, March 21, 2007, H2775– H2776.

139 Rep. Filner of California, speaking in favor of HR 327, *Congressional Record*, October 23, 2007, H11869.

140 "Military Veterans / Suicides," *CBS Evening News*, November 13, 2007, Vanderbilt Television News Archive, www.vanderbilt.edu.

141 Ibid.

142 Ibid.

143 See Ira R. Katz, "A Study of E. Coli Alkaline Phosphate" (PhD diss., Columbia University, 1969). Katz's CV is at "Ira R. Katz," Perelman School of Medicine, University of Pennsylvania, July 31, 2003, www.upenn.edu.

144 Senate Special Committee on Aging, *Suicide and the Elderly: A Population at Risk*, 104th Congress, 2d Session (1996), 113–114.

145 Ibid., 116.

146 Ibid., 118.

147 "Statement of Michael J. Kussman, MD, MS, MACP, Acting Under Secretary of Health, Department of Veterans Affairs," Department of Veterans Affairs, March 8, 2007, www.va.gov.

148 "Military Veterans / Suicides," *CBS Evening News*, November 13, 2007.

149 House Committee on Veterans' Affairs, *Stopping Suicides*, 49.

150 Two VA officials with whom I spoke confirmed this, although both requested to remain anonymous because they were not authorized by the VA to conduct interviews.

151 Sen. Murray of Washington, speaking on Veterans Suicides, *Congressional Record*, November 14, 2007, S14363.

152 Ibid., S14362.
153 Ibid.
154 Ibid.
155 Ibid.
156 Ibid.
157 Ibid., S14363.
158 "Military Veterans / Suicides," *CBS Evening News*, November 14, 2007.
159 Ibid.
160 House Committee on Veterans' Affairs, *Stopping Suicides*, 20.
161 Ibid., 50.
162 Ibid., 51–52.
163 Ibid., 54, 56–57.
164 Ibid., 57.
165 Ibid., 61.
166 Ibid.
167 "Iraq War/ Troop Level / Veteran Suicides," *CBS Evening News*, March 20, 2008, Vanderbilt Television News Archive, www.vanderbilt.edu.
168 Ibid.
169 Ibid.
170 "CBS News Investigation (Military: Veterans Suicides)," *CBS Evening News*, April 21, 2008, Vanderbilt Television News Archive, www.vanderbilt.edu; Les Blumenthal, "VA Accused of Lying on Suicides," *Pittsburgh Post-Gazette*, April 24, 2008.
171 "Military / Veterans Suicides," *CBS Evening News*, April 25, 2008, Vanderbilt Television News Archive, www.vanderbilt.edu.
172 "Military / Veterans Suicides," *CBS Evening News*, April 22, 2008, Vanderbilt Television News Archive, www.vanderbilt.edu.; Carl Young, "Congress Sitting on Vets' Health-Care Bill," *Eureka (CA) Times Standard*, May 4, 2008.
173 Blumenthal, "VA Accused of Lying on Suicides."
174 "Udall Calls for VA Doctor's Dismissal," Associated Press, May 2, 2008.
175 "Front Lines of Care," *Spokane (WA) Spokesman-Review*, May 1, 2008.
176 "Time For Honesty," *Spokane (WA) Spokesman-Review*, April 27, 2008; Neil McFarquhar, "In Federal Suit, 2 Views of Veterans Health Care," *New York Times*, April 22, 2008; and "In Our View: It's a VA Cover-up," *Vancouver (WA) Columbian*, April 25, 2008.
177 House Committee on Veterans Affairs, *The Truth about Veterans' Suicides*, 110th Congress, 2d Session (2008), 2.
178 Ibid., 4–5.
179 Ibid., 18–19.
180 Ibid., 19.
181 Ibid.
182 Ibid., 22.
183 Ibid.

184 Terri Tanielian and Lisa H. Jaycox, eds., *Invisible Wounds of War: Psychological and Cognitive Injuries, Their Consequences, and Services to Assist Recovery* (Santa Monica, CA: RAND Corp., 2008), 303, 267, 272–273.

185 Ibid., 291.

186 Ibid., 265.

187 *"Complaint for Declaratory and Injunctive Relief Under United States Constitution and Rehabilitation Act," Veterans For Common Sense v. R. James Nicholson, et al., July 23, 2007,* 3, in Case files, *Veterans For Common Sense et al. v. James Peake et al.*, No. 08-16728 (9th Cir. 2011), U.S. District Court for the Northern District of California.

188 Ibid., 7. For a critique of the suit, see Finley, *Fields of Combat*, 128; and Glantz, *The War Comes Home*, 126–134.

189 *Veterans For Common Sense*, 5.

190 Ibid., 22.

191 McFarquhar, "In Federal Suit, 2 Views of Veterans Health Care."

192 Ibid.

193 Neil MacFarquhar, "Closing Arguments in Suit on Veterans' Mental Care," *New York Times*, May 1, 2008.

194 "Tell Truth on Vets' Health," *Palm Beach (FL) Post*, May 2, 2008.

195 "The Suffering of Soldiers," *New York Times*, May 11, 2008.

196 "Wars Take a Deadly Toll, Even After Vets Return Home," *USA Today*, May 23, 2008.

197 Gordon P. Erspamer to Samuel Conti, May 28, 2008, *Veterans For Common Sense v. Peake* case files.

198 Finley, *Fields of Combat*, 130.

199 Erspamer to Samuel Conti, May 28, 2008.

200 Arturo J. Gonzalez to Daniel Bensing, May 15, 2008, *Veterans For Common Sense v. Peake* case files.

201 Christopher Lee, "Official Urged Fewer Diagnoses of PTSD," *Washington Post*, May 16, 2008.

202 Rep. Rush of Illinois, speaking on "VA PTSD Diagnoses," *Congressional Record*, May 23, 2008, E1067.

203 "Obama Demands VA Investigation Into PTSD Diagnoses," *CBS News*, May 16, 2008, www.cbsnews.com.

204 "Failing the Troops," *Copley News Service, May 29, 2008.*

205 "Veterans, Cheated by Washington," *Charleston (WV) Gazette*, May 28, 2008.

206 James J. Schwartz to Arturo J. Gonzalez, May 22, 2008, *Veterans For Common Sense et al. v. James Peake et al.*

207 James J. Schwartz to Samuel Conti, June 4, 2008, *Veterans For Common Sense et al. v. James Peake et al.*

208 Senate Committee on Veterans' Affairs, *Oversight Hearing on Systemic Indifference to Invisible Wounds*, 110th Congress, 2d Session (2008), 42, 10, 47.

209 Daniel W. Riley, "Obama Demands VA Investigation into PTSD Diagnoses," *Politico*, May 16, 2008, www.politico.com.

210 Senate Committee on Veterans' Affairs, *Oversight Hearing on Systemic Indifference to Invisible Wounds*, 2. See also Sen. Sanders's (I-VT) questioning of Katz over the emails; ibid., 45–46.

211 Ibid., 5.

212 James J. Schwartz to Samuel Conti, June 4, 2008, *Veterans For Common Sense et al. v. James Peake et al.*

213 Finley, *Fields of Combat*, 129.

214 Senate Committee on Veterans' Affairs, *Oversight Hearing on Systemic Indifference to Invisible Wounds*, 11.

215 Ibid., 11–12.

216 Ibid., 42–43.

217 Ibid., 12.

218 Ibid., 51.

219 Ibid., 57.

220 Ibid, 5.

221 Ibid.

CHAPTER 4. "THE CULTURE OF THE ARMY WASN'T READY"

1 On Casey's father's death, see David Cloud and Greg Jaffe, *The Fourth Star: Four Generals and Their Epic Struggle for the Future of the United States Army* (New York: Three Rivers Press, 2009), 1–9.

2 Senate Committee on Armed Services, *Nominations before the Senate Committee on Armed Services*, 110th Congress, 1st Session (2007), 193–194.

3 Ibid., 288. MHAT = Mental Health Assessment Teams.

4 George Casey, interview with the author, October 13, 2017.

5 Ibid.

6 George Casey, interview with the author, April 6, 2016.

7 Ibid.

8 Ibid.

9 "Information Paper: Post-Traumatic Stress Disorder," September 15, 2008, FOIA. In this volume, papers accessed by the Freedom of Information are noted with the abbreviation FOIA. Copies remain in the author's possession.

10 Ibid. These numbers deserve a gloss, however, because it is important to disentangle the number who have the condition from the number being diagnosed. Thus, the increase was potentially what the Army would call a "good news story" in that it could have reflected better diagnostics, a reduction in stigma, and more people getting care. However, it could also have reflected that more soldiers were suffering.

11 As Finley notes, "The military as an institution was responsive to public pressure and could be made to act quickly and dramatically if subject to the appropriate

leverage"; see Erin Finley, *Fields of Combat: Understanding PTSD among Iraq and Afghanistan Veterans* (Ithaca, NY: Cornell University Press, 2011), 102.

12 Ibid., 104.

13 Jonathan Kors, "How Specialist Town Lost His Benefits," *Nation*, March 29, 2007, www.thenation.com.

14 On the personality debate in general, but not on the Army's response to it, see Aaron Glantz, *The War Comes Home: Washington's Battle with America's Veterans* (Berkeley: University of California Press, 2008), 95–104.

15 American Psychiatric Association, "Personality Disorders," in *Diagnostic and Statistical Manual of Mental Disorders*, 5th ed. (*DSM*-5; Arlington, VA: American Psychiatric Association, 2013); and Aaron T. Beck, "Theory of Personality Disorders," in *Cognitive Therapy of Personality Disorders*, 3rd ed., ed. Aaron T. Beck, Denise D. Davis, and Arthur Freeman (New York: Guilford Press, 2015), 20–21, 24–25.

16 Jerry M. Goffman and E. K. Eric Gunderson, "Prognosis for Personality Disorders in the Naval Service," Report no. 75-25 (San Diego, CA: Naval Health Research Center, 1975), 2; and Jeanne M. Erickson, D. Edwards, and S. F. Bucky, "The Disposition and Organizational Effectiveness of Personality Disorders in a Military Setting," Report no. 73-19 (San Diego, CA: Naval Health Research Center, 1973), 1.

17 Erickson et al., "The Disposition and Organizational Effectiveness of Personality Disorders in a Military Setting," 1.

18 Mark A. Schuckit and E. K. Eric Gunderson, "The Clinical Value of Personality Disorder Subtypes in Naval Service," Report no. 75-22 (San Diego, CA: Naval Health Research Center, 1975), 4.

19 Elspeth Cameron Ritchie, interview with the author, December 15, 2015.

20 Ibid.

21 Ibid.

22 Erin Emery, "Shattered Iraq GIs Say They Face a New Battle at Home," *Denver Post*, December 26, 2007.

23 Matthew D. LaPlante, "Were Pleas of Marine Ignored?" *Salt Lake Tribune*, December 31, 2006; "A Slap in the Face," *St. Louis Post-Dispatch*, January 9, 2007; Dana Priest and Anne Hull, "The War Inside," *Washington Post*, June 17, 2007.

24 House Committee on Veterans Affairs, *Post-traumatic Stress Disorder and Personality Disorder: Challenges for the United States Department of Veterans Affairs*, 110th Congress, 1st Session (2007), 1–2.

25 Ibid., 8.

26 Ibid.

27 Ibid., 9.

28 Ibid., 15.

29 Ibid., 16, 28, 40–41.

30 Ibid., 38–39.

31 Robert Cardona and Elspeth Cameron Ritchie, "Psychological Screening of Recruits prior to Accession in the U.S. Military," in *Recruit Medicine*, ed. Bernard L. DeKoning (Washington, DC: U.S. Department of Defense, 2006), 305.

32 Ibid.

33 According to the Mayo Clinic, "Personality disorders usually begin in the teenage years or early adulthood"; see Mayo Clinic, "Personality Disorders," 2018, www.mayoclinic.org. However, other research has shown that there may often be early evidence of disorders; see Thomas N. Crawford et al., "The Course and Psychosocial Correlates of Personality Disorder Symptoms in Adolescence: Erikson's Developmental Theory Revisited," *Journal of Youth and Adolescence* 33, no. 5 (2005): 373. See also J. G. Johnson et al., "Age-Related Change in Personality Disorder Trait Levels between early Adolescence and Adulthood: A Community-Based Longitudinal Investigation," *Acta Psychiatrica Scandinavica* 102 (2000): 265–275; and Benjamin B. Lahey et al., "Predicting Future Antisocial Personality Disorder in Males from a Clinical Assessment in Childhood," *Journal of Consulting and Clinical Psychology* 73, no. 3 (2005): 389–399.

34 Cardona and Ritchie, "Psychological Screening of Recruits," 306.

35 House Committee on Veterans Affairs, *Post-Traumatic Stress Disorder and Personality Disorder*, 46.

36 Ibid., 48.

37 Defense Health Board Task Force on Mental Health, *An Achievable Vision: Report of the Department of Defense Task Force on Mental Health* (Falls Church, VA: Defense Health Board, 2007), E-4, 22.

38 Senate Committee on Armed Services, *Nominations before the Senate Committee on Armed Services*, 803.

39 "Military Bars Soldiers With Personality Disorders," National Public Radio, October 16, 2007; Daniel Zwerdling, "Army Dismissals for Mental Health, Misconduct Rise," National Public Radio, November 15, 2007, www.npr.org.

40 House Committee on Veterans' Affairs, *Findings of the President's Commission on Care for America's Returning Wounded Warriors*, 110th Congress, 1st Session (2007), 15.

41 Beth Bailey, *America's Army: Making the All-Volunteer Force* (Cambridge, MA: Harvard University Press, 2009), 253.

42 Gale Pollock, interview with the author, November 19, 2015.

43 Ibid.

44 Elspeth Cameron Ritchie, interview with the author, December 15, 2015.

45 Ibid.

46 Ibid.

47 Gale Pollock, interview with the author, November 19, 2015.

48 Rebecca Porter, interview with the author, December 17, 2015.

49 Gale Pollock, interview with the author, November 19, 2015; Jonathan Jaffin, interview with the author, January 30, 2016. Jonathan Jaffin, interview with the author, January 30, 2016.

50 Gale Pollock, interview with the author, November 19, 2015.

51 Gale S. Pollock to Commanders, MEDCOM Regional Medical Command, "Review of Personality Disorder (Chapter 5-13) Separations," August 6, 2007, FOIA.

52 Eric B. Schoomaker to Commanders, MEDCOM Regional Medical Command, "Screening for Post-traumatic Stress Disorder (PTSD) and Mild Traumatic Brain Injury (mTBI) prior to Administrative Separation," May 1, 2008, FOIA.

53 "Information Paper: Post-Traumatic Stress Disorder (PTSD) Screening and Soldiers," July 14, 2008, FOIA.

54 Ronald J. James to Deputy Chief of Staff, G-1, "Enlisted Separations on the Basis of Personality Disorder POLICY MEMORANDUM," February 10, 2009, 1, FOIA.

55 Ibid., 2.

56 For information on these two categories of "Chapter 5 administrative discharges," see "What You Should Know About Chapter 5, AR 635-200 Separation For The Convenience Of The Government," Trial Defense Service, Fort Knox Field Office, n.d., 1. Available at www.knox.army.mil/garrison/supportoffices/tds/docs/chapter_5.pdf.

57 Elspeth Cameron Ritchie, interview with the author, December 15, 2015.

58 Ibid.

59 ANON to See Distribution, "Guidance for Administrative Separation for Personality Disorder and Other Mental Conditions," May 11, 2009, 2–3, FOIA.

60 Ibid., 2.

61 Thomas R. Lamont to Undersecretary of Defense (Personnel and Readiness), "Army Personality Disorder Separation Compliance Report for Fiscal Year 10," March 31, 2011, 2, FOIA.

62 Ibid., 3.

63 Ibid., 4.

64 In psychology, these categories dicussed below are "public stigma," "self-stigma," and "structural stigma"; see Tiffany M. Greene-Shortridge et al., "The Stigma of Mental Health Problems in the Military," *Military Medicine* 152, no. 2 (2007): 157–158, figure at 158. See also Finley, *Fields of Combat*, 83.

65 George Casey, interview with the author, October 13, 2017.

66 Sheena M. Eagan Chamberlin, "Emasculated by Trauma: A Social History of Post-traumatic Stress Disorder, Stigma, and Masculinity," *Journal of American Culture* 35, no. 4 (2012): 358, 364; and Finley, *Fields of Combat*, 7–8, 87, 107–110.

67 Charles Hoge et al., "Combat Duty in Iraq and Afghanistan, Mental Health Problems, and Barriers to Care," *New England Journal of Medicine* 351, no. 1 (2004): 20–21. See also Victoria Langston et al., "Culture: What Is the Effect of Stress in the Military?" *Military Medicine* 172, no. 9 (2007): 932–933; and Defense Health Board Task Force on Mental Health, *An Achievable Vision*, 17, 25.

68 Langston et al., "Culture: What Is the Effect of Stress in the Military?" 933. See also Tracy Stecker et al., "An Assessment of Beliefs about Mental Health Care among Veterans Who Served in Iraq," *Psychiatric Services* 58, no. 10 (2007): 1359.

69 On the stigma-promoting climate at Fort Carson more generally, see Yochi Dreazen, *The Invisible Front: Love and Loss in an Era of Endless War* (New York: Crown, 2014), esp. 229–230.

70 Emery, "Shattered Iraq GIs Say They Face a New Battle at Home"; and Dan Frosch, "Soldiers' Anguish Sometimes Belittled," *New York Times*, May 13, 2007.

71 Finley, *Fields of Combat*, 74, 107–108; and Dreazen, *The Invisible Front*, 194.

72 Senate Committee on Veterans' Affairs, *Seattle Field Hearing: Coming Home from Combat—Are Veterans Getting the Help They Need?* 109th Congress, 1st Session (2005), 56–57. For another example, see House Committee on Veterans' Affairs, Subcommittee on Economic Opportunity, *Transition Assistance for Members of the National Guard*, 109th Congress, 1st Session (2005), 31.
73 Jimmie Keenan, interview with the author, December 1, 2015.
74 Ibid.
75 Peter Roper, "Task Force Studies Combat Stress Cases," *Pueblo (CO) Chieftain*, January 26, 2007.
76 Defense Health Board Task Force on Mental Health, *An Achievable Vision*, 21.
77 Ibid., 23, 8.
78 Ibid., 18.
79 Department of the Army, "Interim Guidance—Army Mild Traumatic Brain Injury (MTBI) and Post-traumatic Stress Disorder (PTSD) Awareness and Response Program," July 15, 2007, FOIA.
80 Ibid. 2; Elisabeth Bumiller, "We Have Met the Enemy and He Is PowerPoint," *New York Times*, April 26, 2010.
81 Department of the Army, "Interim Guidance," 2; Army News Service, "Army Launches Chain Teaching Program for PTSD," July 16, 2007, www. Army.mil..
82 Department of the Army, "Interim Guidance—Army Mild Traumatic Brain Injury (MTBI) and Post-traumatic Stress Disorder (PTSD) Awareness and Response Program," July 15, 2007, 2, FOIA.
83 "Medical Care Available to Soldiers with Post-traumatic Stress Disorder (PTSD)," Information paper, March 18, 2009, FOIA.
84 Finley, *Fields of Combat*, 107.
85 George Casey, interview with the author, April 6, 2016.
86 Ibid.
87 Gale Pollock, interview with the author, November 19, 2015.
88 Ibid.
89 Dreazen, *The Invisible Front*, 195.
90 Ibid., 202.
91 Edmund Sanders, "Attack on U.S. Base Kills 22 at Mess Tent," *San Francisco Chronicle*, December 22, 2004.
92 George Casey, interview with the author, October 13, 2017.
93 Tom Vanden Brook, "General's Story Puts Focus on Stress from Combat," *USA Today*, November 25, 2008.
94 Ibid.
95 Rebecca Porter, interview with the author, December 17, 2015.
96 Ibid. Porter ultimately sought private treatment.
97 Christopher Ivany, interview with the author, December 16, 2015.
98 Gale Pollock, interview with the author, November 29, 2015.
99 See, e.g., "Names of the Dead" items in the *New York Times* for February 1, 2008; February 26, 2008; March 21, 2008; and April 11, 2008, as well as "U.S. Death

Toll," *USA Today*, January 31, 2008. Also see Christopher Ivany, interview with the author, December 16, 2015.

100 Christopher Ivany, interview with the author, December 16, 2015.

101 Ibid.

102 Ibid.

103 Office of the Surgeon, Multi-national Force–Iraq and Office of the Surgeon General, United States Army Medical Command, *Mental Health Advisory Team (MHAT) V: Operation Iraqi Freedom 06-08, 14 February 2008* (MHAT-V; Falls Church, VA: Office of the U.S. Surgeon General, 2008), 68, 71.

104 Christopher Ivany, interview with the author, December 16, 2015.

105 *The Hurt Locker*, directed by Katherine Bigelow (Universal City, CA: Summit Entertainment, 2008; DVD, 2010).

106 Christopher Ivany, interview with the author, December 16, 2015.

107 Michael Bell and Amy Milliken, "Epidemiologic Consultation no. 14-HK-OB1U-09: Investigation of Homicides at Fort Carson, Colorado, November 2008–May 2009" U.S. Army Center for Health Promotion and Preventive Medicine (Aberdeen Proving Ground, MD: U.S. Army Center for Health Promotion and Preventive Medicine, July 2009), 2.

108 Ibid., 11; and Eric Schoomaker, interview with the author, December 17, 2015.

109 Michael Bell, interview with the author, January 19, 2016.

110 Ibid.

111 Ibid.

112 Bell and Milliken, "Epidemiologic Consultation no. 14-HK-OB1U-09," 3.

113 Ibid., 6.

114 Ibid., 3.

115 Michael Bell, interview with the author, January 19, 2016; On Graham's family history, see Dreazen, *The Invisible Front*.

116 Michael Bell, interview with the author, January 19, 2016.

117 Ibid.

118 Ibid.

119 Ibid.

120 Bell and Milliken, "Epidemiologic Consultation no. 14-HK-OB1U-09," 8–9.

121 Elspeth Cameron Ritchie, interview with the author, December 15, 2015; and Eric Schoomaker, interview with the author, April 6, 2016.

122 Bell and Milliken, "Epidemiologic Consultation no. 14-HK-OB1U-09," 11.

123 Ibid., 18.

124 Eric Schoomaker, interview with the author, December 16, 2015.

125 Bell and Milliken, "Epidemiologic Consultation no. 14-HK-OB1U-09," 15.

126 Ibid., 16.

127 Ibid., ES-4.

128 Eric Schoomaker, interview with the author, December 16, 2015.

129 Bell and Milliken, "Epidemiologic Consultation no. 14-HK-OB1U-09," 16, E-7.

130 Ibid., B-6.

131 Ibid., 22–23.

132 Dreazen, *The Invisible Front*, 235, 240–244, 252. The EPICON team had noted this, pointing to the behavioral health staff that had been hired and the treatment programs that had been implemented there; see Bell and Milliken, "Epidemiologic Consultation no. 14-HK-OB1U-09," 17.

133 Rebecca Porter, interview with the author, December 17, 2015. See also Charles W. Hoge et al., "Transformation of Mental Health Care for U.S. Soldiers and Families during the Iraq and Afghanistan Wars: Where Science and Politics Intersect," *American Journal of Psychiatry* 173, no. 4 (2016): 337–338.

134 Hoge et al., "Transformation of Mental Health Care," 337.

135 Christopher Ivany, interview with the author, December 16, 2015; and Bell and Milliken, "Epidemiologic Consultation no. 14-HK-OB1U-09," E-9.

136 Christopher Ivany, interview with the author, December 16, 2015.

137 Ibid.

138 Bell and Milliken, "Epidemiologic Consultation no. 14-HK-OB1U-09," E-9.

139 Dreazen, *The Invisible Front*, 252–253.

140 Christopher Ivany, interview with the author, December 16, 2015.

141 Rebecca Porter, interview with the author, December 17, 2015.

142 Christopher Ivany, interview with the author, December 16, 2015.

143 Ibid.

144 Ibid.

145 Bell and Milliken, "Epidemiologic Consultation no. 14-HK-OB1U-09," E-9.

146 Senior Defense Officials, "Defense Department Background Briefing: Mental Helath," March 22, 2012, FOIA.

147 Christopher Ivany, interview with the author, December 16, 2015.

148 Hoge et al., "Transformation of Mental Health Care," 337.

149 Rebecca Porter, interview with the author, December 17, 2015. There had been some expansions beyond Fort Carson earlier. See Shannon Carabajal, "Army Expanding Successful Embedded Behavioral Health Program," *Fort Hood (TX) Sentinel*, November 24, 2011, www.forthoodsentinel.com.

150 Rebecca Porter, interviews with the author, December 17, 2015, and May 6, 2016; and Hoge et al., "Transformation of Mental Health Care," 338.

151 Hoge et al., "Transformation of Mental Health Care," 338.

152 Charles Engel, interview with the author, January 18, 2016.

153 David Brown, "Army Takes New Tack in Fighting the Unseen Enemy," *Washington Post*, November 23, 1998. See also Hoge et al., "Transformation of Mental Health Care," 338.

154 Charles Engel, interview with the author, January 18, 2016.

155 Ibid.

156 Ibid.

157 Ibid.

158 Ibid.
159 David Brown, "Soldiering on in the Face of Pain," *Washington Post*, November 24, 1998.
160 Charles Engel, interview with the author, January 18, 2016.
161 David Brown, "Soldiering on in the Face of Pain."
162 Charles Engel, interview with the author, January 18, 2016; and Charles C. Engel, Jr., et al., "Relationship of Physical Symptoms to Posttraumatic Stress Disorder among Veterans Seeking Care for Gulf War–Related Health Concerns," *Psychosomatic Medicine* 62, no. 6 (2000): 743, 739.
163 Charles Engel, interview with the author, January 18, 2016.
164 House Committee on Veterans' Affairs, *The Department of Defense and Department of Veterans Affairs: The Continuum of Care for Post Traumatic Stress Disorder*, 109th Congress, 1st Session (2005), 9.
165 Ibid., 27.
166 Ibid., 28.
167 Charles C. Engel et al., "RESPECT-Mil: Feasibility of a Systems-Level Collaborative Care Approach to Depression and Post-traumatic Stress Disorder in Military Primary Care," *Military Medicine* 173, no. 10 (2008): 937–938.
168 Ibid., 938; and Thomas Oxman et al., "RESPECT-MIL Clinician Education Manual," n.pag., FOIA.
169 Thomas Oxman et al., "RESPECT-MIL Clinician Education Manual," n.pag., FOIA.
170 Engel et al., "RESPECT-Mil," 938; and Thomas Oxman et al., "RESPECT-MIL Clinician Education Manual," n.pag., FOIA.
171 Engel et al., "RESPECT-Mil," 938.
172 Ibid.
173 Ibid.
174 Ibid., 939.
175 Ibid.
176 Charles Engel, interview with the author, January 18, 2016.
177 Ibid.
178 Ibid.
179 Ibid.
180 Charles Engel, interview with the author, January 18, 2016.
181 George Casey, interview with the author, December 16, 2015.
182 George Casey, interview with the author, October 13, 2017.
183 Amy Adler, interview with the author, April 21, 2016.
184 Ibid.
185 Charles Engel, interview with the author, January 18, 2016.
186 Thomas Oxman et al., "RESPECT-MIL Clinician Education Manual," n.pag., FOIA.
187 Ibid.
188 Gale Pollock, interview with the author, November 19, 2015.

189 Charles Engel, interview with the author, January 18, 2016.

190 Ibid.

191 Ibid.; and Peter Chiarelli, interview with the author, December 20, 2015.

192 Charles Engel, interview with the author, January 18, 2016.

193 Ibid.

194 Eunice C. Wong et. al., "Evaluating the Implementation of the Re-engineering Systems of Primary Care Treatment in the Military (RESPECT-Mil)," *Rand Health Quarterly* 5, no. 2 November 30, 2015), www.rand.org.

195 Seth Robson, "Soldiers Seeking Routine Medical Care Now Get PTSD Screening as Well," *Stars and Stripes*, July 5, 2012, www.stripes.com.

196 Charles Engel, interview with the author, January 18, 2016.

197 Ibid.

198 Ibid.

199 Senate Committee on Armed Services, *Nominations before the Senate Committee on Armed Services*, 110th Congress, 1st Session (2007), 713.

200 David S. Cloud, "Army Brigade, Long a Symbol of Readiness, Is Stretched Thin," *New York Times*, March 20, 2007; Ann Scott Tyson, "Military Is Ill-Prepared for Other Conflicts," *Washington Post*, March 19, 2007; and Neil Abercrombie and Solomon Ortiz, "How to Fuel Up the Out-of-Gas U.S. Military Machine," *Christian Science Monitor*, March 19, 2007. Erin Finley points out that "by the middle of 2006 the media was raising concerns about the ability of U.S. armed forces to keep up with the logistical requirements of a ground war on two fronts," which while true underestimates the extent to which leaders inside the Army increasingly saw the wars as an existential threat to the All-Volunteer Force; see Finley, *Fields of Combat*, 103.

201 [Authors redacted], "The Association between Number of Deployments to Iraq and Mental Health Screening Outcomes in U.S. Army Soldiers," n.d., 11, FOIA.

202 Paul Bliese, interview with the author, December 3, 2015.

203 Office of the Surgeon, Multi-national Force–Iraq and Office of the Surgeon General, United States Army Medical Command, "Mental Health Advisory Team (MHAT) V," 41, 87, 44.

204 Ibid., 42.

205 Ibid., 47.

206 Gregg Zoroya, "A Fifth of Soldiers at PTSD Risk," *USA Today*, March 7, 2008.

207 George Casey, interview with the author, April 6, 2016.

208 Ibid. Casey's comment reflects the thinking of Army Chief of Staff Eric Shinseki, who testified before Congress in 2002 that invading and occupying Iraq would require a much larger force than the Bush administration had proposed. See Thomas E. Ricks, *Fiasco: The American Military Adventure in Iraq* (New York: Penguin, 2006), 11, 96–97.

209 David S. Cloud, "Gates Favors Faster Expansion of the Army," *New York Times*, September 28, 2007; and House Committee on Armed Services, *Examination of* *the Force Requirements Determination Process*, 110th Congress, 1st Session (2007), 8.

210 Senate Committee on Armed Services, *Department of Defense Authorization for Appropriations for Fiscal Year 2008*, 110th Congress, 1st Session (2007), 245.

211 Senate Committee on Armed Services, *Department of Defense Authorization for Fiscal Year 2009*, 110th Congress, 2nd Session (2008), 82.

212 Senate Committee on Armed Services, *Department of Defense Authorization for Appropriations for Fiscal Year 2008*, 110th Congress, 1st Session (2007), 245; and Senate Armed Services Committee, *Department of Defense Authorization for Appropriations for Fiscal year 2009*, 110th Congress, 2nd Session (2008), 72.

213 Senate Committee on Armed Services, Subcommittee on Readiness and Management Support, *The Current Status of U.S. Ground Forces*, 111th Congress, 1st Session (2009), 5.

214 Ibid., 7.

215 Ibid., 7, 27.

216 Eric B. Schoomaker to Peter Chiarelli, "RE: Other Significant Media Today Not in Early Bird," April 13, 2010, Eric B. Schoomaker personal papers; copy in author's possession.

217 Ibid.

218 Ibid.

219 Senate Committee on Armed Services, *The State of the United States Army*, 110th Congress, 1st Session (2007), 4. On the "Hollow Army," see Bailey, *America's Army*, 173, 228. In his interview with me (May 31, 2016), former Secretary of the Army Pete Geren disputed that the All-Volunteer Force was ever on the brink of collapse. His comments at the time, of course, reflect a greater anxiety.

220 Peter Chiarelli, interview with the author, September 16, 2015.

221 Office of the Surgeon, Multi-national Corps–Iraq and Office of the Surgeon General, United States Army Medical Command, *Mental Health Advisory Team VI (MHAT) VI: Operation Iraqi Freedom 07–09, 8 May 2009* (MHAT-VI; Falls Church, VA: Office of the U.S. Surgeon General, 2009), 36–37.

222 Office of the Surgeon General, United States Army Medical Command; Office of the Command Surgeon, HQ, USCENTCOM; and Office of the Command Surgeon, US Forces, Afghanistan (USFOR-A), *Joint Mental Health Advisory Team 7 (J-MHAT 7): Operation Enduring Freedom 2010, Afghanistan* (J-MHAT-7; Falls Church, VA: Office of the U.S. Surgeon General, 2010), 32–34.

223 M. Shayne Gallaway, C. Lagana-Riordan, A. Milliken, and the Behavioral and Social Health Outcomes Program, "Epidemiological Consultation No. 14-HK-0DHW-10: Reintegration and Risk Assessment at Joint Base Lewis-McChord, Washington, September 2010–January 2011" (Edgewood, MD: Government Printing Office, 2011), ES-2.

224 Ibid., E-29, E-36.

225 Hal Bernton, "Army Is Reviewing Madigan's Reversal of PTSD Diagnoses," *Seattle Times*, January 26, 2012.

226 Rebecca Porter, interview with the author, December 17, 2015; and Adam Ashton, "Army Investigation Finds Colonel Homas Did Not Use Position to Influence PTSD Diagnoses," McClatchyDC News Agency, December 12, 2012.

227 Hal Bernton, "40% of PTSD Diagnoses at Madigan Were Reversed," *Seattle Times*, March 20, 2012; and Sen. Murray of Washington, speaking on S. 3340, *Congressional Record*, June 25, 2012, S4461.

228 Bernton, "40% of PTSD Diagnoses at Madigan Were Reversed."

229 Rebecca Porter, interview with the author, December 17, 2015.

230 Ibid.

231 Ibid.

232 Sen. Murray of Washington, speaking on S. 3340, *Congressional Record*, June 25, 2012, S4462.

CHAPTER 5. "MILITARY FAMILIES ARE QUIETLY COMING APART AT THE SEAMS"

1 Anthony DePalma, "A Nation at War: An Overview: March 21, 2003," *New York Times*, March 22, 2003.

2 Larry Copeland and Valerie Alvord, "In Military Towns, War a Constant Presence," *USA Today*, March 21, 2003.

3 George W. Bush, "The President's Weekly Radio Address, September 29, 2001," *Public Papers of the Presidents of the United States: George W. Bush, 2001, Book 2, July 1–December 31, 2001* (Washington, DC: Government Printing Office, 2005), 1176. See also Mary Dudziak, *War Time: An Idea, Its History, and Its Consequences* (New York: Oxford University Press, 2012), 130–131; and Andrew Bacevich, *Breach of Trust: How Americans Failed Their Soldiers and Their Country* (New York: Metropolitan Books, 2013), 14.

4 Copeland and Alvord, "In Military Towns, War a Constant Presence."

5 Jennifer Mittelstadt, *The Rise of the Military Welfare State* (Cambridge, MA: Harvard University Press, 2016), 175, 183.

6 Karen M. Pavlicin, *Surviving Deployment: A Guide for Military Families* (St. Paul, MN: Elva Resa Publishing, 2003), 12.

7 Meredith Leyva, *Married to the Military: A Survival Guide for Military Wives, Girlfriends, and Women in Uniform* (New York: Fireside, 2003), 160.

8 Shellie Vandevoorde, *Separated by Duty, United in Love: A Guide to Long-Distance Relationships for Military Couples* (2003; reprint, New York: Citadel Press, 2006), 10–11, 12. The 2006 edition is somewhat updated, noting that "we at home may find ourselves feeling gripped with a fear of the unknown or lost in despair or overwhelmed with anxiety or all of these" (119).

9 See, e.g., Curtis Krueger and Chris Tisch, "TV Reports Blindside Soldiers' Families," *St. Petersburg (FL) Times*, March 25, 2003; and Elizabethe Holland, "Hole on Home Front," *St. Louis Post-Dispatch*, April 16, 2003.

10 Gregg Zoroya, "Spouses, Kids Endure Own Agonies of War," *USA Today*, July 11, 2003.

11 Ibid.

12 Andrew Jacobs, "With Breadwinners Overseas, Guard Families Face Struggle," *New York Times*, April 25, 2004.

13 Sarah Williams, "Love and War: Separation, Stress Take Toll on Military Families," *Brownsville (TX) Herald*, March 26, 2006.

14 Ibid.

15 Bob von Sternberg, "Preparing for Soldiers' Homecoming," *Minneapolis (MN) Star Tribune*, September 22, 2006.

16 Angela J. Huebner and Jay A. Mancini, "Adjustment among Adolescents in Military Families When a Parent Is Deployed: A Final Report Submitted to the Military Family Research Institute and the Department of Defense Quality of Life Office" (Falls Church: Virginia Polytechnic Institute and State University, Northern Virginia Center, Department of Human Development, 2005), 30, 19, 11.

17 Ibid., 20, 22, 32.

18 Rachel E. Stassen-Berger, "When Mom and Dad Go to War," *St. Paul (MN) Pioneer Press*, November 5, 2007.

19 Betsy Flagler, "Deployment: Teach Children to Express Their Feelings," *El Paso (TX) Times*, November 11, 2007; Karen Jowers, "Strained Bonds: A Culture of Deployment Is Stressing Families Left Behind," *Marine Corps Times*, September 10, 2007; and Lizette Alvarez, "Long Iraq Tours Can Make Home a Trying Front," *New York Times*, February 23, 2007.

20 "'Heroes at Home' Puts Focus on Troops Families," *Contra Costa (CA) Times*, March 12, 2006.

21 Elizabeth, "Needed to Share . . . ," *MilitarySOS*, June 25, 2005, www.militarysos.com.

22 Oneandonlyval, "Scary Thoughts," *MilitarySOS*, November 11, 2006, www.militarysos.com.

23 LaneyBug, "Nightmares," *MilitarySOS*, January 31, 2006, www.militarysos.com.

24 Jamie+Brandon, "Can't Sleep," *MilitarySOS*, May 27, 2009, www.militarysos.com.

25 Onerosgirl, "How do you cope?" *MilitarySOS*, January 17, 2008, www.militarysos.com.

26 CindyO, "Why Is This Deployment So Much Harder?" *MilitarySOS*, July 26, 2008, www.militarysos.com.

27 BabySealEyes904, "This Sucks . . . ," *MilitarySOS*, June 16, 2009, www.militarysos.com.

28 Williams, "Love and War."

29 Ibid.

30 Gallup poll, "Iraq," 2015, www.gallup.com.

31 Letitia Stein, "Kids in Time of War," *St. Petersburg (FL) Times*, May 9, 2005.

32 Jennifer Levitz, "She Can't Pretend She's Not Lonely," *Providence (RI) Journal*, December 25, 2005.

33 Jonathan Finer, "As N.H. Guardsmen Answer Call of Duty, Families Grow Anxious," *Washington Post*, April 10, 2004.

34 Anna Mulrine, "Uncle Sam Wants You . . . to Reach Out to Help America's Military Families," *Christian Science Monitor*, April 12, 2001; Anita Chandra et al. "Understanding the Impact of Deployment on Military Families," working paper (RAND Corp., 2008), 35; and Ron Harris, "Stress to Families Is Growing Reason for Leaving Military," *St. Louis Post-Dispatch*, June 28, 2004.

35 Thom Shanker and Michael R. Gordon, "Strained, Army Looks to Guard for More Relief," *New York Times*, September 22, 2006.

36 Michael Waterhouse and JoAnne O'Bryant, "National Guard Personnel and Deployment: Fact Sheet" (Washington DC: Congressional Research Service, 2007), 5; Defense Science Board, "Deployment of Members of the National Guard and Reserve in the Global War on Terrorism" (Arlington, VA: Office of the Undersecretary of Defense for Acquisition, Technology, and Logistics, 2007), 8.

37 Defense Science Board, "Deployment of Members of the National Guard and Reserve in the Global War on Terrorism," 17.

38 Harris, "Stress to Families Is Growing Reason for Leaving Military."

39 For an overview, see Mittelstadt, *The Rise of the Military Welfare State*, 148–170. See also Richard Halloran, "Army Acts to Help Families Cope with Pressures of Modern Military Life," *New York Times*, August 17, 1985; and see Community and Family Policy Division, Human Resources Development Directorate, Office of the Deputy Chief of Staff for Personnel, The Army Family Action Plan: Total Army – Unit, Soldier, Family – Readiness (Arlington, VA: United States Army, 1984), DTIC.

40 Delores Johnson, interview with the author, October 10, 2018.

41 Code of Support Foundation, "Executive Director Kristy Kaufmann Briefs Congress," July 2, 2014, www.codeofsupport.org; and Kristy Kaufmann, interview with the author, March 4, 2016.

42 Code of Support Foundation, "Executive Director Kristy Kaufmann Briefs Congress."

43 Kristy Kaufmann, interview with the author, March 4, 2016.

44 Rebekah Sanderlin, "Strong Women Slip through the Cracks," August 10, 2009, and Laura Arenschiled, "Deaths Ruled Murder-Suicide," November 30, 2006, *Fayetteville (NC) Observer*, www.fayobserver.com. See also Kristy Kaufmann, interview with the author, March 4, 2016.

45 Bryan Mims and Ken Smith, "Source: Mother Found Dead with Children Was Depressed," *WRAL.com* (Raleigh, NC), November 30, 3006, www.wral.com; and Arenschiled, "Deaths Ruled Murder-Suicide."

46 On the history of these groups, see Mittelstadt, *The Rise of the Military Welfare State*, 173–176. For a critique of the "established culture of [spouse] volunteerism," see Rebecca A. Adelman, *Figuring Violence: Affective Investments in Perpetual War* (New York: Fordham University Press, 2019), 100. For an account of one FRG's activities, see Kenneth T. MacLeish, *Making War at Fort Hood: Life and Uncertainty in a Military Community* (Princeton: Princeton University Press, 2013), 134–141.

47 Kristy Kaufmann, interview with the author, March 4, 2016.

48 Ibid.
49 Ibid.
50 Ibid.
51 Ibid.
52 Ibid.
53 Ibid.
54 Ibid.
55 Deborah Mancini, *U.S. Army FRG Leader's Handbook*, 3rd ed. (Ithaca, NY: Cornell University, 2006), 37, posted at Fort Leonard Wood, www.wood.army.mil.
56 Ibid., 45.
57 Kristy Kaufmann, interview with the author, March 4, 2016.
58 Kristy Kaufmann Turner, "Family Readiness Groups: Problems and Suggested Solutions," April 2007, 1, copy in author's possession.
59 Ibid., 2–5.
60 Ibid., 8.
61 Ibid., 5.
62 Ibid., 2–3.
63 Ibid., 3.
64 Kristy Kaufmann, email to the author, March 6, 2016.
65 Kristy Kaufmann, interview with the author, March 4, 2016.
66 Ibid.
67 Ibid.
68 Ibid.
69 Adelman, *Figuring Violence*, 104.
70 Kristy Kaufmann, interview with the author, March 4, 2016.
71 Robert Gates, *Duty: Memoirs of a Secretary at War* (New York: Alfred A. Knopf, 2014), 45.
72 Delores Johnson, interview with the author, October 10, 2018.
73 George Casey, interview with the author, April 6, 2016.
74 George Casey, interview with the author, October 13, 2017.
75 George Casey, interview with the author, April 6, 2016.
76 Ibid.
77 Kristy Kaufmann, interview with the author, March 4, 2016.
78 Ibid.
79 Pete Geren, interview with the author, May 31, 2016.
80 Senate Armed Services Committee, *Nominations before the Senate Armed Services Committee*, 110th Congress, 1st Session (2007), 752, 249.
81 Senate Armed Services Committee, *The State of the United States Army*, 110th Congress, 1st Session (2007), 11.
82 Ibid.
83 Ibid., 68.
84 Ibid., 68.

85 Ibid., 74.
86 Pete Geren, interview with the author, May 31, 2016.
87 George W. Casey and Pete Geren, "Memorandum for See Distribution," October 9, 2007, FOIA.
88 Delores Johnson, interview with the author, October 5, 2018.
89 Ibid.
90 Ibid.
91 Ibid.
92 "Information Paper: The Soldier and Family Action Plan (SFAP) Senior Review Group (SRG)," June 7, 2010, FOIA.
93 Donald J. James to Secretary of the Army, "Proposed Charter for the Army Soldier-Family Action Plan (ASFAP)," September 20, 2007, FOIA.
94 Ibid.
95 Laura Avery, "Implementing the Army Family Covenant: How Well Is the Army Doing?" U.S. Army War College Strategy Research Project (Carlisle Barracks, PA: U.S. Army War College, 2009), 1–3, 9.
96 [Redacted to Redacted], "AFC—Year in Review (May 09)," May 26, 2009, FOIA; Pete Geren, interview with the author, May 31, 2016; and "Futures Brief," November 16, 2011, FOIA.
97 "Information Papers: Family Readiness Support Assistance Program," *2008 Army Posture Statement*, 2007, www.army.mil.
98 "Army Family Covenant: Building Resilient Soldiers and Families," September 30, 2009, 5, FOIA; and [Redacted to Redacted], "AFC—Year in Review (May 09)," May 26, 2009, FOIA.
99 Army Community Service, *U.S. Army Family Readiness Support Assistant: FRSA Resource Guide* (Ithaca, NY: Cornell University, Family Life Development Center for the Army Family and Morale, Welfare and Recreation Command, 2007), 11, 5, 9, www.myarmyonesource.com.
100 On the affective function of such programs, see MacLeish, Making War at Fort Hood, 150.
101 U.S. Army Northern Command, "Information Paper in Support of the Secretary of the Army (SA) Two Year Review," July 22, 2009, FOIA; Carter F. Ham to Secretary of the Army, "Information Papers in Support of the Secretary of the Army Two-Year Family Review," July 23, 2009, 1, FOIA; and "Information Paper: Secretary of the Army Two Year Review," July 22, 2009, FOIA.
102 [Redacted to Redacted], "Update to 2 Year AFC Brochure," June 15, 2010, FOIA. See also "Army Family Covenant: Building Resilient Soldiers and Families," 6.
103 [Redacted to Redacted], "Update to 2 Year AFC Brochure," June 15, 2010, FOIA.
104 Carter F. Ham to Secretary of the Army, "Information Papers in Support of the Secretary of the Army Two-Year Family Review," July 23, 2009, 2, FOIA.
105 Ibid., 18.
106 [Redacted to Redacted], "AFC—Year in Review (May 09)," May 26, 2009.

107 Ibid.

108 Avery, "Implementing the Army Family Covenant," 13. The Chaplain Corps had long been at the center of Army family programs. See Mittelstadt, *The Rise of the Military Welfare State*, 148–170. For an overview of the pre-2007 programs, see Helena Olviero, "A Happier Home Front," *Atlanta Journal Constitution*, May 29, 2005.

109 Dolores Johnson, interview with the author, October 5, 2018.

110 Donna Alvah, *Unofficial Ambassadors: American Military Families Overseas and the Cold War, 1946–1965* (New York: New York University Press, 2007), 66. Also see U.S. Army, Family and Morale, Welfare and Recreation Command, "Army Family Covenant: Strategic Action and Marketing Plan" (Fort Sam Houston, TX: U.S. Army Installation Management Command, n.d.), posted at the Association of the United States Army, www.ausa.org. Images of advertisements are at expertinfantry, "Army Posters," Flickr, n.d., www.flikr.com/photos/expertinfantry/albums/72157625923671875.

111 [Redacted to Redacted], "FW: S1NET Message Summary—22 May 09," May 26, 2009, FOIA.

112 Mittelstadt, *The Rise of the Military Welfare State*, 224, 227–228.

113 Carl A. Castro, "Impact of Combat on the Mental Health and Well-Being of Soldiers and Marines: 7 Things I Think I Know," presentation at Smith College, School for Social Work, North Hampton, MA, June 26–28, 2008.

114 Dennis McGurk interview with the author, March 2, 2016.

115 Ibid.

116 Ibid.

117 Ibid.

118 Ibid. The same is true of the Family Readiness Groups, which likewise encourage spouses to become docile subjects who uncritically facilitate wartime deployments. The Army's *Family Readiness Group Handbook*, e.g., begins by explaining that, beyond "helping soldiers and families cope with the stresses of military life . . . in a very real sense, FRGs can help soldiers and leaders with the military mission, too"; see Mancini, *U.S. Army FRG Leader's Handbook*, "Operation Ready," 7).

119 Adelman, Figuring Violence, 104.

120 Cynthia Enloe, *The Curious Feminist: Searching for Women in a New Age of Empire* (Berkeley: University of California Press, 2004), 6, 146. See also Denise M. Horn, "Boots and Bedsheets: Constructing the Military Support System in Time of War," in *Gender, War, and Militarism: Feminist Perspectives*, ed. Laura Sjoberg and Sandra Via (Santa Barbara, CA: Praeger, 2010), 60, 62, 63–64.

121 Mittelstadt, *The Rise of the Military Welfare State*, 42.

122 Adelman, *Figuring Violence*, 104.

123 Dennis McGurk interview with the author, March 2, 2016.

124 Delores Johnson, interview with the author, October 5, 2018.

125 Jimmie Keenan, interview with the author, December 1, 2015.

126 Ibid.

127 Ibid.

128 Ibid.

129 Ibid.

130 "1AD Support of Iron Soldiers and Army Families," 2009, FOIA.

131 William B. Garrett III, "Two-Year Review of the Army Family Covenant (AFC)—United States Army Southern European Task Force (USASETAF)," July 22, 2009, FOIA.

132 [Redacted] to Secretary of the Army, "Command Initiatives in Support of Families—Information Memorandum," July 23, 2009, FOIA.

133 Jimmie Keenan, interview with the author, December 1, 2015.

134 Kristy Kaufmann, interview with the author, March 4, 2016.

135 Ibid.

136 Dennis McGurk interview with the author, March 2, 2016.

137 Ibid.

138 Amy Adler, interview with the author, April 21, 2016; and Dennis McGurk interview with the author, March 2, 2016.

139 Gordon Lubold, "Soldier Rampage Hints at Stress of Repeated Deployments," *Christian Science Monitor*, May 13, 2009.

140 Carolyn Davis, "The Hidden Home Front," *Philadelphia Inquirer*, July 7, 2009.

141 Lou Michel, "Torn by 2 Wars, Home Life Suffers," *Buffalo (NY) News*, June 11, 2008.

142 Richard Danielson, "They're Waging a War Far from the Battlefield," *St. Petersburg (FL) Times*.

143 Kristina Kaufmann, "Addressing Military Families Challenges by Starting the Conversation . . . ," PowerPoint presentation, [2009]); copy in author's possession, courtesy of Kristina Kaufmann.

144 Kristina Kaufmann, interview with the author, March 4, 2016.

145 Ibid.

146 Kristina Kaufmann, "Army Families under Fire," *Washington Post*, May 11, 2009.

147 Ibid.

148 Ibid.

149 Kristy Kaufmann, interview with the author, March 4, 2016.

150 Kristina Kaufmann Turner, "E-Mail Responses to *Washington Post* Op-Ed 'Army Families under Fire,' May 11, 2009," copy in author's possession.

151 Ibid.

152 Ibid.

153 Ibid.

154 MacLeish, *Making War at Fort Hood*, 152.

155 Pete Geren, email to Kristina Kaufmann, May 11, 2009, courtesy of Kristina Kaufmann, copy in author's possession.

156 Pete Geren, email to Kristina Kaufmann, n.d., and Kaufmann, email to Geren, n.d., courtesy of Kristina Kaufmann, copy in author's possession. Geren is likely

referring to the Soldier-Family Action Plan, which the Army issued alongside the Family Covenant (GOMO [General Officer Management Office], email to [Redacted], "SecArmy and CSA Sends: One-Year Anniversary of the Army Family Covenant (UNCLASSIFIED)," November 25, 2008, FOIA.

157 Kristina Kauffman, email to Pete Geren and Kathleen Y. Marin, June 2, 2009, courtesy of Kristina Kaufmann, copy in author's possession.

158 Kristina Kauffmann, email to the author, March 12, 2016.

159 Kaufmann, "Addressing Military Families Challenges by Starting the Conversation," 5, 8.

160 Ibid., 6, 9.

161 Ibid., 11; and Kristina Kaufmann, "Addressing Military Family Challenges Position Paper," [2009], courtesy of Kristina Kaufmann, copy in author's possession.

162 Kaufmann, "Addressing Military Families Challenges by Starting the Conversation," 13.

163 Ibid., 16.

164 The White House, "Executive Order—Improving Access to Mental Health Services for Veterans, Service Members, and Military Families," August 31, 2012, https://obamawhitehouse.archives.gov.

165 Delores Johnson, interview with the author, October 10, 2018.

166 Ibid.

167 See, e.g., the White House, *Strengthening Our Military Families: Meeting America's Commitment* (Washington, DC: White House, 2011), 7, available at the National Center on Domestic and Sexual Violence, www.ncdsv.org; Committee on the Assessment of Readjustment Needs of Military Personnel, Veterans, and Their Families; Board on the Health of Select Populations; Institute of Medicine, *Returning Home from Iraq and Afghanistan: Assessment of Readjustment Needs of Veterans, Servicemembers and Their Families* (Washington, DC: National Academies Press, 2013), 279, www.nationalacademies.org; Blue Star Families, *2014 Military Family Lifestyle Survey Comprehensive Report* (Encinitas, CA: Blue Star Families; Syracuse, NY: Institute for Veterans and Military Families, Syracuse University, 2014), 90, 51, posted at the Institute for Veterans and Military Families website, https://ivmf.syracuse.edu.

168 Committee on the Assessment of Readjustment Needs of Military Personnel, Veterans, and Their Families, *Returning Home from Iraq and Afghanistan*, 257.

169 National Public Radio, "The Conversations about Infidelity on the Homefront," November 26, 2012, www.npr.org.

CHAPTER 6. "THE LIMITED SCIENCE OF THE BRAIN"

1 On Walter Reed's reputation, see James Dao, "Taps Play for a Hospital That Tended All Ranks," *New York Times*, August 26, 2005. For an in-depth ethnography of OIF/OEF veterans receiving treatment there, see Zoe H. Wool, *After War: The Weight of Life at Walter Reed* (Durham: Duke University Press, 2015).

2 Mark Landler, "The Wounded Find Respite at U.S. Unit in Germany," *New York Times*, April 1, 2003; and Neela Bannerjee, "Rebuilding Bodies, and Lives, Maimed by War," *New York Times*, November 13, 2003.

3 Kenneth T. MacLeish, *Making War at Fort Hood: Life and Uncertainty in a Military Community* (Princeton: Princeton University Press, 2013), 12; Tamara Jones and Anne Hull, "Moving Forward, One Step at a Time," *Washington Post*, July 21, 2003; Dave Moniz, "Injured Troops Survive More," *USA Today*, October 26, 2003; Bannerjee, "Rebuilding Bodies"; and Sara Corbett, "The Permanent Scars of Iraq," *New York Times*, February 15, 2004.

4 Steve Vogel, "Doctors Seeing More Brain Injuries from Iraq," *Washington Post*, December 14, 2003.

5 Ibid.

6 Jonathan Jaffin, interview with the author, January 30, 2016.

7 Ibid.

8 Leanne Young, interview with the author, April 27, 2017.

9 Ibid. See also Caroline Alexander, "Blast Force: The Invisible War on the Brain," National Geographic, January 2015. www.nationalgeographic.com.

10 Leanne Young, interview with the author, April 27, 2017.

11 Christian J. Macedonia, interview with the author, October 10, 2018.

12 Macedonia relates this revelation in Jon Hamilton, "How a Team of Elite Doctors Changed the Military's Stance on Brain Trauma," NPR, June 10, 2016, www.npr.org.

13 Christian J. Macedonia, interview with the author, October 10, 2018.

14 Ibid.

15 Christian Macedonia, interview with the author, November 3, 2018. For Macedonia's biography, see Christian Macedonia, "Brain Injuries: Hitches: What Holds Back Human Potential and How Can Systems Science Help?" (Hanover, NH: Dartmouth University, n.d.), www.dartmouth.edu.

16 Hamilton, "How a Team of Elite Doctors Changed the Military's Stance on Brain Trauma."

17 Jerry Lembcke, *PTSD: Diagnosis and Identity in Post-empire America* (Lanham, MD: Lexington Books, 2013), 151, 154, 157.

18 Ibid., 153–154, 146.

19 Ibid., 158; Alexander, "Blast Force."

20 For clinical definitions, see Eric M. Umile et al., "Dynamic Imaging in Mild Traumatic Brain Injury: Support for the Theory of Medial Temporal Vulnerability," *Archives of Physical Medicine and Rehabilitation* 83, no. 11 (2002): 1511; and Nigel A. Shaw, "The Neurophysiology of Concussion," *Progress in Neurobiology* 67 (2002): 281.

21 Mark Faul, et. al., Traumatic Brain Injury in the United States: Emergency Department Visits, Hospitalizations, and Deaths, 2002–2006 (Washington, DC: Centers for Disease Control and Prevention, 2010), 7. www.cdc.gov.

22 Michael P. Poirier, "Concussions: Assessment, Management, and Recommendations for Return to Activity," *Clinical Pediatric Emergency Medicine* 4 (2003): 181; Faul et. al., Traumatic Brain Injury in the United States, 7.

23 Vogel, "Doctors Seeing More Brain Injuries from Iraq."

24 Douglas B. McKeag, "Understanding Sports-Related Concussion: Coming into Focus but Still Fuzzy," *Journal of the American Medical Association* 290, no. 19 (2003): 2604.

25 Michael A. Kiraly and Steven J. Kiraly, "Traumatic Injury and Delayed Sequelae: A Review," *Scientific World Journal* 7 (2007): 1768; Omar Ghaffar et al., "Randomized Treatment Trial in Mild Traumatic Brain Injury," *Journal of Psychosomatic Research* 61 (2006): 153; Carol Y. Crooks, Jennifer M. Zumsteg, and Kathleen R. Bell, "Traumatic Brain Injury: A Review of Practice Management and Recent Advances," *Physical Medicine and Rehabilitation Clinics* 18 (2007): 683; S. Meares et al., "Mild Traumatic Brain Injury Does Not Predict Acute Postconcussion Syndrome," *Journal of Neurology, Neurosurgery, and Psychiatry* 79 (2008): 300; and M. Stulemeijer et al., "Early Prediction of Favorable Recovery 6 Months after Mild Traumatic Brain Injury," *Journal of Neurology, Neurosurgery, and Psychiatry* 79 (2008): 936.

26 Donald Bradshaw, interview with the author, December 16, 2015. See also "Concussion (Mild Traumatic Brain Injury) and the Team Physician: A Consensus Statement," *Medicine and Science in Sports and Exercise* (2006): 395, www.aafp.org.

27 Crooks et al., "Traumatic Brain Injury," 686.

28 Paul Blostein and Susan J. Jones, "Identification and Evaluation of Patients with Mild Traumatic Brain Injury: Results of a Survey of National Level I Trauma Centers," *Journal of Trauma Injury, Infection, and Critical Care* 55, no. 3 (2003): 450; Jeffrey J. Bazarian et al., "Ethnic and Racial Disparities in Emergency Department Care for Traumatic Brain Injury," *Academy of Emergency Medicine* 10, no. 11 (2003): 1211; John Holcomb, interview with the author, February 7, 2007; Peter Chiarelli, interview with the author, September 16, 2015; and Geoffrey Ling, interview with the author, February 21, 2017.

29 Geoffrey Ling, interview with the author, February 21, 2017.

30 Poirier, "Concussions," 179–180.

31 Ibid., 181; and Michael McCrea et al., "Acute Effects and Recovery Time following Concussion in Collegiate Football Players: The NCAA Concussion Study," *New England Journal of Medicine* 290, no. 19 (2003): 2561.

32 Geoffrey Ling, interview with the author, February 21, 2017.

33 "Practice Parameter: The Management of Concussion in Sports (Summary Statement)," *Neurology* 48 (1997): 582–583.

34 Here I date the emergence of attention to TBI somewhat earlier than Lembcke does. See Lembcke, *PTSD*, 153–155.

35 Jones and Hull, "Moving Forward, One Step at a Time"; and "Signature Wounds," *St. Louis Post-Dispatch*, March 5, 2005.

36 Gregg Zoroya, "Brain Injuries Range from Loss of Coordination to Loss of Self," *USA Today*, March 4, 2005.

37 Ibid.

38 Gregg Zoroya, "Key Iraq Wound: Brain Trauma," *USA Today*, March 5, 2005.

39 Lembcke, *PTSD*, 157, 146.

40 Moni Basu, "Unseen Injury," *Atlanta Journal Constitution*, November 19, 2006; Richard Lardner, "The War Comes Home," *Tampa (FL) Tribune*, February 18, 2007; Dana Hull, "Town Rallies behind Injured Vet and Family," *Contra Costa (CA) Times*, December 26, 2006; and Neil Shea, "The Heroes, the Healing: Military Medicine from the Front Lines to the Home Front," *National Geographic*, December 2006, www.nationalgeographic.com.

41 Gregg Zoroya, "Even Mild Brain Injuries Add Up," *USA Today*, June 7, 2006. See also Mark Emmons, "The Signature Wound," *San Jose (CA) Mercury-News*, December 14, 2006.

42 Hull, "Town Rallies behind Injured Vet and Family"; and Shea, "The Heroes, the Healing."

43 "Iraq Losses Should Be Defined," *Buffalo (NY) News*, October 23, 2006.

44 Rep. Honda of California, speaking on "Iraq War Resolution," *Congressional Record*, February 13, 2007, H1539.

45 Rep. Jackson-Lee of Texas, speaking on "Immigration Concerns," *Congressional Record*, February 28, 2007, H2035.

46 Rep. Kilpatrick of Michigan, speaking on HR 1591, *Congressional Record*, April 25, 2007, H4149.

47 See, e.g., Stuart Hoffman, "Fight for Brain Injured Troops," *Atlanta Journal Constitution*, August 21, 2006; and Suzie Kinzie Gays Mills, "Bush Funding Cut for Iraq War Vets with Brain Injuries Is a Betrayal," *Madison (WI) Capital Times*, October 7, 2006.

48 Senate Committee on Appropriations, *Department of Defense Appropriations, Fiscal Year 2007*, Part 2, 109th Congress, 2d Session (2006), 665–666.

49 Geoffrey Ling, interview with the author, February 21, 2017.

50 "DoD Numbers for Traumatic Brain Injury Worldwide—Totals, 2005" (Silver Spring, MD: Defense and Veterans Brain Injury Center, n.d.), and "DoD Numbers for Traumatic Brain Injury Worldwide—Totals, 2007" (Silver Spring, MD: Defense and Veterans Brain Injury Center, n.d.), https://dvbic.dcoe.mil.

51 Ling, interview with the author, February 21, 2017.

52 Senate Committee on Appropriations, *Department of Defense Appropriations, Fiscal Year 2007*, 666; and Louis French, Michael McCrea, and Mark Baggett, "The Military Acute Concussion Evaluation," *Journal of Special Operations Medicine* 8, no. 1 (2008): 69.

53 U.S. Army Office of the Surgeon General, "ALARACT: Concussion in Soldiers on the Battlefield," July 28, 2006, FOIA.

54 Donald Bradshaw, interview with the author, December 16, 2015.

55 Ibid., Alana Foxwell Barajas, "Hooah*," *Wheaton College Alumni Magazine*, Autumn 2012, 32.

56 Donald Bradshaw, interview with the author, December 16, 2015.

57 Traumatic Brain Injury Task Force, "Report to the Surgeon General" (Bradshaw report, Rosslyn, VA: Defense Centers of Excellence for Psychological Health and

Traumatic Brain Injury, January 2008), 12–14, posted at the Homeland Security Digital Library, www.hsdl.org.

58 Ibid., 17, 19, 42.

59 Ibid., 51.

60 Ibid., 14.

61 Donald Bradshaw, interview with the author, December 16, 2015. For a critique, see Aaron Glantz, *The War Comes Home: Washington's Battle With America's Veterans* (Berkeley: University of California Press, 2009), 43-44.

62 Senate Committee on Armed Services, *Department of Defense Authorization for Appropriations for Fiscal Year 2008, First Session on S. 1547, Part 1*, 110th Congress, 1st Session (2007), 217–218.

63 House Committee on Veterans' Affairs, Subcommittee on Health, *The U.S. Department of Veterans Affairs Fiscal Year 2008 Budget*, 110th Congress, 1st Session (2007), 10–11.

64 House Committee on Armed Services, Military Personnel Subcommittee, *The State of the Military Health Care System*, 110th Congress, 1st Session (2007), 7–8.

65 Charles W. Hoge et al., "Mild Traumatic Brain Injury in U.S. Soldiers Returning from Iraq," *New England Journal of Medicine* 358 (2008): 459.

66 Ibid., 458–459.

67 Ibid., 457.

68 Ibid., 461.

69 Ibid. See also Alexander, "Blast Force."

70 Charles Hoge, interview with the author, December 31, 2015.

71 Jan E. Kennedy, et al., "Post-Traumatic Stress Disorder and Post-Traumatic Stress Disorder-Like Symptoms and Mild Traumatic Brain Injury," *Journal of Rehabilitation Research and Development* 44, no. 7 (2007): 895–920. For a brief, useful overview of these debates, see Manfred F. Greiffenstein and W. John Baker, "Validity Testing in Dually-Diagnosed Post-Traumatic Stress Disorder and Mild Closed Head Injury," *Clinical Neuropsychologist* 22, no. 3 (2008): 565–567.

72 Dean Baker, "The Price of PTSD," *Vancouver (WA) Columbian*, November 11, 2007.

73 See, e.g., Marty Toohey, "Blasts Leave U.S. Troops Scarred by Brain Injury," *Austin (TX) American-Statesman*, September 16, 2007; Ann Scott Tyson, "Troops' Mental Healthcare 'Woefully Inadequate,' Pentagon Finds," *Washington Post*, June 17, 2007; Bart Jansen, "Battlefield Brain Injuries Defy Easy Detection," *Portland (ME) Press Herald*, May 13, 2007. Ritchie is quoted at House Committee on Armed Services, Military Personnel Subcommittee, *The State of the Military Health Care System*, 4.

74 Kennedy et al., "Post-Traumatic Stress Disorder and Post-Traumatic Stress Disorder-Like Symptoms and Mild Traumatic Brain Injury," 896, 913; Michael Koenigs et al., "Focal Brain Damage Protects against Post-Traumatic Stress Disorder in Combat Veterans," *Nature Neuroscience* 11, no. 2 (2008): 236; and Edward Kim et al., "Neuropsychiatric Complications of Traumatic Brain Injury: A Critical Review of the Literature," *Journal of Neuropsychiatry and Clinical Neuroscience* 19:2 (2007), 120.

75 Jeffrey M. Rogers and Christina A. Reed, "Psychiatric Comorbidity following Traumatic Brain Injury," *Brain Injury* 21, no. 13–14 (2007): 1322; Alexander, "Blast Force."

76 Peter Chiarelli, interview with the author, December 9, 2015.

77 Geoffrey Ling, interview with the author, February 21, 2017.

78 Traumatic Brain Injury Task Force, "Report to the Surgeon General," 41, 52; Senate Committee on Armed Services, *Department of Defense Authorization for Appropriations for Fiscal Year 2008, First Session on S. 1547, Part 1*, 217.

79 Jonathan Jaffin, interview with the author, January 30, 2016.

80 Eric Schoomaker, interview with the author, December 16, 2015.

81 Ibid.

82 Charles Hoge, email to the author, February 28, 2018.

83 Charles Hoge, interview with the author, December 31, 2015; emphasis mine, based on recording.

84 Hoge, email to the author, February 28, 2018.

85 Ibid. Moreover, as Hoge explains, the symptoms of persistent post-concussion syndrome are equally difficult to tease out from other conditions like PTS and depression.

86 Schoomaker, email to the author, February 20, 2018; and Hoge et al., "Mild Traumatic Brain Injury in U.S. Soldiers Returning from Iraq," 461.

87 Geoffrey Ling, interview with the author, February 21, 2017.

88 In fact, a 2011 article by Heidi Terrio and colleagues stated that the "structured clinical interview . . . is the gold standard for retrospective TBI assessment"; see Heidi P. Terrio et al., "Postdeployment Traumatic Brain Injury Screening Questions: Sensitivity, Specificity, and Predictive Values in Returning Soldiers," *Rehabilitation Psychology* 56, no. 1 (2011): 27.

89 Anne Hull and Dana Priest, "Little Relief on Ward 53," *Washington Post*, June 18, 2007. See also Gregg Zoroya, "Scientists: Brain Injuries Far Worse than Thought," September 24, 2007, and "Marine Didn't Recognize Signs of Brain Injury," November 23, 2007, both in *USA Today*.

90 Sen. Murray of Washington, speaking on "Dignified Treatment of Wounded Warriors Act," *Congressional Record*, July 25, 2007, S9863.

91 Office of the Surgeon General, "ALARACT: Documenting Blast Exposure/Injury in Theater Medical Records," June 15, 2007, FOIA.

92 Office of the Surgeon General, "ALARACT 173/2008," July, 2008, FOIA.

93 Terrio et al., "Postdeployment Traumatic Brain Injury Screening Questions," 27; Karen A. Schwab et al., "Screening for Traumatic Brain Injury in Troops Returning from Deployment in Afghanistan and Iraq: Initial Investigation of the Usefulness of a Short Screening Tool for Traumatic Brain Injury," *Journal of Head Trauma and Rehabilitation* 22, no. 6 (2007): 377–389; and Charles Hoge, interview with the author, December 31, 2015.

94 Charles Hoge, interview with the author, December 31, 2015.

95 Ibid.

96 Terrio et al., "Postdeployment Traumatic Brain Injury Screening Questions," 30.

97 Ibid.
98 Charles Hoge, interview with the author, December 31, 2015.
99 Leanne Young, interview with the author, April 27, 2017.
100 Ibid.
101 Christian J. Macedonia, interview with the author, October 10, 2018.
102 Ibid. See also Alexander, "Blast Force."
103 Eric Schoomaker, email to the author, February 21, 2018; and Christian J. Macedonia, email to the author, November 3, 2018.
104 Yochi Dreazen, *The Invisible Front: Love and Loss in an Era of Endless War* (New York: Crown, 2014), 218.
105 Greg Cloud and David Jaffe, *The Fourth Star: Four Generals and Their Epic Struggle for the Future of the United States Army* (New York: Crown, 2011), 152. See also Dreazen, *The Invisible Front*, 219.
106 Robert Gates, "Promotion and Swearing-In of General Peter Chiarelli (Arlington, VA), August 4, 2008," United States Department of Defense, n.d., www.defense.gov.
107 Christopher Philbrick, interview with the author, June 15, 2016.
108 David Jaffe, "Army's Vice Chief of Staff, Gen. Peter W. Chiarelli, Gives Closing Words of Advocacy," *Washington Post*, January 28, 2012.
109 Bruce Shahbaz, interview with the author, February 24, 2016.
110 Peter Chiarelli, interview with the author, September 15, 2015.
111 Ibid. This revelatory moment is one that Chiarelli has related elsewhere; see, e.g., Howard Schultz and Rajeev Chandrasekaran, "General Chiarelli's Brain Crusade," *Politico*, December 27, 2014, www.politico.com.
112 Peter Chiarelli, interview with the author, September 15, 2015.
113 Ibid.
114 Ibid.
115 Cloud and Jaffee, *The Fourth Star*, 58, 249, 272, 294; and Dreazen, *The Invisible Front*, 218.
116 Rebecca Porter, interview with the author, May 6, 2016.
117 Pete Geren, interview with the author, May 31, 2016.
118 Eric Schoomaker, interview with the author, December 15, 2015. Shoomaker's then deputy and subsequent successor Patricia Horoho made almost the same point: "We had to make sure that we were disciplined enough . . . to make sure that what we were adopting for the Army was going to be evidence-based" because, for "every hypothesis, every idea that was out there, there was a competing, opposite hypothesis," she explains, and "if we just adopted something . . . and then we found out years later that what we adopted was actually hurting our soldiers, that's unethical"; see Patricia Horoho, interview with the author, January 14, 2016.
119 Ibid.
120 Peter Chiarelli, interview with the author, December 9, 2015.
121 Daniel Zwerdling, "NPR Interview with the Surgeon General, Lt. Gen. Eric Schoomaker," transcript (Washington, DC: National Public Radio, May 11, 2010), 5, Eric B. Schoomaker personal papers (hereafter EBS Papers). Schoomaker used almost the

same example in his December 16, 2015, interview as well as in a 2011 speech; see Eric B. Schoomaker, "Keynote Address: 5th Annual Defense and Veteran's Brain Injury Center Summit," August 22, 2011, EBS Papers; copy is in the author's possession.

122 Yuri Rassovsky et al., "Functional Outcome in TBI II: Verbal Memory and Information Processing Speed Mediators," *Journal of Clinical and Experimental Neuropsychology* 28, no. 4 (2007): 588; and Judith Marcoux et al., "Persistent Metabolic Crisis as Measured by Elevated Cerebral Microdialysis Lactate-Pyruvate Ratio Predicts Chronic Frontal Lobe Brain Atrophy after Traumatic Brain Injury," *Critical Care Medicine* 36, no. 10 (2008): 2871.

123 Peter Chiarelli, interview with the author, December 9, 2015. Greg Jaffe, "Military Reckons with the Mental Wounds of War," *Washington Post*, July 18, 2010; Charles Hoge, interview with the author, December 31, 2015; Alexander, "Blast Force."

124 Peter Chiarelli, interview with the author, December 9, 2015.

125 Ibid.; and Jaffe, "Military Reckons with the Mental Wounds of War."

126 Peter Chiarelli, interview with the author, December 9, 2015.

127 Jaffe, "Military Reckons with the Mental Wounds of War"; and Schultz and Chandrasekaran, "General Chiarelli's Brain Crusade."

128 Peter Chiarelli, interview with the author, December 9, 2015.

129 Eric Schoomaker, email to the author, April 28, 2016.

130 Christian J. Macedonia, interview with the author, October 10, 2018.

131 Schoomaker, email to the author, April 28, 2016.

132 Rebecca Porter, interview with the author, May 6, 2016. Elsewhere, Chiarelli is quoted as telling his doctors, "I come from 'take a hill by 5 p.m.' You are 'take as long as you want.'" See David Finkel, *Thank You for Your Service* (New York: Sarah Crighton Books, 2013), 78.

133 Leanne Young, interview with the author, April 27, 2017.

134 Peter Chiarelli, interview with the author, September 16, 2015.

135 Ibid.

136 Rebecca Porter, interview with the author, May 6, 2016; and Amy Adler, interview with the author, April 1, 2016.

137 Donald Bradshaw, interview with the author, December 16, 2015.

138 Amy Adler, interview with the author, April 1, 2016; and Rebecca Porter, interview with the author, May 6, 2016.

139 Eric Schoomaker, interview with the author, April 6, 2016.

140 Russ Zajtchuck, ed. *Textbook of Military Medicine*, pt. 1 (Washington, DC: Office of the Surgeon General at TMM Publications, Center of Excellence in Military Medical Research and Education, 1990), 5:242.

141 Ibid., 5:262.

142 Ibid., 5:242, 246, 249.

143 Ibid., 5:244–245, 249. See also Alexander, "Blast Force."

144 Zajtchuck, ed. *Textbook of Military Medicine*, 5:246.

145 Most of the research that had been done as part of the Vietnam Head Injury Study, which the army conducted longitudinally from 1967 through the 1980s, examined veterans who had suffered penetrating head wounds. Only one article—published in 1998—suggested that post-traumatic stress and TBI may often have been co-occurring and that Vietnam veterans' TBIs had potentially been overlooked in ways that affected their ability to recover from PTSD. See Jordan Grafman et al., "Frontal Lobe Injuries, Violence, and Aggression: A Report from the Vietnam Head Injury Study," *Neurology* 46, no. 5 (1996): 1231; Jane K. Sweeney and Michael A. Smutok, "Vietnam Head Injury Study: Preliminary Analysis of the Functional and Anatomical Sequelae of Penetrating Head Trauma," *Physical Therapy* 63, no. 12 (1983), 2018–2025; Zeev Grosswasser et al., "Quantitative Imaging in Late TBI, Part II: Cognition and Work after Closed and Penetrating Head Injury: A Report of the Vietnam Head Injury Study," *Brain Injury* 16, no. 8 (2002): 687, 684; and David L. Trudeau et al., "Findings of Mild Traumatic Brain Injury in Combat Veterans with PTSD and a History of Blast Concussion," 10, no. 3 (1998): 312, Alexander, "Blast Force."

146 Geoffrey Ling, interview with the author, February 21, 2017.

147 Ibid.

148 Ibid.

149 The description of this experiment was described by Leanne Young in an interview with the author, April 27, 2017. It is also described in Richard A. Bauman et al., "An Introductory Characterization of a Combat-Casualty-Care Relevant Swine Model of Closed Head Injury Resulting from Exposure to Explosive Blast," *Journal of Neurotrauma* 26 (2009): 842, 848–849.

150 Bauman et al., "An Introductory Characterization of a Combat-Casualty-Care Relevant Swine Model of Closed Head Injury," 843–846.

151 Leanne Young, interview with the author, April 27, 2017.

152 Geoffrey Ling, interview with the author, February 21, 2017.

153 Leanne Young, interview with the author, April 27, 2017.

154 Bauman et al., "An Introductory Characterization of a Combat-Casualty-Care Relevant Swine Model of Closed Head Injury," 852.

155 Ibid., 851–852.

156 Geoffrey Ling, interview with the author, February 21, 2017.

157 The Bauman study did determine that after the explosion the swine struggled to maneuver in a manner that suggested they had been injured by the blast, but the study stopped short of concluding that the injury was to the brain and not muscular. See Bauman et al., "An Introductory Characterization of a Combat-Casualty-Care Relevant Swine Model of Closed Head Injury," 857–858.

158 Leanne Young, interview with the author, April 27, 2017.

159 Ibid.; Alexander, "Blast Force."

160 Walter Carr et al., "Repeated Low-Level Blast Exposure: A Descriptive Human Subjects Study," *Military Medicine* 181, no. 5 (2016): 28.

161 Ibid., 29.

162 Ibid., 38. This study is also described in Alexander, "Blast Force."

163 Geoffrey Ling, interview with the author, February 21, 2017.

164 Ibid. See also Alexander, "Blast Force."

165 "New Helmet Sensors to Measure Blast Impact," States News Service, January 7, 2008; and Gregg Zoroya, "Army Device Will Gauge Blast Hits," *USA Today*, July 19, 2011.

166 Geoffrey Ling, interview with the author, February 21, 2017.

167 John Holcomb, interview with the author, February 7, 2017. Holcomb and his colleagues also made this point in a 2010 article in which they determined that the most common form of blast injury was the ruptured eardrum. The only discussion of primary blast injury to the brain comes as skepticism: "If there is primary blast-induced brain injury, the group with tympanic membrane injury is likely to have the highest incidence." See Amber E. Ritenour et al., "Incidence of Primary Blast Injury in US Military Overseas Contingency Operations," *Annals of Surgery* 251, no. 6 (2010): 1143.

168 Hoge, email to the author, February 28, 2018.

169 Ibid.

170 Ibid.

171 Christian J. Macedonia, interview with the author, October 10, 2018.

172 This point is also made in Robert F. Worth, "What If PTSD Is More Physical than Psychological?" *New York Times Magazine*, June 10, 2016. However, Worth does not discuss the debate over how close to an explosion a soldier had to be in order to receive mandatory screening, a policy colloquially known as the fifty-meter rule, described in detail below.

173 See, e.g., Rosemary Scolaro Moser et al., "Neurological Evaluation in the Diagnosis and Management of Sports-Related Concussion," *Archives of Clinical Neuropsychology* 22 (2007): 909–916; Louis DeBeaumont et al., "Long-Term and Cumulative Effects of Sports Concussion on Motor Cortex Inhibition," *Neurosurgery* 61, no. 2 (2007): 329–337; and S. E. Wall et al., "Neuropsychological Dysfunction following Repeat Concussion in Jockeys," *Journal of Neurology, Neurosurgery, and Psychiatry* 77 (2006): 518–520. For popular press stories, see Alan Schwarz, "12 Athletes Leaving Brains to Researchers," September 24, 2008, "Sixth N.F.L. Player's Brain Found to Have Damage," January 28, 2009, and "N.F.L. Acknowledges Long-Term Concussion Effects," December 21, 2009, all in the *New York Times*; and Greg Bishop, "As Signs of Concern Spread, N.F.L. Revisits Concussions," *New York Times*, November 30, 2009.

174 Lizette Alvarez, "Home from War, Veterans Say Head Injuries Go Unrecognized," *New York Times*, August 26, 2008.

175 Ibid.

176 Peter Chiarelli, interview with the author, September 16, 2015. Chiarelli's encounter and embrace of the slide of the brain of the hit football player is also recounted, with a nearly identical quote from Chiarelli, in Schultz and Chandrasekaran, "General Chiarelli's Brain Crusade."

177 Peter Chiarelli, interview with the author, September 16, 2015.

178 Peter Chiarelli, interview with the author, December 9, 2015. A version of this story also appears in Schultz and Chandrasekaran, "General Chiarelli's Brain Crusade"; and in Worth, "What If PTSD Is More Physical than Psychological?"

179 Christian J. Macedonia, interview with the author, October 10, 2018.

180 Hamilton, "How a Team of Elite Doctors Changed the Military's Stance on Brain Trauma"; and Christian J. Macedonia, interview with the author, October 10, 2018.

181 Christian J. Macedonia, email to the author, November 3, 2018.

182 Christian J. Macedonia, interview with the author, October 10, 2018.

183 Ibid.

184 Ibid.; Christian J. Macedonia, email to the author, November 3, 2018; Hamilton, "How a Team of Elite Doctors Changed the Military's Stance on Brain Trauma." On "Red Teaming," see "University of Foreign Military and Cultural Studies / Red Teaming" (Fort Leavenworth, KS: United States Army Combined Arms Center, September 26, 2018), https://usacac.army.mil.

185 Geoffrey Ling, interview with the author, February 21, 2017.

186 Ibid.

187 Christian J. Macedonia, email to the author, November 3, 2018.

188 Peter Chiarelli, interview with the author, December 5, 2015.

189 Ibid.

190 Christian J. Macedonia, interview with the author, October 10, 2018; Hamilton, "How a Team of Elite Doctors Changed the Military's Stance on Brain Trauma." In 2016, the DOD discontinued use of the blast sensors over concerns over their effectiveness. See Jon Hamilton, "Pentagon Shelves Blast Gauges Meant to Detect Battlefield Brain Injuries," National Public Radio, December 20, 2016, www.npr.org.

191 Christian J. Macedonia, interview with the author, October 10, 2018.

192 Ibid.; Geoffrey Ling, interview with the author, February 21, 2017. See also Hamilton, "How a Team of Elite Doctors Changed the Military's Stance on Brain Trauma"; and William J. Lynn III to Secretaries of the Military Departments et al., "Directive-Type Memorandum (DTM) 09-033: 'Policy Guidance for Management of Concussion / Mild Traumatic Brain Injury in the Deployed Setting" (Washington, DC: Office of the Undersecretary of Defense, June 21, 2010), posted at the *Government Executive* website, www.govexec.com.

193 Office of the Surgeon General, "HQDA EXORD 253–10: Management of Concussion / Mild Traumatic Brain Injury (MBTI) in the Deployed Setting" (Falls Church, VA: Office of the Surgeon General, June 21, 2010). See also Hamilton, "How a Team of Elite Doctors Changed the Military's Stance on Brain Trauma."

194 Geoffrey Ling, interview with the author, February 21, 2017.

195 Ibid.

196 Ibid.

197 Peter Chiarelli, interview with the author, December 9, 2015.

198 Eric Schoomaker, interview with the author, December 16, 2015.

199 Charles Hoge, interview with the author, December 31, 2015.

200 Macedonia, email to the author, November 3, 2018.

201 John Holcomb, interview with the author, February 7, 2017.

202 Eric Schoomaker, interview with the author, December 16, 2015.

203 Ibid.

204 Jonathan Jaffin, interview with the author, January 30, 2016.

205 William J. Lynn III, "Memorandum for Secretaries of the Military Departments, et al.," June 21, 2010, 7–8, FOIA.

206 Jonathan Jaffin, interview with the author, January 30, 2016.

207 William J. Lynn III, "Memorandum for Secretaries of the Military Departments, et al.," June 21, 2010, 7–8, FOIA.

208 Schultz and Chandrasekaran, "General Chiarelli's Brain Crusade."

209 Eric Schoomaker, email to the author, April 28, 2016.

210 "History of the ANAM Program Prepared for TSG," November 2, 2011, EBS Papers.

211 Ibid.; Traumatic Brain Injury Task Force, "Report to the Surgeon General," 5, 64.

212 Senate Committee on Armed Services, *Current and Future Readiness of the Army and Marine Corps*, 110th Congress, 1st Session (2007), 67–68.

213 S.1065, *The Heroes at Home Act*, 110th Congress, 1st Session (2007), 1; H.R. 4986, *The National Defense Authorization Act of 2008*, 110th Congress, 1st Session (2008), 449–450.

214 "History of the ANAM Program Prepared for TSG," November 2, 2011, EBS Papers.

215 Ibid.

216 Ibid.; Daniel Zwerdling, "Military's Brain-Testing Program a Debacle," National Public Radio, November 28, 2011.

217 Traumatic Brain Injury Task Force, "Report to the Surgeon General," 40.

218 Ibid.

219 Charles Hoge, interview with the author, December 31, 2015. See also Zwerdling, "Military's Brain-Testing Program a Debacle."

220 "History of the ANAM Program Prepared for TSG," November 2, 2011, EBS Papers.

221 Ibid.

222 Charles Hoge, interview with the author, December 31, 2015.

223 Traumatic Brain Injury Task Force, "Report to the Surgeon General," 40.

224 Ibid.

225 Headquarters, US Army Medical Command, "Operations Order 08-65 (Predeployment Neurocognitive Testing)," August 2008, FOIA.

226 Ibid.

227 Ibid., 2.

228 Gregg Zoroya, "Military Fails on Brain-Test Follow-Ups," *USA Today*, June 14, 2010.

229 Leanne Young, interview with the author, April 27, 2017.

230 Ibid.; Zoroya, "Military Fails on Brain-Test Follow-Ups."

231 "Questions for LTG Schoomaker Ref: ANAM Tool Selection Process Reporters: NPR-Daniel Zwerdling / ProPublica—Joaquin Sapien," November 1, 2011, EBS Papers.

232 Rodney L. Coldren et al., "The ANAM Lacks Utility as a Diagnostic or Screening Tool for Concussion More than Ten Days following Injury," *Military Medicine* 177, no. 2 (2012): 182.

233 Center for the Study of Human Operator Performance, "Summary of Pre- and Post-deployment Cognitive Testing in the U.S. Military," PowerPoint Presentation, 2009, FOIA.

234 Ibid.

235 "ANAM Program Review," June 16, 2010, 1, FOIA.

236 Office of the Surgeon General, "Pre-deployment and Appropriate In-Theater Brain Function Testing for Concussion," July 16, 2008, FOIA.

237 Ibid.

238 Ibid., 1–2; "Information Paper: ANAM Program Review," June 16, 2010, FOIA.

239 Zwerdling, "Military's Brain-Testing Program a Debacle."

240 Ibid.

241 Eric B. Schoomaker to NPR, December 1, 2011, EBS Papers.

242 Ibid.

243 Ibid.

244 Ibid.

245 Christian J. Macedonia, interview with the author, October 10, 2018.

CHAPTER 7. "LEADERS CAN ONCE AGAIN DETERMINE THE KIND OF CULTURE THE ARMY IS BUILDING"

1 Nathaniel Fick, *One Bullet Away: The Making of a Marine Officer* (New York: Houghton Mifflin, 2005), 363; and Carl A. Castro, "Battlemind Training I: Transitioning from Combat to Home," PowerPoint presentation (Silver Spring, MD: Walter Reed Army Institute of Research, n.d.), 15, www.army.mil.

2 Suicide Grief Support Forum, "Sgt. Douglas E. Hale, 08/19/1983–07/07/10," July 2, 2011, www.suicidediscussionboard.com.

3 Gregg Zoroya, "Thousands Strain Post's Mental Health System," August 23, 2010, and "Army Efforts Don't Stem Suicides at Fort Hood," January 6, 2011, both at *USA Today*.

4 David Tarrant, "Fort Hood Reflects Army's Increase in Suicides," *Dallas Morning News*, June 12, 2012, www.dallasnews.com; James C. McKinley, Jr., "Four Suicides in a Week Take Toll on Fort Hood," *New York Times*, September 29, 2010; and Michael Hoffman, "Guard, Reserve Suicide Rate Sees Big Spike," *Army Times*, January 19, 2011, www.armytimes.com. Kenneth T. MacLeish, *Making War at Fort Hood: Life and Uncertainty in a Military Community* (Princeton: Princeton University Press, 2013), 225.

5 U.S. Army, *Health Promotion, Risk Reduction, Suicide Prevention Report, 2010* (Red Book; Arlington, VA: U.S. Army, 2010), i, www.army.mil.

6 Gregg Zoroya, "June Was Worst Month on Record for Army Suicides," *USA Today*, July 16, 2010.

7 Greg Jaffe, "Army Suicides Hit Monthly High in June," *Washington Post*, July 17, 2010; as MacLeish points out, for the Army, "suicide is a public health problem

and a threat to the organizational body" and it became "a highly visible scandal for the Army" (*Making War at Fort Hood*, 226–27).

8 Beth Bailey, *America's Army: Making the All-Volunteer Force* (Cambridge, MA: Harvard University Press, 2009), 206.

9 Howard Schultz and Rajeev Chandrasekaran, "General Chiarelli's Brain Crusade," *Politico*, December 27, 2014, www.politico.com.

10 On the increasing rate, see Erin Finley, *Fields of Combat: Understanding PTSD among Veterans of Iraq and Afghanistan* (Ithaca, NY: Cornell University Press, 2011), 69.

11 James Martin, interview with the author, July 17, 2017.

12 On Wickham's conversion and vision for the Army, see Randall Herbert Balmer, *Encyclopedia of Evangelicalism* (Louisville: Westminster John Knox Press, 2002), 621; Seth Dowland, *Family Values and the Rise of the Christian Right* (Philadelphia: University of Pennsylvania Press, 2015), 199; Anne C. Loveland, *American Evangelicals and the U.S. Military, 1942–1993* (Baton Rouge: Louisiana State University Press, 1996), 276–277; and Jennifer Mittelstadt, *The Rise of the Military Welfare State* (Cambridge, MA: Harvard University Press, 2015), 152–153, 155.

13 Robert M. Wildzunas, "Suicide Prevention," *Countermeasure: Army Ground Risk-Management Publication* 21, no. 1 (2000): 3.

14 John J. South, "The Army Suicide Prevention Program" (MA thesis, Army War College, Carlisle Barracks, PA, 2003), 17–18. On evangelicals' "zeal and cohesiveness" in the military, see Loveland, *American Evangelicals and the U.S. Military*, 340.

15 South, "The Army Suicide Prevention Program," 19.

16 Gale Pollock, interview with the author, November 19, 2015.

17 U.S. Army Center for Health Promotion and Preventive Medicine, "Suicide Prevention: A Resource Manual for the United States Army" (Aberdeen Proving Ground, MD: U.S. Army Center for Health Promotion and Preventive Medicine, 2007), 3, www.army.mil.

18 Ibid., 6–8.

19 Ibid., 12.

20 Ibid., 15–16.

21 Eric B. Schoomaker, interview with the author, December 16, 2015.

22 U.S. Army, *Health Promotion, Risk Reduction, Suicide Prevention Report, 2010*, 16.

23 Dana Priest, "Soldier Suicides at Record Level," *Washington Post*, January 31, 2008.

24 Ibid.

25 Sen. Murray of Washington, speaking on Army Suicides, Congressional Record, January 31, 2008, S528.

26 Ibid.

27 "New Reason for a Draft?" *Providence (RI) Journal*, April 4, 2008.

28 Rep. Wasserman Schultz of Florida, speaking on 30-Something Working Group, *Congressional Record*, April 9, 2008, H2138. The 30-Something Working Group was "a loose collection of House Democrats in their 30s and 40s" who, the *Washington Post* reported in 2006, "played an important role in helping their party take control

of Congress" during the 2006 midterms. See Lyndsey Layton, "In New Congress Seniority Takes Back Seat to Spirit," *Washington Post*, December 11, 2006.

29 Rob Nixon, *Slow Violence and the Environmentalism of the Poor* (Cambridge, MA: Harvard University Press, 2011), 200.

30 Ibid.

31 Ibid., 2.

32 U.S. Army Center for Health Promotion and Preventive Medicine, "Suicide Prevention: A Resource Manual for the United States Army," 23–25.

33 Senate Armed Services Committee, Subcommittee on Personnel, *The Findings and Recommendations of the Department of Defense Task Force on Mental Health, the Army's Mental Health Advisory Team Reports, and Department of Defense and Service-Wide Improvements in Mental Health Resources, Including Suicide Prevention for Servicemembers and Their Families*, 110th Congress, 2d Session (2008), 128.

34 The report was also likely the source for the *Post*'s article on Whiteside; see Priest, "Soldier Suicides at Record Level."

35 Suicide Risk Management and Surveillance Office, *Army Suicide Event Report (ASER), Calendar Year 2007* (Tacoma, WA: Madigan Army Medical Center, 2008), 2.

36 Ibid., 15, 13.

37 Ibid., 19.

38 Ibid., 21, 57.

39 Greg Zoroya, "Army Tallies Record Number of Suicides among Soldiers," *USA Today*, May 30, 2008.

40 "Suicide Rate for Soldiers Rose in '07," *New York Times*, May 30, 2008.

41 "The Strain on Our Soldiers," *Anniston (AL) Star*, June 2, 2008.

42 "The War Raging Within," *Quincy (MA) Patriot Ledger*, June 30, 2008. See also "Rising Army Suicides Demand Attention," *San Gabriel Valley (CA) Tribune*, June 4, 2008.

43 Elizabeth Mohr, "Soldier's Death a Reminder of War's Stress and the Toll It Takes," *St. Paul (MN) Pioneer Press*, June 24, 2008.

44 Jim Carney, "Akron, Ohio, Soldier Ends War's Pain with His Suicide," *Akron (OH) Beacon Journal*, August 15, 2008.

45 "Army to Report Record Suicide Rate," CNN, January 30, 2009, www.cnn.com.

46 Gregg Zoroya, "Soldier's Death Reflects Impact of Stress in Ranks," *USA Today*, January 13, 2009.

47 Ibid.

48 Tom Infield, "Another Casualty of War, after the Fact," *Philadelphia Inquirer*, February 7, 2009.

49 Cathy Kelly, "Santa Cruz Memorial Grows at Spot of Soldier's Suicide," *San Jose (CA) Mercury News*, May 27, 2009.

50 Erica Goode, "After Combat, Victims of an Inner War," *New York Times*, August 2, 2009.

51 Steve Vogel, "Families Affected by Suicide Feel Sting on Memorial Day," *Washington Post*, May 25, 2009.

52 Matthew D. LaPlante, "Family Pressures, Military Stress Add to Suicide Risk for Guard, Reserve Troops," *Salt Lake (UT) Tribune*, February 12, 2008; and Gregg Zoroya, "Army Tallies Record Number of Suicides Among Soldiers," *USA Today*, May 30, 2008.

53 Diana Washington Valdez, "Suspected Suicides Increase at Fort Bliss," *El Paso (TX) Times*, May 30, 2008.

54 Gordon Lubold, "US Army Suicides on Track to Hit New High in 2009," *Christian Science Monitor*, November 17, 2009. See also "Editorial: Suicide Rates Are Unacceptably High," *Reading (PA) Eagle*, February 15, 2009; John Simerman, "Army Suicides Continue Troubling Climb," *Contra Costa (CA) Times*, May 7, 2009.

55 Nancy A. Youseff, "Army Suicides Are Expected to Set Records," *St. Paul (MN) Pioneer Press*, November 17, 2009.

56 Eric B. Schoomaker, interview with the author, December 16, 2015.

57 Elspeth Cameron Ritchie, interview with the author, December 15, 2016.

58 Ibid.

59 Eric B. Schoomaker, interview with the author, December 16, 2015.

60 Ibid.

61 Ibid.; and Bruce Shahbaz, interview with the author, February 24, 2016.

62 Elspeth Cameron Ritchie, interview with the author, December 15, 2015.

63 Ibid.

64 Jeffrey Castro, "First Woman Becomes Army Provost Marshall General," U.S. Army News Service, January 15, 2010, www.army.mil.

65 Ibid.

66 Ibid.

67 Ibid.; and Bruce Shahbaz, interview with the author, February 24, 2016.

68 Colleen McGuire, interview with the author, March 7, 2018.

69 Bruce Shahbaz, interview with the author, February 24, 2016.

70 Ibid.

71 Bruce Shahbaz, interview with the author, February 24, 2016.

72 Charles Hoge, interview with the author, December 31, 2015.

73 Eric B. Schoomaker, interview with the author, December 17, 2015.

74 Bruce Shahbaz, interview with the author, February 24, 2016.

75 Ibid.

76 Senate Committee on Armed Services, Subcommittee on Personnel, *The Incidence of Suicides of United States Servicemembers and Initiatives within the Department of Defense to Prevent Military Suicides*, 111th Congress, 1st Session (2009), 10, 35–36.

77 House Armed Services Committee, Military Personnel Subcommittee, *Psychological Stress in the Military: What Steps Are Leaders Taking?* 111th Congress, 1st Session (2009), 9.

78 Ibid., 19.

79 Peter Chiarelli, interview with the author, September 18, 2015.

80 Eric B. Schoomaker, interview with the author, December 17, 2015.

81 U.S. Army, *Health Promotion, Risk Reduction, Suicide Prevention Report, 2010*, 39.

82 Ibid., 40.

83 Ibid.

84 Ibid., 11.

85 Ibid.

86 Ibid., 37.

87 Ibid., 24, 76, 100, 106.

88 Ibid., 40, 53.

89 Thomas E. White and Eric K. Shinseki, *A Statement on the Posture of the United States Army, 2001, Presented to the Committees and Subcommittees of the United States Senate and the House of Representatives*, 107th Congress, 1st Session (Arlington, VA: Office of the Chief of Staff, 2001) , 3, www.army.mil.

90 Bailey, *America's Army*, 206.

91 Lauren Berlant, *Cruel Optimism* (Durham, NC: Duke University Press, 2011), 11. Also see Andrew Bacevich, *Breach of Trust: How Americans Failed Their Soldiers and Their Country* (New York: Metropolitan Books, 2013), 11; and Gallup poll, "Confidence in Institutions," n.d., www.gallup.com.

92 Bailey, *America's Army*, 252, 249.

93 Brad Knickerbocker, "Sergeant Seen as 'Kill Team' Leader Found Guilty in Afghanistan Atrocities," *Christian Science Monitor*, November 10, 2011; Jon Krakauer, *Where Men Win Glory: The Odyssey of Pat Tillman* (New York: Doubleday, 2009); U.S. Army, *Army 2020: Generating Health and Discipline in the Force ahead of the Strategic Reset: Report, 2012* (Gold Book; Arlington, VA: Department of the Army, 2012), 90–91.

94 Berlant, *Cruel Optimism*, ix.

95 U.S. Army, *Health Promotion, Risk Reduction, Suicide Prevention Report, 2010*, 36. See also U.S. Army, *Army 2020*, 7.

96 Bailey, *America's Army*, 253.

97 Michael Schoenbaum et al., "Predictors of Suicide and Accident Death in the Army Study to Assess Risk and Resilience in Servicemembers (Army STARRS): Results from the Army Study to Assess Risk and Resilience in Servicemembers (Army STARRS)," *JAMA Psychiatry* 71, no. 5 (2014): 499.

98 U.S. Army, Health Promotion, Risk Reduction, Suicide Prevention Report, 2010, 68–69.

99 U.S. Army, *Army 2020*, 90.

100 Elisabeth Bumiller, "Pentagon Report Places Blame," *New York Times*, July 29, 2010; and Bruce Shahbaz, email to the author, February 22, 2016.

101 Raymond Palumbo, interview with the author, April 24, 2016.

102 Rebecca Porter, interview with the author, December 17, 2015.

103 Ibid.

104 Colleen McGuire, interview with the author, March 7, 2018.

105 Ibid.

106 U.S. Army, Health Promotion, Risk Reduction, Suicide Prevention Report, 2010, iii.

107 Ibid., 35–37.

108 Ibid., 37.

109 Rebecca Porter, interview with the author, December 17, 2015.

110 Ibid.

111 Raymond Palumbo, interview with the author, April 24, 2016.

112 Colleen McGuire, interview with the author, March 7, 2018.

113 U.S. Army, *Health Promotion, Risk Reduction, Suicide Prevention Report, 2010*, 40.

114 Chiarelli, quoted in Nancy A. Youssef, "As Iraq Winds Down, U.S. Army Confronts a Broken Force," McClatchyDC News Agency, September 17, 2010, www.mcclatchydc.com.

115 U.S. Army, Health Promotion, Risk Reduction, Suicide Prevention Report, 2010, iii.

116 Ibid.

117 Ibid., 64.

118 Ibid., 69.

119 Ibid., 40.

120 Ibid, 68–69.

121 U.S. Army, *Army 2020*, 86, 80, 32.

122 Ibid., 80.

123 At the same time, this language opens the door to dismissing, rather than treating, soldiers with mental health issues and raises the question of whether increased discharges of those who develop mental health issues while serving will occur (as they did during the Iraq War under the rubric of "Personality Disorder"), and it also casts the military as unsympathetic to mental health conditions.

124 Greg Barnes, "The Last Battle," *Fayetteville (NC) Observer*, September 25, 2012, www.fayobserver.com.

125 Dave Phillips, "Left Behind: No Break for the Wounded," *Colorado Springs (CO) Gazette*, May 20, 2013, https://gazette.com.

126 Ibid.

127 Daniel Zwerdling, "Thousands of Soldiers with Mental Health Disorders Kicked Out for 'Misconduct,'" National Public Radio, October 28, 2015, www.npr.org. The Army disputed this account, claiming that it had appropriately followed its procedures; see "Senators, Military Specialists Say Army Report on Dismissed Soldiers Is Troubling," National Public Radio, December 1, 2016, www.npr.org.

128 M. A. Fischl, "Combat Stress Decompression" (Washington, DC: Walter Reed Army Institute of Research, March 7, 1991), 4, courtesy of James Martin, copy in author's possession.

129 *Carl Levin and Howard P. "Buck" McKeon National Defense Authorization Act for Fiscal Year 2015*, 113th Congress, 2d Session (2014), 106; and Josh Hicks, "Mental-Health Experts Would Review Discharges," *Washington Post*, December 6, 2014.

130 "In Response to Letter from Baldwin and Others, U.S. Army Updates Regulations Covering Soldiers Discharged for Misconduct While Suffering from PTSD and TBI," website for Sen. Tammy Baldwin (D-WI), September 16, 2016, www.senate.gov.
131 Megan Cerullo, "Mental Woe for 62% of Booted Troops," *New York Daily News*, May 17, 2017.
132 George Casey, interview with the author, October 13, 2017. On this concept, see Bailey, *America's Army*, 73.
133 Christopher Ivany, interview with the author, December 16, 2015.
134 Jimmie Keenan, interview with the author, December 1, 2015.
135 Colleen McGuire, interview with the author, March 7, 2018.
136 Ibid.
137 Bruce Shahbaz, interview with the author, February 24, 2016.
138 Ibid.
139 Christopher Ivany, interview with the author, December 16, 2015.
140 Ibid.
141 Ibid.
142 Ibid.
143 Bruce Shahbaz, interview with the author, February 24, 2016.
144 Ibid.
145 Ibid.
146 Rebecca Porter, interview with the author, May 6, 2016.
147 Rebecca Porter, interview with the author, December 16, 2015.
148 Ibid.
149 Pete Geren, interview with the author, May 31, 2016.
150 Bruce Shahbaz, interview with the author, February 24, 2016.
151 Greg Jaffe, "Honoring the Service of Soldiers Who Commit Suicide," *Washington Post*, July 18, 2010.
152 Ibid.; and "Army Staff Sergeant Thaddeus S. Montgomery, Jr.," *Military Times*, n.d., www.militarytimes.com.
153 Yochi Dreazen, *The Invisible Front: Love and Loss in an Era of Endless War* (New York: Crown, 2014), 235, 259–260.
154 Richard Maurer, "Military Commanders Take On Stigma regarding Suicides," *Anchorage (AK) Daily News*, February 7, 2010, www.mcclatchydc.com, posted at Franklin James Cook, "Alaska Army Post Battles Stigma to Prevent Suicide," *Suicide Prevention News and Comment* (blog), February 7, 2010, https://wordpress.com.
155 Bruce Shahbaz, interview with the author, February 24, 2016.
156 Jaffe, "Honoring the Service of Soldiers Who Commit Suicide"; and Greg Jaffe, "Diagnosis: Battle Wound," *Washington Post*, July 18, 2010.
157 Christopher Philbrick, interview with the author, June 5, 2016.
158 Jaffe, "Diagnosis: Battle Wound"; Dreazen, *The Invisible Front*, 220; and David Finkel, *Thank You for Your Service* (New York: Sarah Crighton Books, 2013), 73.
159 Bruce Shahbaz, interview with the author, February 24, 2016.

160 Christopher Philbrick, interview with the author, June 5, 2016.
161 Peter Chiarelli, interview with the author, December 9, 2015.
162 Ann Scott Tyson and Greg Jaffe, "Generals Find Suicide a Frustrating Enemy," *Washington Post*, May 23, 2009; and Jamie Tarabay, "Suicide Rivals the Battlefield in Toll on U.S. Military," *National Public Radio*, June 17, 2010, www.npr.org.
163 Finkel, *Thank You for Your Service*, 76.
164 Raymond Palumbo, interview with the author, April 24, 2016.
165 Ibid.
166 Christopher Philbrick, interview with the author, June 5, 2016.
167 Rebecca Porter, interview with the author, May 6, 2016, and December 17, 2015.
168 Colleen McGuire, interview with the author, March 7, 2018.
169 Raymond Palumbo, interview with the author, April 24, 2016.
170 See, e.g., Finkel, *Thank You for Your Service*, 72–73.
171 Rebecca Porter, interview with the author, May 6, 2016.
172 Ibid.
173 Eric Schoomaker, interview with the author, April 6, 2016.
174 Rebecca Porter, interview with the author, December 17, 2015, and May 6, 2016.
175 Rebecca Porter, interview with the author, December 17, 2015.
176 Ibid.
177 Colleen McGuire, interview with the author, March 7, 2018.
178 Rebecca Porter, interview with the author, December 17, 2015.
179 Ibid.
180 Ibid.
181 Colleen McGuire, interview with the author, March 7, 2018; and Rebecca Porter, interview with the author, May 6, 2016.
182 Rebecca Porter, interview with the author, May 6, 2016.
183 Senate Committee on Armed Services, *The Progress in Preventing Military Suicides*, 111th Congress, 2d Session (2020), 6.
184 Ibid., 7. In fact, the suicide rate did not decrease, though it did stabilize by 2012; Bruce Shahbaz, email to the author, February 22, 2016.
185 Senate Committee on Armed Services, *The Progress in Preventing Military Suicides*, 7, 47.
186 Ibid., 41, 52.
187 Ibid., 54.
188 Ibid., 56.
189 James C. McKinney, "Despite Army's New Efforts, Suicides Continue at Grim Pace," *New York Times*, October 11, 2010.
190 Richard Sisk, "Mocked to Death," *New York Daily News*, October 8, 2010.
191 Nancy A. Youseff, "Suicides Rise Sharply among Reservists, Guardsmen," *St. Paul (MN) Pioneer Press*, January 19, 2011; and Greg Jaffe, "Army Reports Big Increase in Guard, Reserve Suicides," *Washington Post*, January 20, 2011.
192 Greg Jaffe, "Army Suicides at Record High in July," *Washington Post*, August 13, 2011.

193 This was a point that Chiarelli also made in Stephanie Gaskell, "Peter Chiarelli Continues His Siege on Suicides," *Politico*, September 20, 2012.

194 Peter Chiarelli, interview with the author, September 16, 2015.

CHAPTER 8. "THE CHALLENGE TO THE VA IS EXECUTION AND IMPLEMENTATION"

1 Joshua Omvig Veterans Suicide Prevention Act, S. 479, 110th Congress, 1st Session (February 1, 2007), posted at GovTrack, www.govtrack.us; and Deputy Undersecretary for Health Operations and Management to Network Director, "Mental Health Funding for Suicide Prevention Coordinators," February 8, 2007, FOIA.

2 Deputy Undersecretary for Health Operations and Management to Network Director, "Mental Health Funding for Suicide Prevention Coordinators," February 8, 2007, FOIA.

3 VA Office of the Inspector General, *Healthcare Inspection: Implementing VHA's Mental Health Strategic Plan for Suicide Prevention* (Washington, DC: VA Office of the Inspector General, 2007), 27.

4 Ibid., 29, 33, 37, 38.

5 Ibid., 31.

6 Ibid., 46, 49–50.

7 Ibid., 52.

8 Department of Veterans Affairs, "Veterans Integrated Healthcare Network (VISN) 19 Mental Illness Research, Education and Clinical Center (MIRECC) Reverse Site Visit (2010–2014), 2014, 4, FOIA.

9 Ibid. Also see the Center of Excellence in Suicide Prevention's fifteen-page Self Study (*VISN 2 Center of Excellence for Suicide Prevention [COE]* [Canandaigua, NY: Department of Veterans Affairs, n.d.], 1, FOIA.)

10 Paul Shekelle, Steven Bagley, and Brett Munjas, *Strategies for Suicide Prevention in Veterans* (Washington, DC: Department of Veterans Affairs, January 2009), 1.

11 Ibid., 6, 23, 20–27.

12 Ibid.

13 Ibid.

14 Ibid., 5; and J. John Mann et al., "Suicide Prevention Strategies: A Systematic Review," *Journal of the American Medical Association* 294, no. 16 (2005): 2065.

15 House Committee on Veterans' Affairs, *Stopping Suicides: Mental Health Challenges within the U.S. Department of Veterans Affairs*, 110th Congress, 1st Session, (2007), 85.

16 House Committee on Veterans' Affairs, Subcommittee on Health, *The U.S. Department of Veterans Affairs Suicide Prevention Hotline*, 110th Congress, 2d Session (2008), 14, 15.

17 Ibid., 15, 18, 20, 24.

18 Ibid., 28, 14, 18.

19 Ibid., 9.

20 Ibid., 26, 28.

21 Ibid., 13, 22, 28.
22 Ibid., 18.
23 Ibid., 19.
24 Jay Winter, "Thinking about Silence," in *Shadows of War: A Social History of Silence in the Twentieth Century*, ed. Efrat Ben-Ze'ev, Ruth Ginio, and Jay Winter (New York Cambridge University Press, 2010), 6.
25 House Committee on Veterans' Affairs, Subcommittee on Health, *The U.S. Department of Veterans Affairs Suicide Prevention Hotline*, 3, 26, 29.
26 Ibid., 59, 60.
27 Ibid., 26. Jeremy K. Saucier, "Calls of Duty: The World War II Combat Video Game and the Construction of the 'Next Great Generation,'" in *The War of My Generation: Youth Culture and the War on Terror*, ed. David Kieran (New Brunswick, NJ: Rutgers University Press, 2015), 128, 132–135. On the difference between the two generations, see Evan Wright, *Generation Kill: Devil Dogs, Ice Man, Captain America, and the New Face of American War* (New York: Penguin, 2008), 5.
28 Michael J. Kussman to Veteran, n.d., FOIA.
29 Gerald Cross and William Schoenhard to Network Directors, "Patients at High Risk for Suicide," April 24, 2008, 1, FOIA.
30 Vicki Johnstone-Lawrence et al., "VA Primary Care–Mental Health Integration: Patient Characteristics and Receipt of Mental Health Services, 2008–2010," *Psychiatric Services* 63, no. 11 (2012): 1137, 1139–1140.
31 Gerald Cross and William Schoenhard to Network Directors, "Patients at High Risk for Suicide," April 24, 2008, 2, FOIA.
32 Department of Veterans Affairs, "SP Scorecard (Feb10)," March 18, 2010, FOIA.
33 Department of Veterans Affairs, "FY13 Scorecard (Nov12)," December 14, 2011, FOIA.
34 National Suicide Prevention Coordinator to Facility Suicide Prevention Coordinators and Teams, September 1, 2009, FOIA.
35 Ibid.
36 Jan E. Kemp to VHA Suicide Prevention Coordinators, "Questions," September 15, 2009, FOIA.
37 Robert Bossarte, Ira Katz, and Janet Kemp, "Evaluation of the VA's suicide Event Reporting System," n.d., FOIA; Jan E. Kemp to VHA VISN Directors et al., "Suicide Prevention Coordinator Report Due October 9th," October 9, 2009, FOIA. For an example of the spreadsheet, see files like "SPC Report for Sep09," FOIA.
38 See "SPC Monthly Report (4)," FOIA.
39 "All VISN Suicide Attempts and Death Data," FOIA.
40 Ibid.
41 Sherry L. Murphy et al., "Deaths: Final Data for 2010," *National Vital Statistics Reports* 61, no. 4 (2013): 34.
42 William Schoenhard to Network Directors and Chief Medical Officers, April 18, 2011, FOIA.

43 Frederic C. Blow and Janet Kemp, "Suicide among Individuals Receiving VHA Health Services, FY01–FY09," presentation to OMHO [Office of Mental Health Operations] / OMHS [Office of Mental Health Services] Suicide Workgroup, November 28, 2011, FOIA.

44 Ibid.

45 VA Suicide Prevention Program, "Facts about Veteran Suicide," April, 2010, FOIA.

46 Senate Committee on Veterans Affairs, *The Fiscal Year 2010 Budget for Veterans' Programs*, 111th Congress, 1st Session (2009), 26–27.

47 Joseph L. Galloway, "At Long Last, Sir, Have You Left No Sense of Decency?" *Salt Lake Tribune*, January 9, 2009.

48 Julie Robinson, "More Counseling, Fewer Pills," *Charleston (WV) Gazette*, April 12, 2009.

49 Kevin Graman, "VA Center Wasn't Aware of Many Veteran Suicides," *Spokane (WA) Spokesman Review*, August 9, 2009.

50 Leo Shane III, "A Matter of Degrees," *Stars and Stripes*, June 27, 2012, www.stripes.com.

51 Bill Rautenstrauch, "VA Clinic Stresses Mental Health Care Need," *La Grande (OR) Observer*, April 8, 2009.

52 Kevin Graman, "VA Doctors Protest 'Crisis,'" *Spokane (WA) Spokesman Review*, August 13, 2009; and "Suicide Hotline Referrals for FY 2010 by Site by Month," FOIA.

53 "More than Memories: Give Returning Troops the Specialized Care They Need," *Daytona Beach (FL) News-Journal*, November 11, 2009.

54 Gordon Lubold, "'Surge Home Overwhelms Veterans Affairs Clinics," *Christian Science Monitor*, August 30, 2009; and Clint Van Winkle, *Soft Spots: A Marine's Memoir of Combat and Post-traumatic Stress Disorder* (New York: St. Martin's, 2009).

55 Van Winkle, *Soft Spots*, 32.

56 Ibid., 86, 25.

57 Ibid., 37, 90.

58 Ibid., 90.

59 Ibid., 25–26.

60 Steven M. Silver, Susan Rogers, and Mark Russell, "Eye Movement Desensitization and Reprocessing (EMDR) in the Treatment of War Veterans," *Journal of Clinical Psychology: In Session* 64, no. 8 (2008): 947–948. On the VA's embrace of EMDR, see Matthew J. Friedman, "Posttraumatic Stress Disorder among Military Returnees from Afghanistan and Iraq," *American Journal of Psychiatry* 163, no. 4 (2006): 591; Mark C. Russell et al., "Responding to an Identified Need: A Joint Department of Defense/Department of Veterans Affairs Training Program in Eye Movement Desensitization and Reprocessing (EMDR) for Clinicians Providing Trauma Services," *International Journal of Stress Management* 14, no. 1 (2007): 61–71; and Robert P. Salvatore, "Posttraumatic Stress Disorder: A Treatable Public Health Problem," *Health and Social Work* 34, no. 2 (2009): 153. For an overview of the evidence-based treatments that the VA has embraced, see Erin Finley, *Fields of Combat: Understanding*

PTSD among Iraq and Afghanistan Veterans (Ithaca, NY: Cornell University Press, 2011), 123–130. David J. Morris, *The Evil Hours: A Biography of Post-Traumatic Stress Disorder* (New York: Houghton-Mifflin, 2015), 244–46.

61 Van Winkle, *Soft Spots*, 204–205.

62 Senate Committee on Veterans' Affairs, *Hearing on Mental Health Care and Suicide Prevention for Veterans*, 111th Congress, 2d Session (2010), 57–58.

63 Ibid., 57.

64 Ibid., 29.

65 House Committee on Veterans' Affairs, Subcommittee on Oversight and Investigations, *Examining the Progress of Suicide Prevention Outreach Efforts at the U.S. Department of Veterans Affairs*, 111th Congress, 2d Session (2010), 6.

66 Ibid., 6.

67 Ibid., 8.

68 Ibid., 11.

69 Ibid., 15.

70 Ibid., 11–12.

71 Ibid., 22–23.

72 The Plowshare Group, "Suicide Prevention PSA," October 15, 2008, YouTube, www.youtube.com/watch?v=5VlRY3_4Z1g.

73 House Committee on Veterans' Affairs, Subcommittee on Oversight and Investigations, *Examining the Progress of Suicide Prevention Outreach Efforts at the U.S. Department of Veterans Affairs*, 23.

74 Ibid., 23–24.

75 Senate Committee on Veterans' Affairs, *Hearing on Mental Health and Suicide Prevention for Veterans*, 111th Congress, 2d Session (2010), 63.

76 Class VI items, according to the military's *General Supply and Field Operations* field manual, are "personal demand items," but in everyday parlance the classification signifies alcohol. On Army posts, e.g., you buy your liquor at a "Class Six store." See U.S. Army, *ATP 4-42: General Supply and Field Operations* (Arlington, VA: Department of the Army, 2014), 4-4, www.army.mil. On the role that slang plays in constructing group identity, see Connie C. Elbe, *Slang and Sociability: In-Group Language among College Students* (Chapel Hill: University of North Carolina Press, 1996), 11–12, 119.

77 Janet Kemp and Robert Bossarte, *Suicide Data Report, 2012* (Canandaigua, NY: Department of Veterans Affairs, Mental Health Services Suicide Prevention Program, 2012), www.va.gov.

78 Department of Veterans Affairs, "Suicide Hotline Referrals for FY 2010 by Site by Month," FOIA.

79 Department of Veterans Affairs, "Center of Excellence Mental Health Crisis/Suicide Hotline YTD 2011 Referral Breakdown," May 11, 2011, FOIA.

80 Ibid.

81 Aaron Glantz, "After Service, Veteran Deaths Surge," *New York Times*, October 17, 2010; Joyce Tsai, "A Soldier, a Son, Lost," *Lowell (MA) Sun*, May 29, 2011; and Mary

McCarty, "PTSD—The 'Next Epidemic' for Vets," *Dayton (OH) Daily News*, May 27, 2012.

82 Aaron Glantz, "Suicides Highlight Failures of Veterans' Support System," *New York Times*, March 25, 2012.

83 Graman, "VA Center Wasn't Aware of Many Veteran Suicides."

84 See chapter 3.

85 Jan E. Kemp to Suicide Prevention Coordinators, "Reporting Information—FYI," September 28, 2009, FOIA.

86 Jan C. Kemp to VHA Suicide Prevention Coordinators; VHA CLE BVAC Group, "Comparisons of Suicides Reported in SPAN and NDI," September 1, 2011. FOIA.

87 Heather Shaw and Robert M. Bossarte, "State Mortality Database," presentation to the VISN 2 Center for Excellence for Suicide Prevention Advisory Board, April 28–29, 2011, FOIA.

88 Ibid.

89 Blow and Kemp, "Suicide among Individuals Receiving VHA Health Services, FY01–FY09"; and VA Suicide Prevention Program, "Facts about Veteran Suicide," April 2010, FOIA.

90 VA Suicide Prevention Program, "Facts about Veteran Suicide," May 2011, 3, FOIA.

91 Ibid.

92 Blow and Kemp, "Suicide among Individuals Receiving VHA Health Services, FY01–FY09."

93 VA Suicide Prevention Program, "Facts about Veteran Suicide," April 2010, FOIA.

94 Schoenhard to Network Directors and Chief Medical Officers, April 18, 2011, FOIA.

95 Kemp and Bossarte, *Suicide Data Report, 2012*, 10.

96 Ibid.; and Heather Shaw and Robert M. Bossarte, "State Mortality Database," presentation to the VISN 2 Center for Excellence for Suicide Prevention Advisory Board, April 28–29, 2011, FOIA.

97 On the underreporting of suicide, see Colin Pritchard and Lars Hansen, "Examining Undetermined and Accidental Deaths as Source of 'Under-Reported-Suicide' by Age and Sex in Twenty Western Countries," *Community Mental Health Journal* 51, no. 3 (2015): 365–376.

98 Kemp and Bossarte, *Suicide Data Report, 2012*.

99 Heather Shaw and Robert M. Bossarte, "State Mortality Database," presentation to the VISN 2 Center for Excellence for Suicide Prevention Advisory Board, April 28–29, 2011, FOIA.

100 Kemp and Bossarte, *Suicide Data Report, 2012*, 15.

101 Greg Jaffe, "VA Study Finds More Veterans Committing Suicide," *Washington Post*, February 1, 2013; and Department of Veterans Affairs, "Veterans Integrated Healthcare Network (VISN) 19 Mental Illness Research, Education and Clinical Center (MIRECC) Reverse Site Visit Review (2010–2014)," 2014.

102 Kemp and Bossarte, *Suicide Data Report, 2012*, 15.

103 Jan E. Kemp to Suicide Prevention Coordinators, "Reporting Information—FYI," September 28, 2009, FOIA.

104 Gregg Zoroya, "VA Study: 22 Vets Commit Suicide Every Day," *USA Today*, February 1, 2013.

105 Jaffe, "VA Study Finds More Veterans Committing Suicide."

106 Ibid.

107 James Dao, "As Suicides Rise in U.S., Veterans Are Less of Total," *New York Times*, February 2, 2013.

108 *Veterans For Common Sense v. Shinseki*, 644 F.3d (9th Cir. 2011); this is also quoted in James Dao, "Court Backs Veterans' Complaints on Mental Health Services," *New York Times*, May 11, 2011.

109 Ibid.

110 Dao, "Court Backs Veterans' Complaints on Mental Health Services."

111 "A Victory for Veterans," *New York Times*, May 18, 2011,which is also reprinted in *Deseret Morning News*, May 20, 2011; "Overdue Aid for Vets," *Lancaster (PA) Intelligencer Journal*, May 26, 2011; and "A Disgrace," *Brattleboro (VT) Reformer*, May 13, 2011.

112 Leila Levinson, "Veterans Bitterly Disappointed by Obama," *Huffington Post*, August 26, 2011, www.huffingtonpost.com.

113 *Veterans For Common Sense V. Shinseki*, D.C. No. 3:07-cv-03758-SC, November 16, 2011.

114 Levinson, "Veterans Bitterly Disappointed by Obama."

115 *Veterans For Common Sense v. Shinseki et. al.*, 08-16728, 4829–4830, www.uscourts.gov.

116 Senate Committee on Veterans Affairs, *VA Mental Health Care: Evaluating Access and Accessing Care*, 112th Congress, 2d Session (2012), 1.

117 Department of Veterans Affairs, Office of the Inspector General, *Veterans Health Administration: Review of Veterans' Access to Mental Health Care* (Washington, DC: Department of Veterans Affairs, 2012), i.

118 Ibid., 1.

119 Ibid., iii.

120 Ibid., iii, 6, 7.

121 "VHA Issue Brief: Suicide Risk Performance Measure," May 25, 2012, 1, FOIA.

122 Department of Veterans Affairs, Office of the Inspector General, *Veterans Health Administration: Review of Veterans' Access to Mental Health Care*, 7.

123 Ibid.

124 Ibid., 5.

125 Department of Veterans Affairs, "Access Audit: System-Wide Review of Access: Results of Access Audit Conducted May 12, 2014, through June 3, 2014" (Washington, DC: Department of Veterans Affairs, 2014), 3–4, www.va.gov.

126 Department of Veterans Affairs, Office of the Inspector General, *Veterans Health Administration: Review of Veterans' Access to Mental Health Care*, 8.

127 "VHA Issue Brief: Suicide Risk Performance Measure," May 25, 2012, 1, FOIA.

128 Senate Committee on Veterans Affairs, *VA Mental Health Care*, 33.
129 Ibid., 34.
130 Ibid., 35.
131 Ibid.
132 Ibid., 65–66.
133 Ibid., 66.
134 Mike Scotti, "The V.A.'s Shameful Betrayal," *New York Times*, May 27, 2012.
135 Mike Scotti, *The Blue Cascade: A Memoir of Life after War* (New York: Grand Central Publishing, 2012).
136 Van Winkle, *Soft Spots*; and Roxana Robinson, *Sparta* (New York: Sara Creighton Books, 2013).
137 See chapter 3.
138 Mike Scotti, *The Blue Cascade: A Memoir of Life after War* (New York: Grand Central Publishing, 2012), 12–13.
139 Nathaniel Fick, *One Bullet Away: The Making of a Marine Officer* (Boston: Mariner Books, 2005). Robinson acknowledges having read Fick's memoir, and her debt to him is evident throughout. See, e.g., Robinson, *Sparta*, 24; and Fick, *One Bullet Away*, 4–5.
140 Robinson, *Sparta*, 21.
141 Scotti, *The Blue Cascade*, 4.
142 Robinson, *Sparta*, 95.
143 Ibid., 333.
144 Scotti, *The Blue Cascade*, 24.
145 Ibid., 94–96; and Robinson, *Sparta*, 243. Also see Van Winkle, *Soft Spots*, 111.
146 Robinson, *Sparta*, 33.
147 Ibid., 68.
148 Scotti, *The Blue Cascade*, 17–18 (italics in the original).
149 Robinson, *Sparta*, 347.
150 Scotti, *The Blue Cascade*, 90–91.
151 Ibid., 150.
152 Ibid., 127.
153 Ibid., 318.
154 Scotti, *The Blue Cascade*, 18–19.
155 Ibid., 20.
156 Van Winkle, *Soft Spots*, 25; and Robinson, *Sparta*, 367.
157 Robinson, *Sparta*, 370.
158 Ibid.
159 Scotti, *The Blue Cascade*, 32.
160 Robinson, *Sparta*, 314.
161 Ibid., 55.
162 Scotti, *The Blue Cascade*, 16, 32.
163 Jan E. Kemp to VHA Suicide Prevention Coordinators, "FW: E-Mail to SPCs regarding HBO Documentary," October 23, 2013, FOIA.

164 Dave Boyer, "VA Raises Concerns about Gun Registry with Offer of Free Gun Locks," *Washington Times*, January 6, 2015. FOX News similarly reported that "the Veterans Administration is offering free gun locks to former military members—but wants some key personal information in exchange," including the number of guns that the veteran owns; see "VA Wants Personal Information in Exchange for Free Gun Locks," FOX News, January 8, 2015, www.foxnews.com. See also Alan Zarembo, "Obama Pushes to Extend Gun Background Checks to Social Security," *Los Angeles Times*, July 18, 2015.

165 Robert M. Bossarte, "Evaluation of Implementation of VA's project Child Safe," Presentation to Advisory Board, April 28–29, 2011, FOIA.

166 Jan E. Kemp to Andrea Wicker, VHA Suicide Prevention Coordinators, and Lisa Brenner, "Re: Gun Locks," October 30, 2009, FOIA.

167 Ibid.; Stephanie Alton and Robert M. Bossarte, "Project Child Safe," Presentation to Advisory Board, April 28–29, 2011, FOIA.

168 Jan E. Kemp to Jan E. Kemp and Robert M. Bossarte, "Gun Lock Program," February 4, 2013, FOIA; Michael E. DeBakey VA Medical Center, "Veteran Family Safety: Free Gun Locks," news release (Houston, TX: Michael E. DeBakey VA Medical Center, January 9, 2012), www.va.gov; Jan E. Kemp to Jan E. Kemp and Robert M. Bossarte, "Gun Lock Program," February 4, 2013, FOIA.

169 VA Central Office, "Attempted Hanging from a Door in a Locked Mental Health Unit," October 6, 2009, 1, 2, FOIA; William Schoenhard to Network Directors, "Wireless Access Points on mental Health Units Treating Suicidal Patients," November 15, 2012, FOIA; Department of Veterans Affairs, "Operation S.A.V.E.," May 2012, 25, FOIA; Jan E. Kemp to VHA Suicide Prevention Coordinators et al., "FW: New Posting on NCPS Intranet," November 2, 2011, FOIA; VA Central Office, "Plastic Cards Can Be Used to Disable Locks in Inpatient Mental Health and Emergency Department (ED) Settings," April 9, 2012, FOIA; and Jan E. Kemp to Jan E. Kemp, "FW: Rx Cap Logos," September 19, 2012, FOIA.

170 As Matthew Miller explains, "In one study of near-lethal suicide attempts, for example, 24% of attempters took less than 5 minutes between the decision to kill themselves and the actual attempt (70% took less than one hour)"; see Matthew Miller, "Preventing Suicide by Preventing Lethal Injury: The Need to Act on What We Already Know," *American Journal of Public Health* 102, no. S1 (2012): e1. My reference to the amount of time it takes to smoke a few cigarettes comes from John Hiatt's "Adios to California," a song about his wife's suicide: "Two cigarettes from the package gone / I guess you thought about it just that long."

171 Department of Veterans Affairs, "Safety Plan Quick Guide for Clinicians," March 2012, 3, FOIA. On the need for such a program, see Miller, "Preventing Suicide by Preventing Lethal Injury," e2.

172 *Veterans Second Amendment Protection Act*, HR 577, 113th Congress, 1st Session (2013), 2, www.congress.gov.

173 *Clay Hunt SAV Act*, Public Law 114-2, 2015, www.congress.gov.

174 Senate Committee on Veterans' Affairs, *Mental Health and Suicide among Veterans*, 113th Congress, 2d Session (2014), 48.
175 Ibid., 14–15.
176 Ibid., 22–23.
177 Ibid., 21.
178 Committee on Veterans Affairs, Subcommittee on Health, *Understanding and Preventing Veteran Suicide*, 112th Congress, 1st Session (2011), 11.

CONCLUSION

1 Rhonda Cornum, interview with the author, April 13, 2016.
2 Elaine Sciolino, "Female P.O.W. Is Abused, Kindling Debate," *New York Times*, June 29, 1992.
3 Rhonda Cornum, interview with the author, April 13, 2016.
4 Ibid.
5 Ibid.
6 George Casey, interview with the author, April 6, 2016.
7 Ibid.
8 Eric Schoomaker, interview with the author, April 6, 2016.
9 This chapter does not offer a comprehensive analysis of resiliency training in the Army. For that, see Emily Sogn, "Internal Frontiers: Health, Emotion, and the Rise of Resilience Thinking in the U.S. Military," PhD diss., New School, New York, 2016.
10 Patricia Horoho, interview with the author, January 16, 2016; and Paul B. Lester et al., "Bringing Science to Bear: An Empirical Assessment of the Comprehensive Soldier Fitness Program," *American Psychologist* 66, no. 1 (2011): 77; and Alison Howell, "Resilience, War, and Austerity: The Ethics of Military Enhancement and the Politics of Data," *Security Dialogue* 46, no. 1 (2015): 16.
11 Sogn, "Internal Frontiers," 54; Howell, "Resilience, War, and Austerity," 18-21, 25.
12 Eric Schoomaker, interview with the author, April 6, 2016.
13 Sogn, "Internal Frontiers," 18, 73.
14 Eric Schoomaker, interview with the author, April 6, 2016.
15 Ibid.
16 George Casey, interview with the author, October 13, 2017.
17 Sogn, "Internal Frontiers," 54.
18 Martin E. P. Seligman, "Building Resilience," *Harvard Business Review* (April 2011), https://hbr.org/. As Sogn explains, "In the contemporary United States, resilience . . . is called upon to address a panoply of problems"; it "firmly acknowledges crisis as a potential condition of the future" and "is often cast as a positive-reframing of health that emphasizes the capacities of individuals and communities over traits of the social or economic environment." See Sogn, "Internal Frontiers," 47, 53, 55.

19 Seligman, "Building Resilience."
20 Rhonda Cornum, interview with the author, April 13, 2016.
21 For critiques of both failure and normalcy as cultural constructs, see Judith/Jack Halberstam, *The Queer Art of Failure* (Durham, NC: Duke University Press, 2011); and Lennard Davis, *Enforcing Normalcy: Disability, Deafness, and the Body* (New York: Verso, 1995).
22 Rhonda Cornum, interview with the author, April 13, 2016.
23 Ibid.
24 Amy Adler, interview with the author, April 1, 2016.
25 Karen J. Revich, M. E. Seligman, and S. McBride, "Master Resilience Training in the U.S. Army," *American Psychologist* 66, no. 1 (2011): 25–26; and Sogn, "Internal Frontiers," 75–76.
26 George Casey, interview with the author, October 13, 2017.
27 Paul D. Lester et al., "The Comprehensive Soldier Fitness Program Evaluation, Report #3: Longitudinal Analysis of the Impact of Master Resilience Training on Self-Reported Resilience and Psychological Health Data" (Arlington, VA: Department of the Army, December 2011), 1, posted at the University of Pennsylvania, www.upenn.edu.
28 Ibid.
29 Ibid., 15.
30 Amy Adler, interview with the author, April 1, 2016; and Sogn, "Internal Frontiers," 198–199.
31 Jennifer Mittelstadt, *The Rise of the Military Welfare State* (Cambridge, MA: Harvard University Press, 2015), 224–225; Howell, "Resilience, War, and Austerity," 16.
32 Sogn, "Internal Frontiers," 78; see also Howell, "Resilience, War, and Austerity," 16 and 26.
33 Carl Castro, interview with the author, April 15, 2016.
34 Howell, "Resilience, War, and Austerity," 23.
35 Senate Committee on Armed Services, *The Progress in Preventing Military Suicides and the Challenges in Detection and Care of the Invisible Wounds of War*, 111th Congress, 2d Session (2010), 57.
36 Patricia Horoho, interview with the author, January 16, 2016.
37 Christopher Ivany, interview with the author, December 16, 2015.
38 Rebecca Porter, interview with the author, December 17, 2015.
39 Ibid.
40 Patricia Horoho, interview with the author, January 16, 2016.
41 Ibid.
42 Peter Chiarelli, interview with the author, September 15, 2015.
43 Patricia Horoho, interview with the author, January 16, 2016.
44 Bruce Shahbaz, interview with the author, February 24, 2016.
45 Ibid.
46 Ibid.
47 Ibid.
48 Charles Engel, interview with the author, January 18, 2016.

49 U.S. Army Medical Command, "ARMY PCMH Implementation Manual: Leaders Guide to Army Patient Centered Medical Home Transformation" (San Antonio, TX: U.S. Army Medical Command, January 20, 2014), 21, posted at Uniformed Services Academy of Family Physicians, www.usafp.org.

50 Charles Engel, interview with the author, January 18, 2016.

51 Ibid.

52 Ibid.

53 Ibid.

54 Geoff Ling, interview with the author, February 21, 2016.

55 Ibid.

56 Charles Engel, interview with the author, January 18, 2016.

57 George Casey, interview with the author, October 13, 2017.

58 Pentagon Channel, "General Chiarelli Retires," United States Army, January 31, 2012, www.army.mil.

59 Gale Pollock, interview with the author, December 1, 2015.

60 George Casey, interview with the author, April 6, 2016.

61 John Dyckman, "Exposing the Glosses in Seligman and Fowler's (2011) Straw-Man Arguments," *American Psychologist* 66, no. 7 (October 2011): 644. This critique anticipated Alison Howell's argument that "resilience . . . is a tool for waging unending war"; see Howell, "Resilience, War, and Austerity," 28.

62 Martin E. P. Seligman, "Helping American Soldiers in Time of War: Reply to the Comments on the Comprehensive Soldier Fitness Special Issue," *American Psychologist* 66, no. 7 (October 2011): 647.

63 On permanent war, see Mary Dudziak, *Wartime: An Idea, Its History, and Its Consequences* (New York: Oxford University Press, 2012), 8. Here, I differ from Howell, "Resilience, War, and Austerity," 23.

64 Charles Engel, interview with the author, January 18, 2016.

65 Gale Pollock, interview with the author, December 1, 2015.

66 Ibid.

67 Dave Baiochi, *Measuring Army Deployments to Iraq and Afghanistan* (Santa Monica: RAND Corp., 2012), 1; U.S. Army, *2012 Army Posture Statement* (Arlington, VA: Department of the Army, 2012), www.army.mil; and Charles W. Hoge et al., "Transformation of Mental Health Care for U.S. Soldiers and Families during the Iraq and Afghanistan Wars: Where Science and Politics Intersect," *American Journal of Psychiatry* 173, no. 4 (2016): 334.

INDEX

ABOUT THE AUTHOR

David Kieran is Assistant Professor of History and Director of the American Studies concentration at Washington & Jefferson College in Washington, Pennsylvania. He is the author of *Forever Vietnam: How a Divisive War Changed American Public Memory* (2014), the editor of *The War of My Generation: Youth Culture and the War on Terror* (2015), and the co-editor of *At War: The Military and U.S. Culture in the Twentieth Century and Beyond* (2018).